C. Grondahl

Two-Phase Steam Flow in Turbines and Separators

SERIES IN THERMAL AND FLUIDS ENGINEERING

EDITORS:

JAMES P. HARTNETT and THOMAS F. IRVINE, JR.

Chang	• Control of Flow Separation: Energy Conservation, Operational Efficiency, and Safety
Chi	• Heat Pipe Theory and Practice: A Sourcebook
Eckert and Goldstein	• Measurements in Heat Transfer, 2nd edition
Hsu and Graham	• Transport Processes in Boiling and Two-Phase Systems, Including Near-Critical Fluids
Keairns	• Fluidization Technology
Moore and Sieverding	• Two-Phase Steam Flow in Turbines and Separators: Theory, Instrumentation, Engineering

Two-Phase Steam Flow in Turbines and Separators
theory, instrumentation, engineering

Edited by

M. J. MOORE
Central Electricity Research Laboratory
Leatherhead, England

C. H. SIEVERDING
von Karman Institute for Fluid Dynamics
Rhode-Saint-Genèse, Belgium

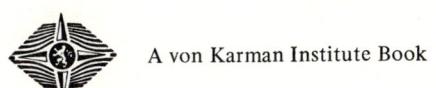
A von Karman Institute Book

HEMISPHERE PUBLISHING CORPORATION
Washington   London

McGRAW-HILL BOOK COMPANY

New York	Johannesburg	Paris
St. Louis	London	São Paulo
San Francisco	Madrid	Singapore
Auckland	Mexico	Sydney
Bogotá	Montreal	Tokyo
Düsseldorf	New Delhi	Toronto
	Panama	

TWO-PHASE STEAM FLOW IN TURBINES AND SEPARATORS: Theory, Instrumentation, Engineering

Copyright © 1976 by Hemisphere Publishing Corporation. All rights reserved. Printed in the United States of America. No part of this publication may be reproduced, stored in a retrieval system, or transmitted, in any form or by any means, electronic, mechanical, photocopying, recording, or otherwise, without prior written permission of the publisher.

1 2 3 4 5 6 7 8 9 0   K P K P   7 8 3 2 1 0 9 8 7 6

This book was set in Press Roman by Hemisphere Publishing Corporation. The editors were Mary A. Phillips and William E. Pettit; the designer was Lilia Guerrero; the production supervisor was Rebekah McKinney; and the compositor was Wayne Hutchins.
The printer and binder was The Kingsport Press, Inc.

Library of Congress Cataloging in Publication Data

Two-phase steam flow in turbines and separators.

  (Series in thermal and fluids engineering)
  "A collection of the edited lecture notes for a course . . . presented at the von Karman Institute for Fluid Dynamics in May 1974."
  Includes bibliographical references.
  CONTENTS: Gyarmathy, G. Basic notions.–Moore, M. J. Gas dynamics of wet steam and energy losses in wet-steam turbines.–Gyarmathy, G. Condensation in flowing steam. [etc.]
    1. Steam-turbines—Addresses, essays, lectures. 2. Steam-separators—Addresses, essays, lectures. 3. Two-phase flow—Addresses, essays, lectures. 4. Steam flow—Addresses, essays, lectures. I. Moore, M. J. II. Sieverding, C. H., date.
TJ737.T9   621.1'65   76-9125
ISBN 0-07-042992-8

Contents

The Authors   ix
Preface   xi

CHAPTER 1   **BASIC NOTIONS**   *G. Gyarmathy*   1

1.1 Brief History of Problems Associated with Steam Wetness   1
1.2 State of the Art of Wet-Steam Turbines   6
     1.2.1 Conventional Low-Pressure Condensing Turbines   6
     1.2.2 Nuclear High-Pressure Wet-Steam Turbines   11
1.3 Properties of Wet Steam   20
     1.3.1 Composition, Vapor Pressure   20
     1.3.2 Physical Data of Steam and Water   21
     1.3.3 Structure of Wet Steam   29
     1.3.4 Supersaturation and Subcooling   31
1.4 Behavior of Wet Steam   35
     1.4.1 Thermodynamics of Steam/Droplet Mixtures   35
     1.4.2 Heat Transfer in Wet Steam   38
     1.4.3 Mechanics of Steam/Droplet Mixtures   43
     1.4.4 Motion of Liquid on Channel Walls   52
Nomenclature   54
     Subscripts   55
     Superscripts   55
     Abbreviations   55
References   56

CHAPTER 2   **GAS DYNAMICS OF WET STEAM AND ENERGY LOSSES IN WET-STEAM TURBINES**   *M. J. Moore*   59

2.1 Equations of State for Steam   59
     2.1.1 Equations for Superheated and Supersaturated Steam   59
     2.1.2 Equations for Wet Steam   62
2.2 Relaxation   64
     2.2.1 General System   64
     2.2.2 Thermal Relaxation of Wet Steam   67
     2.2.3 Inertial Relaxation of Water Droplets   68

   2.2.4 Typical Values of Relaxation Time 69
 2.3 One-Dimensional Equations for Wet-Steam Flow 70
   2.3.1 Continuity Equation 71
   2.3.2 Momentum Equation for the Liquid Phase 73
   2.3.3 Momentum Equation for the Gaseous Phase 74
   2.3.4 Energy Equation 74
   2.3.5 Equations of State 75
   2.3.6 Supplementary Equations 76
   2.3.7 Dimensionless Equations for Steady Flow 77
   2.3.8 Application of the Flow Equations 78
   2.3.9 Examples of Flow Calculations 81
 2.4 One-Dimensional Flow with Spontaneous Condensation 83
   2.4.1 Flow Equations 83
   2.4.2 Transonic Solution 85
   2.4.3 Sudden Heat Release on Condensation 88
 2.5 Calculated Examples of Wet-Steam Flows 91
   2.5.1 Condensation in a Multistage HP Turbine 91
   2.5.2 Transonic LP Blading 91
   2.5.3 Condensation in a Two-Dimensional Flow Field 92
 2.6 Acoustic Velocity in Wet Steam 94
 2.7 Shock Waves in Wet Steam 99
 2.8 Wetness Losses in Turbines 104
   2.8.1 Empirical Approach 104
   2.8.2 Theoretical Approach 105
   2.8.3 Isentropic Wet-Steam Stage 106
   2.8.4 Stage Work-Loss Coefficient 110
   2.8.5 Components of Stage-Work Loss 111
   2.8.6 Losses Due to Supercooling and Nucleation 119
   2.8.7 Example Calculation of Losses 122
Nomenclature 122
  Subscripts 124
  Superscripts 124
  Units 124
References 124

CHAPTER 3 CONDENSATION IN FLOWING STEAM *G. Gyarmathy* 127
 3.1 Introduction 127
   3.1.1 Historical Review 127
   3.1.2 Basic Concepts 129
   3.1.3 Steady Gas Flow in Laval Nozzles 131
 3.2 Experimental Evidence in Nozzles 137
   3.2.1 Test Equipment Used 137
   3.2.2 Phenomenology of Condensation in Nozzles 140
   3.2.3 Synopsis of Experimental Data 145

3.3 Theoretical Description of the Condensation Process   155
    3.3.1 Nucleation Theory   155
    3.3.2 Droplet Growth Theory   163
    3.3.3 One-Dimensional Flow with Condensation   169
3.4 Influence of Foreign Nuclei   177
3.5 Condensation in Cascades and Turbines   179
    3.5.1 Experimental Evidence   180
    3.5.2 Theoretical Explanations   184
Nomenclature   186
    Subscripts   187
    Superscripts   187
References   187

CHAPTER 4 INSTRUMENTATION FOR WET STEAM   191

4.1 A Review of Instrumentation for Wet Steam   *M. J. Moore*   191
    4.1.1 Pressure Measurement   191
    4.1.2 Temperature Measurement   201
    4.1.3 Velocity Measurement   201
    4.1.4 Measurement of Unsteady Flow   212
    4.1.5 Measurement of Wetness Fraction   217
    4.1.6 Measurement of Droplet Size   234
4.2 A Light-Scattering Probe for Droplet Size and Wetness Fraction Measurement in Two-Phase Flows   *A. Ederhof*   249
    4.2.1 The Principle of Measurement of the Light-Scattering Probe   249
    4.2.2 Calibration of the Light-Scattering Probe   250
    4.2.3 Example of Applications   250
Nomenclature   257
    Subscripts   258
References   258

CHAPTER 5 EXPERIMENTAL DEVELOPMENT OF WET-STEAM TURBINES   *A. Smith*   261

5.1 Influence of Moisture on Blading Efficiency   261
5.2 Moisture-Removal Devices   269
5.3 Erosion Protection   282
Nomenclature   288
References   289

CHAPTER 6 OPERATING EXPERIENCE OF WET-STEAM TURBINES   *W. Engelke*   291

6.1 Erosion-Corrosion   291
    6.1.1 Causes of Erosion-Corrosion   291
    6.1.2 Evaluation of Erosion-Corrosion   293
    6.1.3 Agreement of Predictions with Measurements   296

6.2 Erosion of LP Blades   298
    6.2.1 Causes of Erosion   298
    6.2.2 Erosion Coefficient   299
    6.2.3 Examples of Erosion with Different Machines   303
    6.2.4 Experience with Erosion on 750-mm LP Blades   303
    6.2.5 Design Principles   303
6.3 Overspeed after Major Load Rejection   312
    6.3.1 Calculation Method   313
    6.3.2 Test Results   313
    6.3.3 Measurements to Prevent Excessive Overspeed   313
References   315

CHAPTER 7 EXTERNAL WATER SEPARATORS   317

7.1 Performance of Knitted Wire Mesh and Corrugated Plate Separators
    G. C. Gardner   317
    7.1.1 Introduction to Basic Concepts   317
    7.1.2 Fractional Separation Efficiency   318
    7.1.3 Breakthrough Characteristics   329
7.2 External Moisture Separator-Reheaters   *R. L. Coit, P. D. Ritland,*
    T. F. Rabas, and P. W. Viscovich   337
    7.2.1 Physical Features of MSRs   337
    7.2.2 The Evolution of MSR Design   344
    7.2.3 Operating Experience   348
    7.2.4 MSRs of the Future   364
Nomenclature   367
    Subscripts   368
References   368

    Author Index   371
    Subject Index   379

The Authors

R. L. Coit is the Engineering Manager of the Heat Transfer Division of Westinghouse Electric Corp., Lester, Pennsylvania, U.S.A. He is responsible for the design of large external moisture separator-reheaters and, with his coauthors, has contributed part 2 of Chapter 7 on the practical aspects of separator design and operation.

A. Ederhof contributed part 2 of Chapter 4 on a light-scattering probe for measuring droplet distribution in wet-steam flows. Mr. Ederhof is currently a Research Engineer at the Institut für Dampf- und Gasturbinen, RWTH, Aachen, West Germany.

W. Engelke is General Manager of Steam Turbine Design Division, at the Mülheim, West Germany, works of Kraftwerk Union AG. In Chapter 6 he describes some problems experienced in the operation of wet-steam turbines and their solution.

G. C. Gardner has contributed part 1 of Chapter 7 on the theory of water separation. Mr. Gardner is a Research Engineer at the Central Electricity Research Laboratories (CERL) of the CEGB at Leatherhead, England.

G. Gyarmathy is Deputy Head of Turbomachinery Research at BBC Brown Boveri Ltd., Baden, Switzerland. He provides an introduction to wet-steam turbine development in Chapter 1 and a detailed description in Chapter 3 of condensation in flowing steam.

M. J. Moore is Head of the Thermodynamics Section at CERL, Leatherhead, England. In addition to acting as Course Director, Dr. Moore has contributed Chapter 2 on the gas dynamics of wet steam and part 1 of Chapter 4, a review of wet-steam instrumentation.

A. Smith is Manager of the Mechanical Research Department at C.A. Parsons Ltd., Newcastle-upon-Tyne, England. In Chapter 5 he describes experimental development of wet-steam turbines.

Preface

This publication is a collection of the edited lecture notes for a course on two-phase flow in turbines presented at the von Karman Institute for Fluid Dynamics in May 1974. The course, Lecture Series 70, was held under the auspices of the Head of Fluid Dynamics, Professor J. Chauvin. The lectures, contributed by specialists from Europe and the United States, are intended mainly for students, scientists, and practicing engineers involved in design, development, research, and operation of turbine plants. The chapters on condensation processes and gas dynamics of wet steam should, however, be useful in a general way for all researchers working in the domain of two-phase flows. The term "wet-steam" used here and throughout the text refers to a two-phase mixture in which the liquid phase is dispersed in various forms. In a continuous vapor medium the water formed by partial condensation of the steam appears in the low-pressure part of large turbines and also in the high-pressure part of nuclear turbines for water-cooled reactor systems. Its presence is marked by a reduction in thermodynamic efficiency and by various problems of erosion and corrosion. These and other aspects are described from the basic fundamentals of the condensation process to the latest operating experience with modern turbines. In the preparation of the course we wish to acknowledge the valuable suggestions made by Dr. G. Gyarmathy. We are also indebted to Mr. G. C. Gardner, who provided on short notice a contribution on the theory of water separation. We thank the following companies for their cooperation: BBC Brown Boveri Ltd., C. A. Parsons Ltd., Central Electricity Research Laboratories (CEGB), Kraftwerk Union AG, Institut für Dampf- und Gasturbinen, RWTH, Aachen, and Westinghouse Electric Corporation.

M. J. Moore
Course Director

C. H. Sieverding
VKI Coordinator

Two-Phase Steam Flow in Turbines and Separators

CHAPTER 1

Basic Notions

G. Gyarmathy

1.1 BRIEF HISTORY OF PROBLEMS ASSOCIATED WITH STEAM WETNESS

The problems of wet steam are almost as old as steam turbines themselves. Why?

From the early days of steam turbines, designers strived to improve the efficiency of the thermodynamic cycle of steam. The most promising means to do this consisted in raising the pressure and temperature of live steam as high as compatible with the best construction materials available at an acceptable price. It appeared soon that the limits set to the temperature of highly stressed parts were rather strict while the limits pertaining to the pressure in the boiler and at the inlet for the turbine could be raised by improved design.

As illustrated by Figs. 1.1.1a and b, an increase of live-steam pressure p_0 at constant live-steam temperature T_0 raises the mean temperature level of heat addition to the cycle, $\bar{T}$, which means an improvement of Carnot efficiency.

It turned out however that limits were set for p_0, too. Surprisingly, the limits on live-steam pressure were not set by the high-pressure, high-temperature parts of the boiler or of the turbine; they were rather set by the fact that an increase in p_0 tended to increase the wetness fraction $1 - X_E$ at the turbine exit to intolerable levels.

With the increase of exit wetness, two particular problems were encountered. First, severe erosion was found to occur in the blading at the low-pressure end of turbines. Frequently, after a few months or few years of service, the low-pressure rotor blades were damaged, i.e., roughened, pitted, or even mutilated. In order to avoid grave erosion damage, a limitation of exit wetness to about 12% was found to be necessary. This limit remained valid up to the present, even though considerable improvements in design and construction were made as years progressed. Such design measures involve

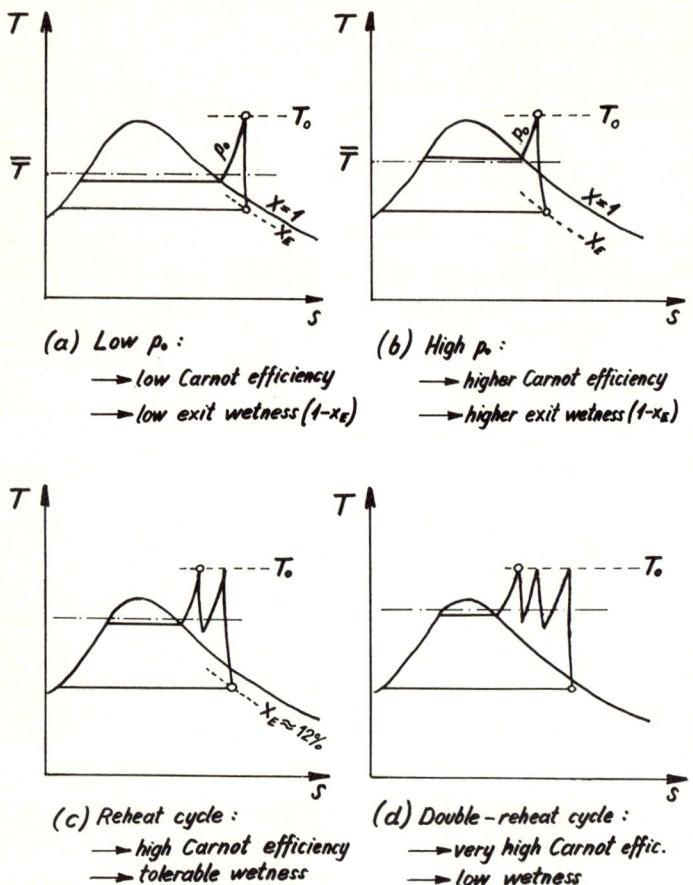

FIG. 1.1.1 Temperature-entropy diagrams of conventional steam turbine cycles with and without reheat.

blade hardening or blade shielding with stellite strips and water drainage devices in the casing walls. Due to the 12% limit set by erosion, wetness became an equally severe limiting factor for the thermodynamic efficiency of steam turbine plants as the temperature resistance of materials.

As a second though less critical nuisance, it was found that the aerodynamic efficiency of turbine stages operating in the wet steam region was considerably lower than that of dry stages. K. Baumann established as early as 1910 that 1% of wetness present in a stage was likely to cause about 1% of efficiency decrease (Baumann's rule). This loss diminished the gain obtained by increasing the live-steam pressure, but fortunately was never able to cancel it.

In some applications, such as marine propulsion where the size of the boiler had to be kept small, *saturated* steam was commonly used to drive the turbine. In geothermal plants live steam was, and is, frequently *wet*. In these instances the erosion problems play an even more decisive role in the design of the turbine.

The great attention devoted to wet steam in the pioneering period is documented in the work of Stodola [1-1]. Later, when the limits set by materials to temperature and to boiler pressure were further raised in the course of metallurgical and design progress, the strict limits to exit wetness required the intermediate reheating of steam (Fig. 1.1.1c), and thus imposed a severe complication on steam plants. However, reheating also meant a further increase of $\bar{T}$ and of Carnot efficiency. In the 1950s plants with double reheat (Fig. 1.1.1d) were constructed and a final solution for the problems of steam wetness seemed to be emerging. Indeed, with double reheat the limits on p_0 were imposed by the design of the high-pressure parts and no longer by wetness. Optimum cycle selection usually led to a mere 6 to 10% wetness at the turbine exit, an amount well controllable with blade materials and blade shields developed in the meantime.

Increase in plant size and the corresponding increase in blade speed caused a renewal of attention toward blade erosion in the early 1960s [1-2]. In the meantime it also became obvious that, due to economic reasons, double-reheat systems would not completely supercede single-reheat ones, and therefore high exit wetness would remain a necessity. The critical question emerged whether the empirical wetness limits pertaining to earlier designs remained applicable to advanced designs having higher tip speed ($\geqslant 550$ m/s) and supersonic flow in the region of blade tips.

Another development that vitalized the interest in wet steam was the development of light-water-cooled nuclear reactors generating saturated, high-pressure steam. While the low-pressure (LP) part of the associated turbines remained essentially similar to conventional designs, the high-pressure (HP) part presented novel problems. Saturated or even slightly wet steam of 45 to 65 bar pressure entered the first high-pressure section and had to be expanded to 15% or more of wetness corresponding to a pressure level of about 5 bar. A further straightforward expansion, leading eventually to about 25% wetness at the vacuum end, was clearly not acceptable. The solution currently used for nuclear plants therefore involves water separators between the HP and LP turbine, the separators usually being combined with reheaters that superheat the steam to a temperature sufficiently high to ensure permissible values of the steam wetness X_E at the LP exit (see Fig. 1.1.2a). In such turbines involving reheaters water separation is essential for high cycle efficiency. In some instances however water separation without superheating is being used (Fig. 1.1.2b). In these cases the wetness problems in the LP turbine require particular attention, and water separators are essential for preventing erosion. There exist several competitive methods to achieve separation, and a final assessment of relative merits is not yet quite easy to make. These topics will be treated in detail in Chap. 7.

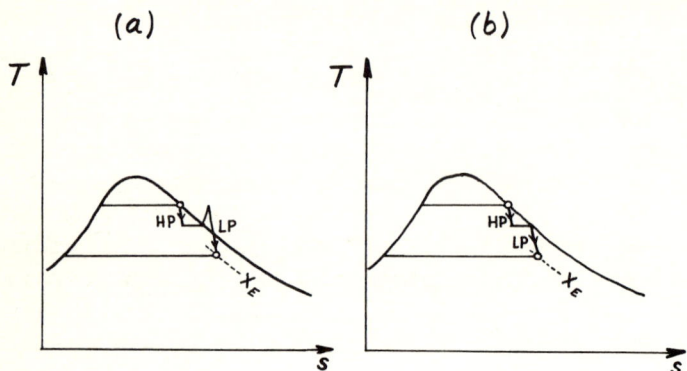

FIG. 1.1.2 Temperature-entropy diagram of nuclear steam turbine cycles. (*a*) Intermediate water separation and steam superheat. (*b*) Intermediate water separation only.

Since the 1960s, a great amount of research has been conducted in various countries in order to shed light on the thermodynamic and flow phenomena associated with wet steam. Our insight has been significantly deepened but we are, despite all endeavor, not yet capable of theoretically explaining and predicting all phenomena observed in wet-steam turbines and connected devices.

Before beginning to deal with specific problems in detail, let us highlight some aspects of economic trends that bear upon the development of steam turbines. Electric power consumption is known to have doubled every ten years in past times, and a further increase at nearly this rate can be foreseen for the next decade. The increase in network output encourages utility companies to take full advantage of the decrease in specific (i.e., per MW)

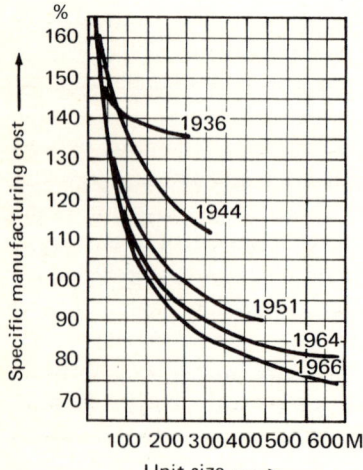

FIG. 1.1.3 Trends in power plant manufacturing cost [1-3].

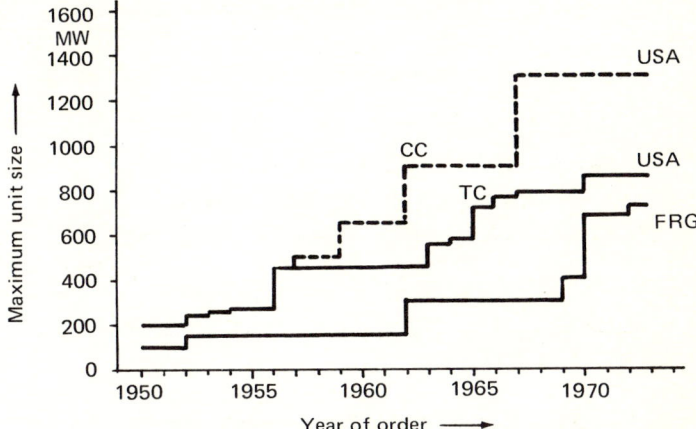

FIG. 1.1.4 Trends in turbine/generator unit size [1-3]. (Note: FRG = Federal Republic of Germany, CC = two-shaft "cross-compound" units, TC = one-shaft "tandem-compound" units.)

plant manufacturing cost with increasing unit size (Fig. 1.1.3). As a result, maximum power of turbine units has increased from 200 MW in 1950 to 800-1200 MW by 1970 (Fig. 1.1.4).

In the turbine the size-limiting element being the last (low-pressure) stage, blade designs have been continually developed toward greater mass flow rates, i.e., toward transonic flow, greater blade length, and higher tip speed. These developments have been closely associated with problems of wet steam.

Besides increasing the size of large turbines, there is also considerable economic interest in improving their efficiency. This may be illustrated with a few figures. It is common nowadays to impose performance penalties on nuclear turbines ranging between $400 and $500 per kilowatt power deficit caused by poor turbine efficiency. Thus an efficiency defect of say 0.1% on a 1000 MW machine has an economic equivalent of $400,000 to $500,000. In fossil-fired units the penalty rates are usually lower (by about half), but the sums are still quite considerable.

In view of these sums the incentive for improvements on turbines must be great. On the other hand, however, development time and development cost for implementing new designs and for maturing new components has risen considerably. Caution is further prompted by the extreme demands posed with respect to reliability of large power production units. Manufacturers are therefore compelled to scrutinize development projects closely and to proceed only with the most promising ones.

Improved theoretical explanation of problems is clearly desirable, because better understanding facilitates the choice between alternative developments, helps to avoid failures, and accelerates progress. It is hoped that the present course will encourage further efforts toward these goals.

1.2 STATE OF THE ART OF WET-STEAM TURBINES

1.2.1 Conventional Low-Pressure Condensing Turbines

1.2.1.1 General Remarks

Low-pressure (LP) condensing turbines, for which manufacturers normally possess a set of standard designs of different exhaust area size, are the most expensive elements of both fossil-fired and nuclear power plant turbines. A typical schematic diagram of a fossil-fired plant is shown in Fig. 1.2.1, together with the pertinent expansion line. While the high-pressure (HP) and intermediate-pressure (IP) turbines expand superheated steam, the last few stages of LP turbines are working with wet steam.

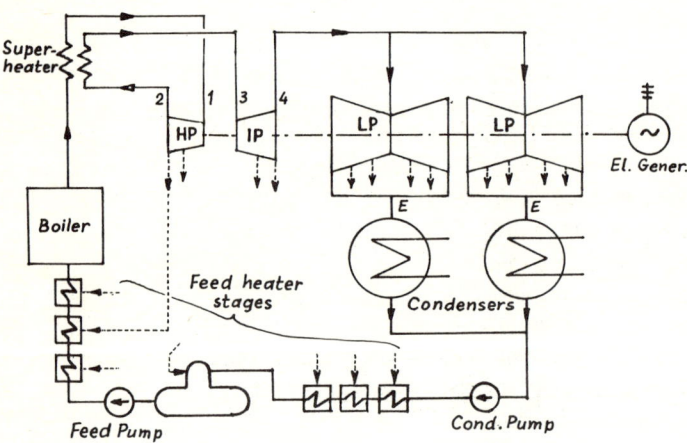

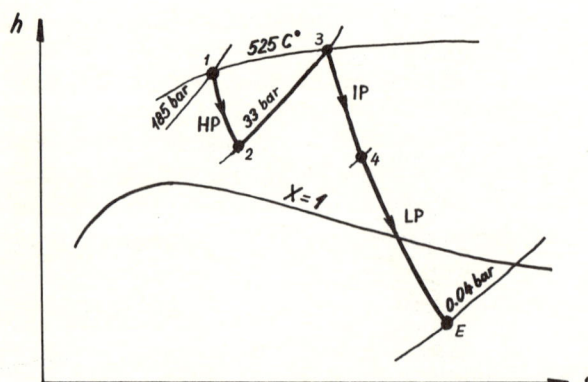

FIG. 1.2.1 Schematic diagram and expansion line of a typical conventional steam turbine power plant.

Basic Notions 7

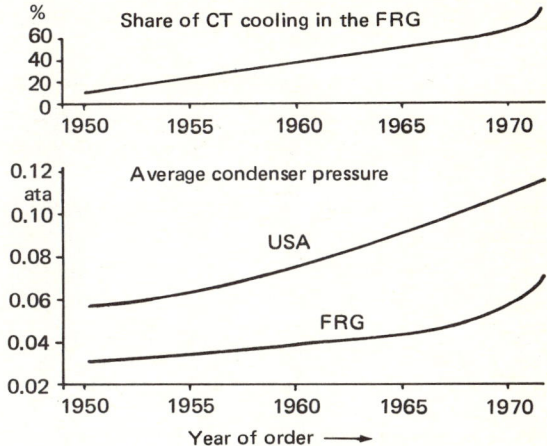

FIG. 1.2.2 Trends of back pressure [1-3].

Several LP flows in parallel are required to handle the enormous volume flow rates typical of last stages. Apart from end-stage design considerations, the number of flows is essentially dictated by two factors: unit size and back pressure. The current transition from through-flow cooling (TFC) of condensers to cooling towers (CT) brings about a steady increase of mean condenser pressures as shown by Fig. 1.2.2. At medium back pressures (say, 0.08 bar), one square meter of exhaust area is required per 25 to 30 MW. The number and size of LP flows are shown in Fig. 1.2.3 in terms of condenser pressure and turbine power output.

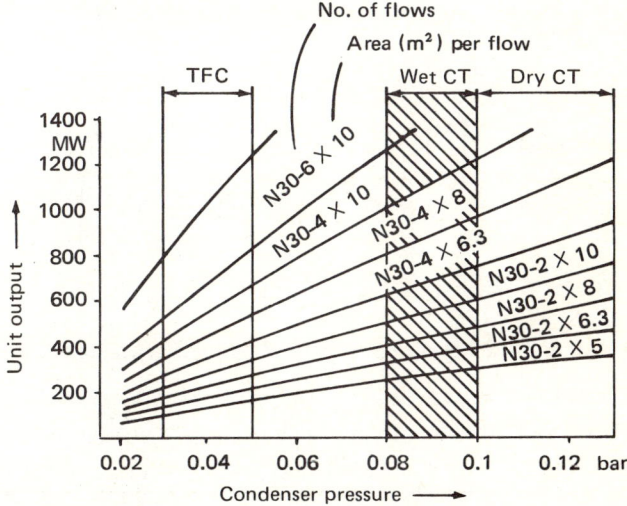

FIG. 1.2.3 Number and size of LP flows typically required [1-3].

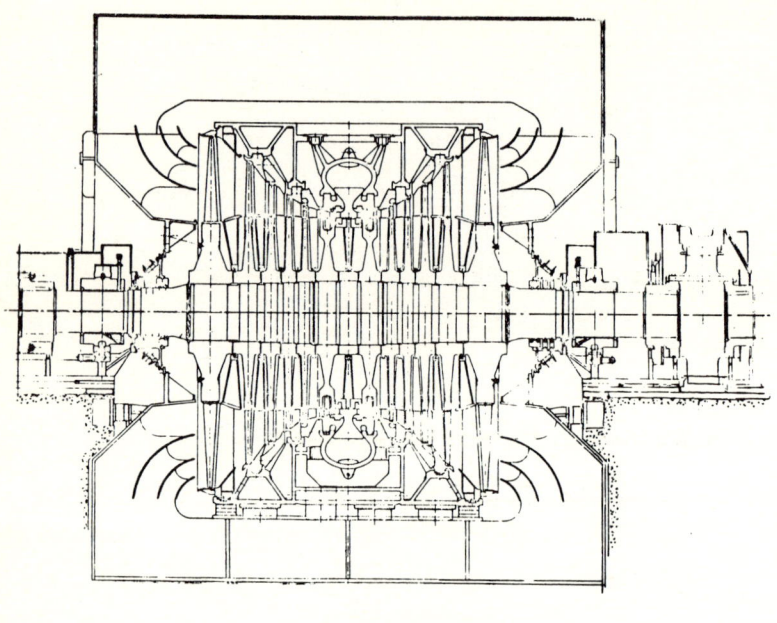

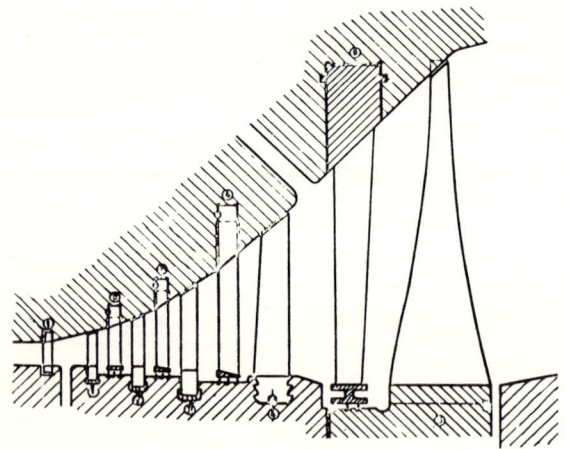

FIG. 1.2.4 Typical cross sections of LP turbines [1-5, 1-6].

1.2.1.2 Design of the LP Turbine; Expansion Line

Cross sections of two LP turbines are shown in Fig. 1.2.4. Typical data of some standard designs of LP last stages currently used on 3000-rpm turbines are shown in Table 1.2.1.

A correlation between last-stage isentropic heat drops and blade lengths may be seen in Fig. 1.2.5. A typical expansion line in a modern five-stage LP

Basic Notions

TABLE 1.2.1 Typical data of LP last stages

Type	A	B	C	D
Blade length (mm)	670	790	870	960
Root diameter (mm)	1680	1590	1740	1900
Exhaust area (m^2)	5	6	7	8.5
Tip speed (m/s)	470	500	545	600

turbine is shown in Fig. 1.2.6. It is seen that typically the last two stages are working entirely in the wet region and the last-stage rotor blades receive steam of about 8% wetness.

1.2.1.3 Effects and Handling of Moisture in LP Turbines

Due to the high tip speed and high local wetness values, erosion in the tip region (Fig. 1.2.7) is a common occurrence, unless protective measures are taken. Manufacturers generally harden tip leading edges or shield them with stellite strips. Other protective measures consist of removal of water through water drainage arrangements in the casing walls (Fig. 1.2.8) or through suction slots made in hollow stator blades (Fig. 1.2.9). An unusual antierosion device is the so-called Baumann stage depicted in Fig. 1.2.10.

Turbines of nuclear power plants are often designed for half speed (i.e., 1500 rpm in Europe and 1800 rpm in America). In these large-diameter LP machines peripheral speeds can be chosen more conservatively than in full-speed ones, and therefore erosion problems are somewhat less severe.

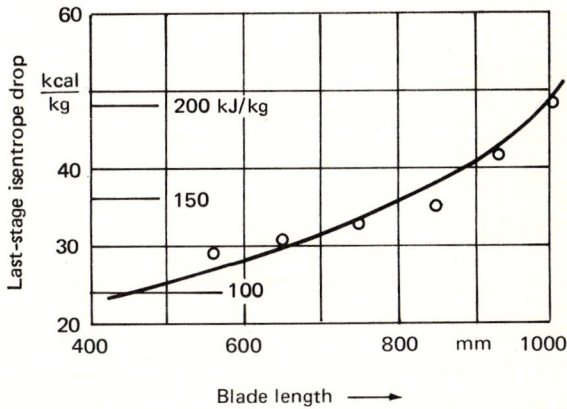

FIG. 1.2.5 Variation of last-stage heat drop with blade length [1-4].

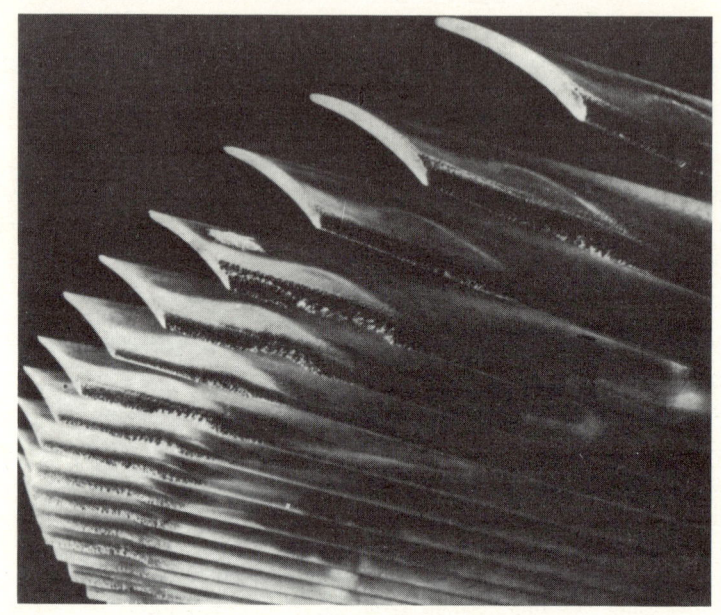

FIG. 1.2.7 Eroded LP blades [1-7].

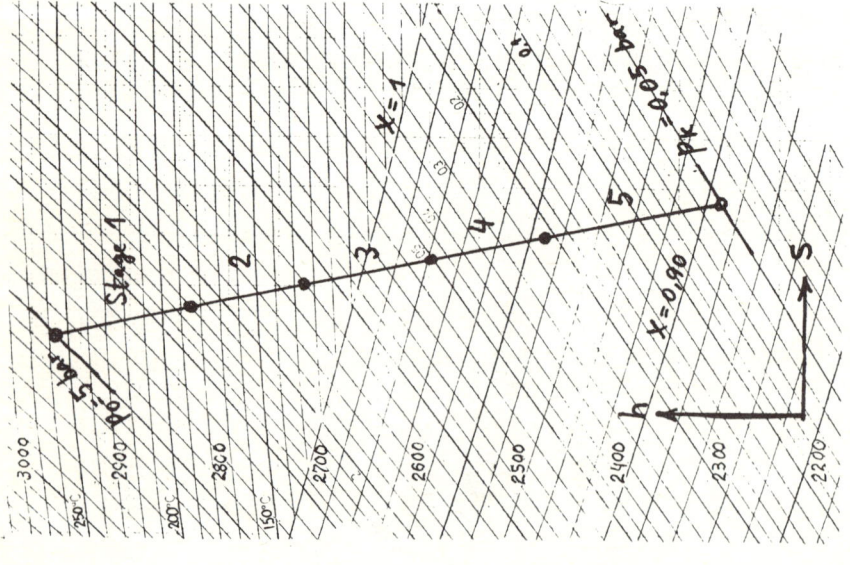

FIG. 1.2.6 Expansion line of typical LP turbine.

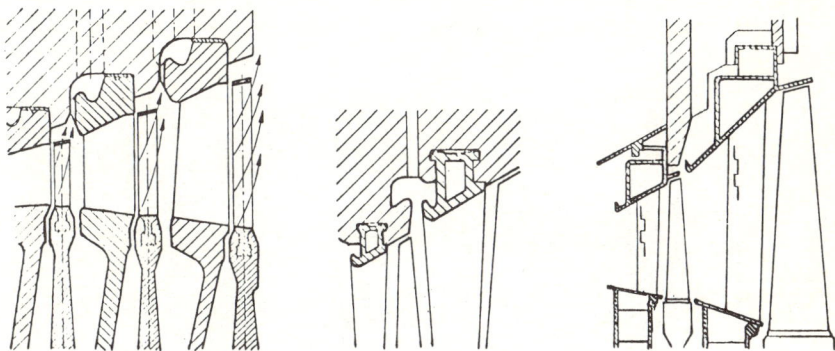

FIG. 1.2.8 Various LP drainage devices [1-4].

1.2.2 Nuclear High-Pressure Wet-Steam Turbines

1.2.2.1 General Remarks

Present-day nuclear power plants mostly involve light-water-cooled nuclear reactors which are either of the pressurized-water (PWR) or of the boiling-water (BWR) type. The reactor cycles are shown in Fig. 1.2.11. In both cases *saturated steam* of low wetness (ca. 0.2%) and of 50–70 bar pressure is being

FIG. 1.2.9 Hollow-blade suction slots [1-7].

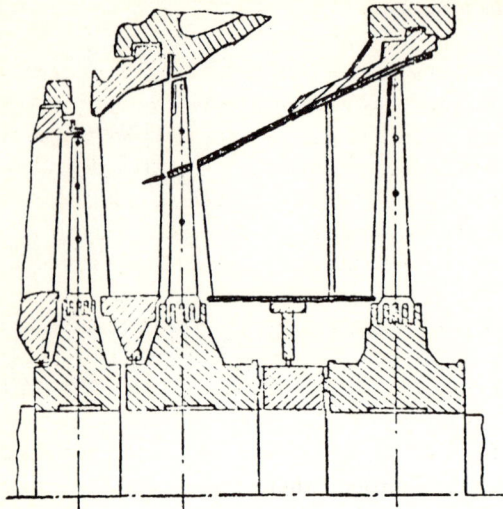

FIG. 1.2.10 Exhaust with Baumann-type penultimate stage.

(a) PWR Plant

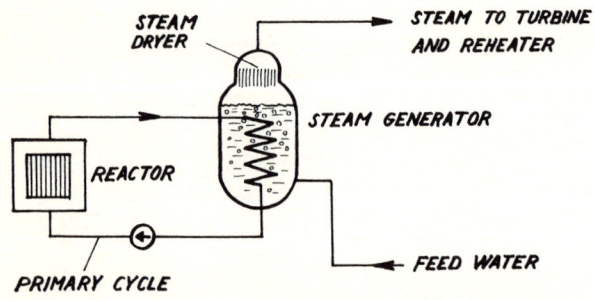

(b) BWR Plant

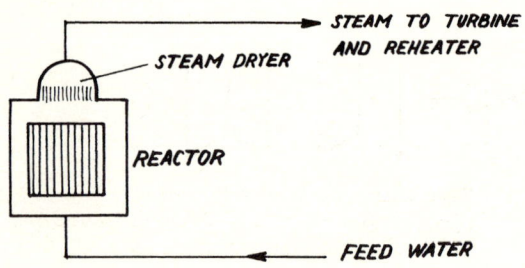

FIG. 1.2.11 Principles of PWR and BWR plants.

generated. The main difference affecting the operation of the turbine stems from the fact that the PWR cycle produces nonradioactive steam while the BWR cycle produces radioactively contaminated steam. In BWR plants reduced accessibility stresses the importance of reliable operation. A second difference consists in the chemical nature of contaminants present in the feed water which manifests itself as different corrosion problems in each plant. In some PWR systems the steam generator is designed for generating dry steam of slight initial *superheat*.

Unit sizes of present-day nuclear power plants range between 300 and 1300 MW rated electrical output. The combination of huge output and of relatively low live-steam parameters (60–70 bar, 260–290°C) results in enormous mass flow rates (e.g., nearly 2 tons of steam per second in a 1200 MW plant). At the low-pressure end very large flow sections are required to deal with the exhaust volume flow rates involved. In Europe, where turbines of 3000 rpm are commonly used, quite high outputs are feasible without requiring absurdly large numbers of LP flows or the use of more expensive blade materials, like titanium [1-6]. In America, however, the 60-cycle system dictates a rotor speed of 3600 rpm, and the more severe diameter limits imposed at this speed lead at large outputs to worm-shaped machines like the one sketched in Fig. 1.2.12*a*. Economics dictate here the transition to half speed (1800 rpm) (see Fig. 1.2.12*c*). Reference lists of manufacturing companies (e.g., [1-8]) show that all nuclear turbines installed in the United States since about 1970 run at 1800 rpm, have a single shaft with 4 or 6 exhaust flows, and the majority are equipped with steam reheat. Typical last-stage bucket lengths range from 900 to 1100 mm.

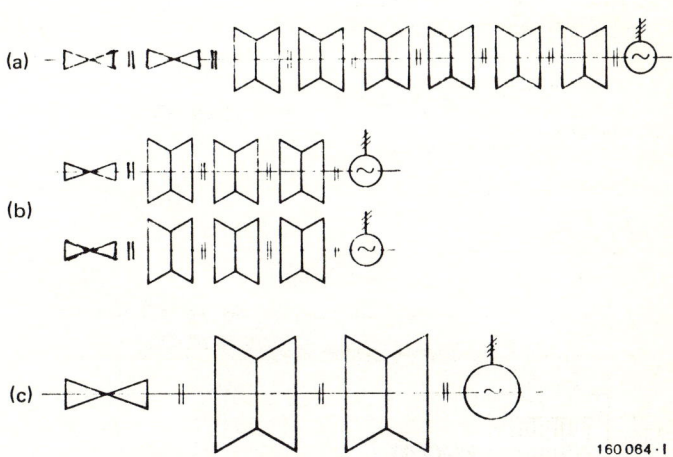

FIG. 1.2.12 Basic arrangements of very large (>1000 MW) nuclear steam turbines [1-9]. (*a*) One turbine of 3600 rpm for the full output. (*b*) Two turbines of 3600 rpm for half output each. (*c*) One turbine of 1800 rpm for the full output.

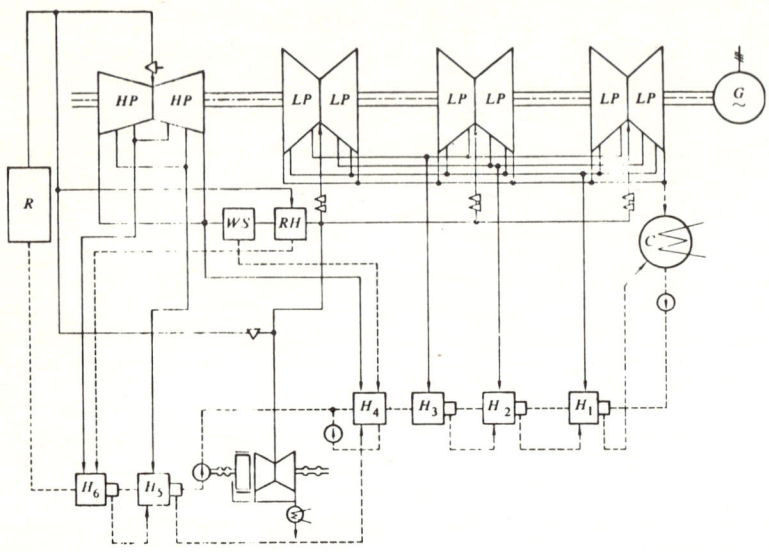

FIG. 1.2.13 Schematic arrangement of an 1160 MW nuclear steam turbine [1-9].

The schematic cycle layout of an 1160 MW nuclear steam turbine operating at 1800 rpm [1-9] is shown in Fig. 1.2.13. The unit comprises a double-flow HP turbine and three double-flow LP turbines. Between the HP and the LP part there is a water separator (WS) and steam reheater (RH). As a result of reheating, the LP turbines operate under conventional conditions. The feed-water heating system comprises six stages, two of which are using steam

FIG. 1.2.14 Model view of an 1160 MW, 1800 rpm nuclear steam turbine [1-9].

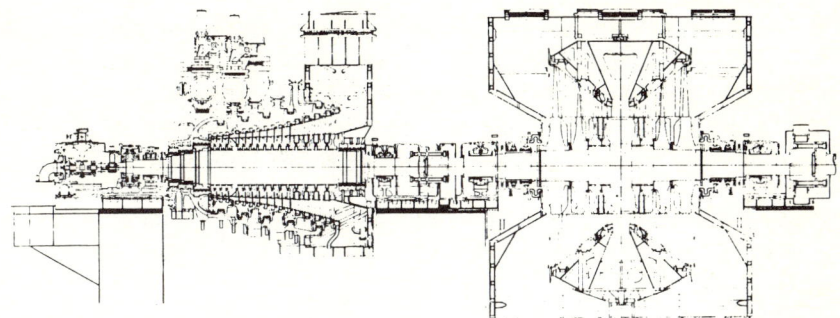

FIG. 1.2.15 HP and LP turbine of the nuclear plant Gundremmingen (237 MW, 70 bar/286°C, no external water separation, no reheat) [1-11].

extracted from the HP turbine. A scale model of the turbine/generator unit is shown in Fig. 1.2.14. Steam admission and HP turbine are seen on the left, three LP casings in the middle, and the electrical generator at the far extreme. In this particular arrangement separators and reheaters are contained in two horizontal cylindrical vessels seen on both sides of the turbine. The HP exit crossover pipes are omitted from the model; they would enter the separator vessel from the bottom.

1.2.2.2 Design of the HP Turbine

While in early units of smaller output the HP turbine usually was of single-flow type (Fig. 1.2.15), at present most HP casings feature a double-flow arrangement (smaller dimensions, compensated thrust). As a first example, Fig. 1.2.16 shows the longitudinal cross section of the 10-stage, double-flow HP cylinder of the 1160 MW turbine shown in Fig. 1.2.14 [1-9]. Moisture is removed through the steam tappings situated after the fourth and

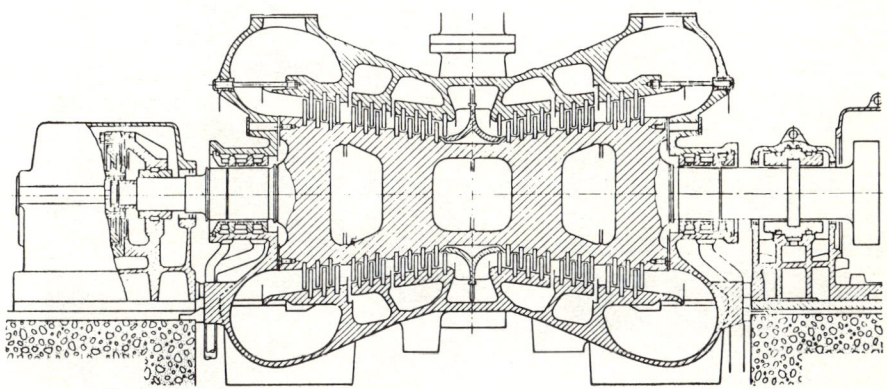

FIG. 1.2.16 Double-flow HP turbine of an 1160 MW, 1800 rpm unit, with 50% reaction blading [1-9].

the seventh stages, and through drainage devices provided at the rotor tips of the three last stages. The blading is of 50%-reaction type. The second example concerns an HP turbine with low-reaction blading (Fig. 1.2.17). The design is of the wheel-and-diaphragm type. A moisture drainage belt is provided after each rotor blade row, and two of the drainage belts are combined with steam tappings for feed-water heating. In some turbines the rotor blades of the last few HP stages have radial grooves near the leading edge in order to facilitate removal of water [1-8].

Peripheral speeds in HP stages vary between 100 and 200 m/s, depending on unit size, rotor speed, and type of blading used.

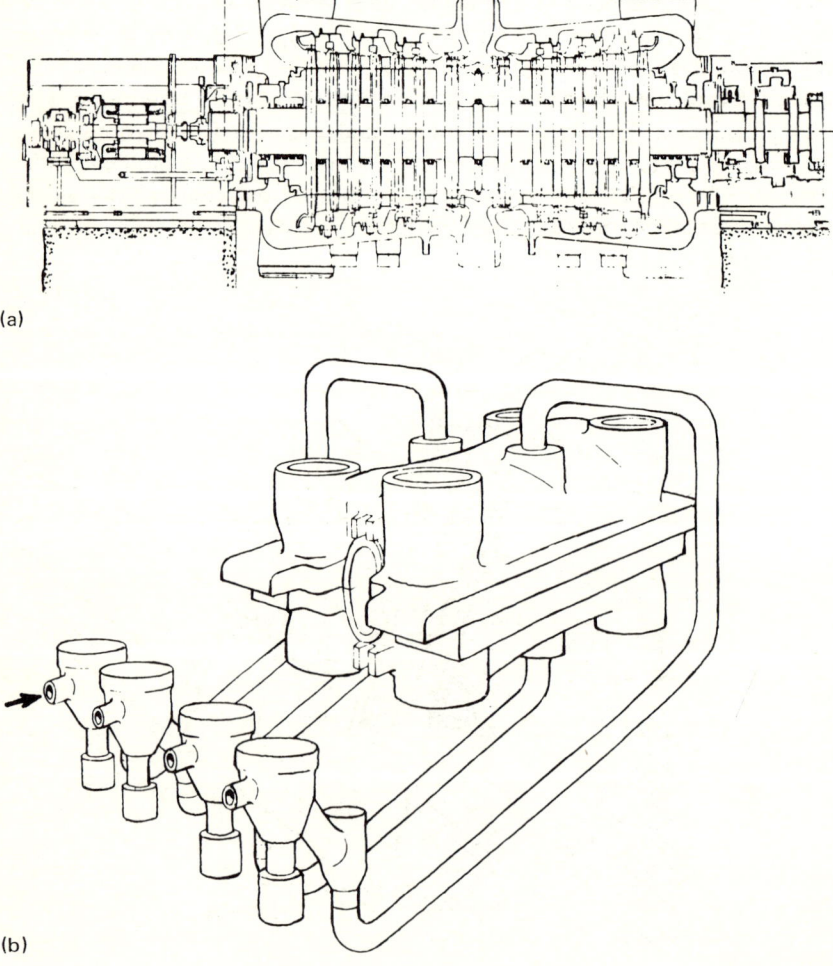

FIG. 1.2.17 Double-flow HP turbine of a 940 MW, 1500 rpm unit, with low-reaction blading [1-5].

1.2.2.3 Process of Expansion in HP Turbines

The evolution of steam conditions in the HP turbine is influenced by numerous factors:

1 inlet pressure p_0 (usually 50–70 bar)
2 inlet superheat (mostly zero) or inlet wetness (usually $< 0.3\%$)
3 presence of secondary steam inlet (unusual)
4 efficiency of HP stages (usually 80–90%)
5 drainage devices and steam tappings
6 exit (cross-over) pressure p_E

Factors 1, 2, and 3 depend on the choice of the reactor system; 4 and 5 are a matter of the turbine manufacturer (and of physics); and 6 is set by economic optimization. In Fig. 1.2.18 a number of typical expansion lines are sketched. For sake of simplicity, moisture removal is ignored in most lines shown (lines a, b, c); its effect is illustrated by the steps indicated in line d. Except for line b all lines begin with dry (or slightly wet) saturated conditions, while line b corresponds to 15°K inlet superheat. Line c shows the case where a considerable amount of secondary (saturated) steam is injected into the

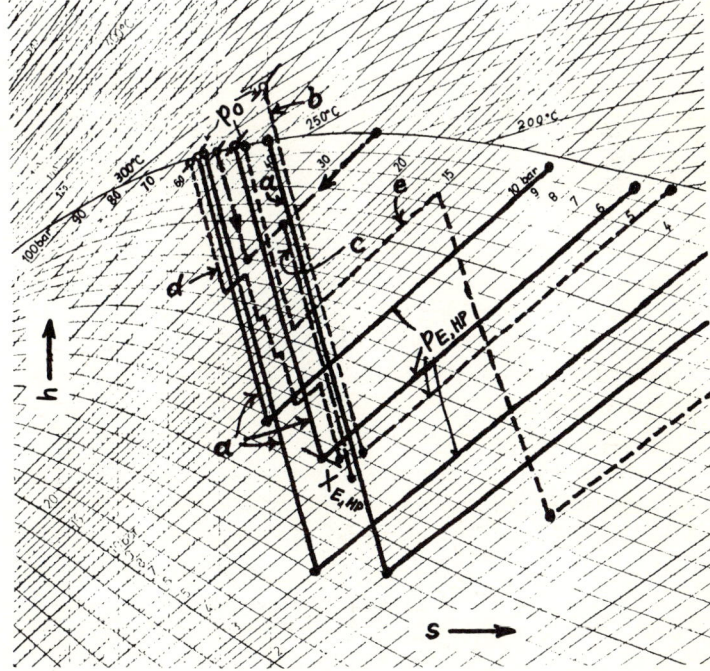

FIG. 1.2.18 Various typical expansion lines of HP wet-steam turbines in the Mollier enthalpy/entropy diagram.

turbine at a pressure level inferior to p_0. The exit pressure p_E can vary over a wide range; its choice depends on many factors like unit size, the kind of water separators used, presence or absence of reheat, etc. The exit wetness fraction $1 - X_E$ is seen to lie between 13 and 17%; real values are somewhat lower, due to water removal through drainage devices and steam tappings.

For HP turbines that are equipped with an internal moisture separator, the expansion line has the shape shown by line e.

The expansion in the HP turbine is virtually always followed by separation of steam and water at constant pressure p_E in an external water separator. The steam quality after separation is close to unity ($X > 0.97$, mostly 0.99 and above). External separation is dealt with in Chap. 7.

1.2.2.4 Effects and Handling of Moisture in HP Turbines

High-pressure wet steam flowing at high speed is an unusually aggressive medium. The steam path and all leakage routes have to be protected against erosion and corrosion. Classical (impact) erosion of the moving blades is fortunately insignificant in high-pressure turbines, because peripheral speeds remain well under dangerous levels. Protective measures other than the choice of highly alloyed (12% Cr) blade steel are not necessary [1-5].

Casing walls however are usually made of low-chromium steel and have to be protected against water sprayed off rotor blade tips in stages where moisture is abundant [1-5, 1-6, 1-8]. This is usually achieved by high-chromium shielding of these parts (Fig. 1.2.19a).

Leakage paths, if not protected by highly alloyed steel, can suffer serious damage as a result of the erosive and corrosive action of the high-speed wet steam blowing through gaps (Fig. 1.2.19b). Manufacturers usually protect exposed surfaces by inlays and coatings of nobler material and pay particular attention to good sealing at all joints. Similar destruction can occur in drainage and extraction piping where flow speed of wet steam becomes locally excessive. Criteria for avoiding corrosion-erosion are presented in Chap. 6.

The efficiency of the HP turbine is considerably affected by the presence of wetness. Figure 1.2.20 shows test results obtained on a laboratory turbine [1-10]. The data suggest that wetness losses amount to several percent in HP turbines. Countermeasures consist in moisture-removal devices in some or all of the HP stages. These devices can comprise peripheral drainage slots, rotor blades with leading-edge grooves, or hollow stator blades equipped with slots for removal of water films by boundary-layer suction. In Chap. 5, wetness losses and moisture removal will be discussed in detail.

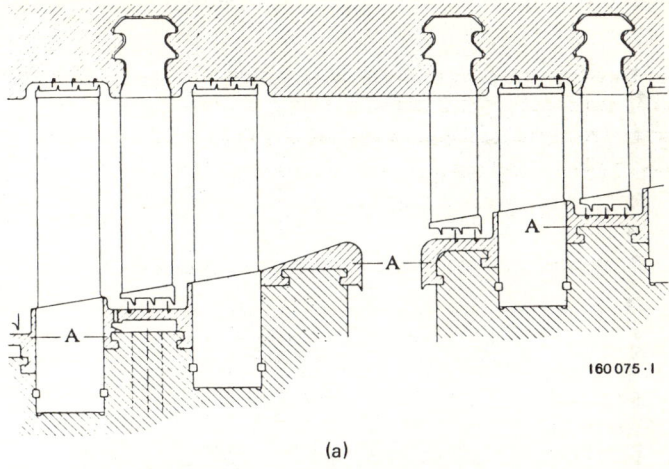

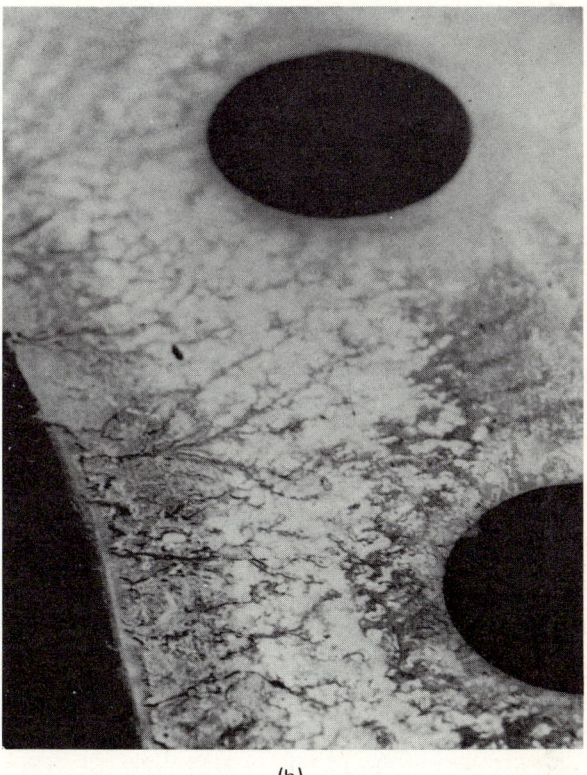

FIG. 1.2.19 Erosion and erosion protection in the HP turbine. (a) Chromium-steel shield rings A for protecting the casing walls [1-9]. (b) Erosion-corrosion due to steam leakage in a carbon-steel flange [1-6].

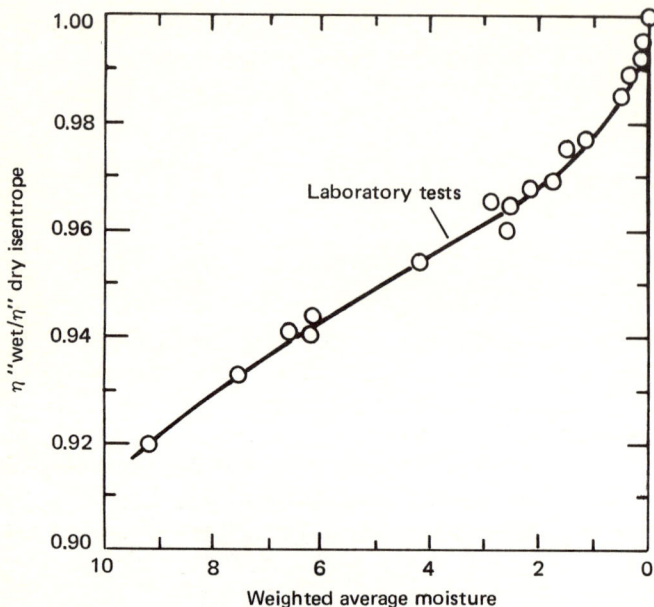

FIG. 1.2.20 Wetness loss in a laboratory HP turbine, as a function of inlet superheat [1-10].

1.3 PROPERTIES OF WET STEAM

1.3.1 Composition, Vapor Pressure

Wet steam is a mixture of pure steam and water. If the total mass of the mixture is m_t and is composed of m_g parts of steam and $m_f = m_t - m_g$ parts of water (index g and f denoting gaseous and liquid phase, respectively), we define:

$$\text{Dryness fraction} \quad X \equiv \frac{m_g}{m_t} \quad (1.3.1)$$

$$\text{Wetness fraction} \quad Y \equiv \frac{m_f}{m_t} = 1 - X \quad (1.3.2)$$

If the mixture is enclosed adiabatically and kept at constant volume, the phases will (by partial evaporation or condensation) tend to reach *thermodynamic equilibrium*, in which both phases have the same pressure and temperature. In this condition, steam and water are said to be *saturated*. Quantities referring to saturated steam and water will be denoted by $''$ and $'$, respectively. It is assumed in this definition that all phase interfaces are plane surfaces, i.e., the vapor is in equilibrium with its "bulk" liquid (and not with droplets).

In the state of saturation, pressure p is uniquely determined by temperature T and vice versa. The relationship

$$p = p_s = p_s(T) \qquad (1.3.3)$$

is called the *vapor pressure* function. The inverse relationship

$$T = T_s = T_s(p) \qquad (1.3.4)$$

gives the *saturation temperature* T_s at given p.

It will be shown later that the conditions of equilibrium expressed by Eqs. (1.3.3) and (1.3.4) are not valid if the liquid is present in form of droplets. Deviations, however, become important only if droplet sizes are below 0.1 μm. In a steam environment that is in equilibrium with respect to bulk liquid, droplets will tend to evaporate. Therefore an adiabatically enclosed mixture consisting of bulk liquid and droplets of various sizes will reach complete equilibrium only when all droplets have evaporated (and the amount of bulk liquid has correspondingly increased).

1.3.2 Physical Data of Steam and Water

1.3.2.1 General Data

Water is a chemically stable[*] substance and is characterized by the following basic data:

Chemical formula: H_2O
Molecular weight: $M_m = 18,015$
Triple point: $t_T = 0.01°C$ ($p_T = 0.0006112$ bar)
Critical point: $t_K = 374.15°C$ ($p_K = 221.20$ bar)

Physical constants derived from the molecular weight are:

Avogadro number: $N_m = 3.34 \cdot 10^{25}$ molecules/kg
Mass of one molecule: $m_m = 1/N_m = 2.99 \cdot 10^{-26}$ kg
Gas constant: $R = 461.5$ kJ/kg · K

In the liquid state (at density $\rho_f = 10^3$ kg/m³) one molecule occupies, on the average, a volume equivalent to a sphere having the

Molecular radius: $r_m = (3 m_m / 4\pi \rho_f)^{1/3} = 1.93 \cdot 10^{-10}$ m

---

[*] Its degree of dissociation is $2 \cdot 10^{-9}$ at room temperature.

1.3.2.2 Equations of State

Of all substances, water is the one with the most accurately and most completely known thermodynamic properties. The equations of state currently in widespread use (International Formulation "IFC 1967") are summarized, tabulated, and charted in [1-12]. A more recent equation of state "MIT 1969," which has a more satisfactory though less practical form, is tabulated in [1-12a].

An overall view of the various phase domains can be obtained from Fig. 1.3.1. Besides the vapor, liquid, and wet-steam regions that are of interest in turbines, the upper diagram also includes the solid phase and its transitional domains. Wet-steam turbines usually operate within the region between the enthalpies $h = 2000$ and $h = 3000$ kJ/kg. Of this portion of the Mollier chart an enlarged view is given in Fig. 1.3.2. The vapor pressure function $p_s(T)$ and the latent heat $\Delta h_{fg} = h'' - h'$ are related by the Clausius-Clapeyron equation

$$\frac{dp_s}{dT} = \frac{\Delta h_{fg}}{(v'' - v')T} \tag{1.3.9}$$

For approximate calculations at low pressures, simplified versions of the equations of state may be resorted to. In the crudest approximation steam is considered as an *ideal gas* and water as an *incompressible liquid*, both with constant specific heats. This leads to the following equations

for steam:
$$pv_g = RT$$
$$h_g = h_{g0} + c_p(T - T_0)$$
$$s_g = s_{g0} + c_p \ln\left(\frac{T}{T_0}\right) - R \ln\left(\frac{p}{p_0}\right)$$

for water:
$$v_f = \frac{1}{\rho_f}$$
$$h_f = h_{f0} + c_f(T - T_0)$$
$$s_f = s_{f0} + c_f \ln\left(\frac{T}{T_0}\right)$$

(1.3.10)

where the constants are chosen, e.g., as follows:

$$T_0 = 273.16°K$$
$$p_0 = 611.2 \text{ N/m}^2$$
$$h_{f0} = s_{f0} = 0$$
$$h_{g0} = \Delta h_{fg0} = 2501.6 \text{ kJ/kg}$$

$$s_{g0} = \frac{\Delta h_{fg0}}{T_0} = 9.1575 \text{ kJ/kg} \cdot \text{K}$$

$$c_f = 4200 \text{ kJ/kg} \cdot \text{K}$$

$$c_p = 1880 \text{ kJ/kg} \cdot \text{K}$$

$$\rho_f = 1000 \text{ kg/m}^3$$

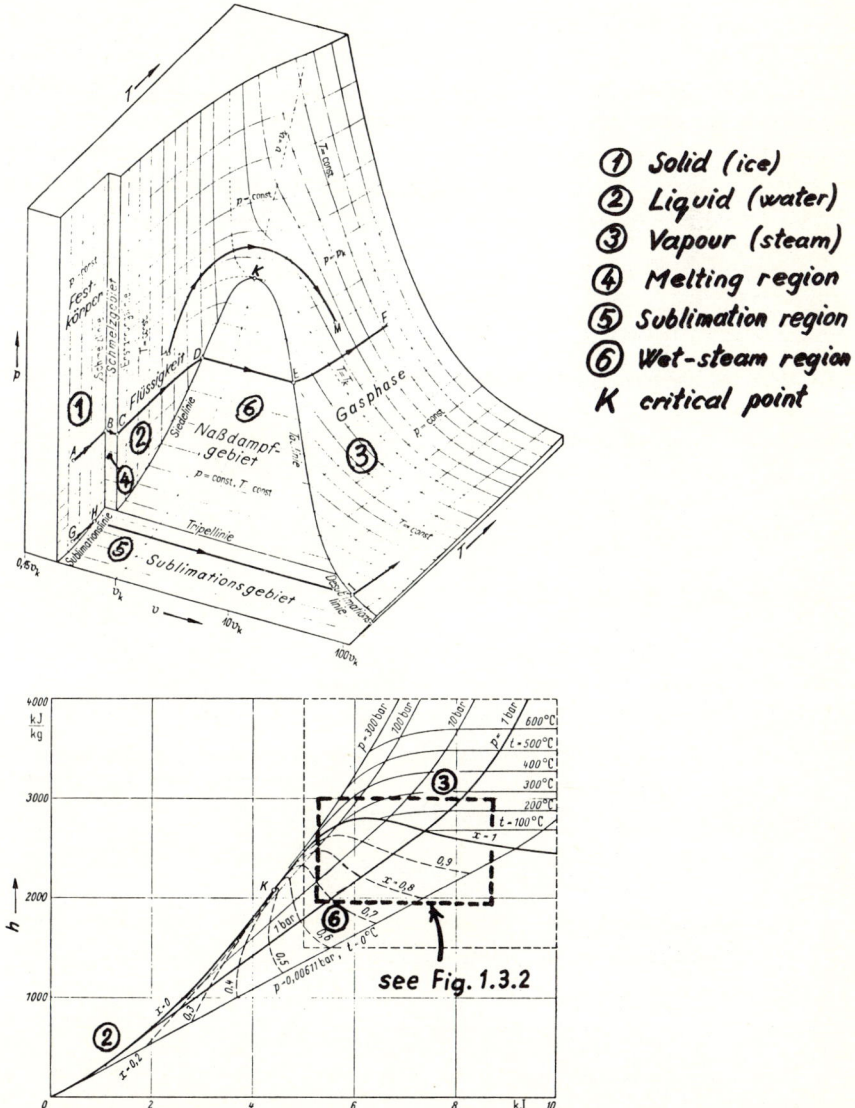

FIG. 1.3.1 Graphical representation of the equation of state of water, from [1-16]. Top: $(p, \log v, T)$ surface (qualitatively drawn). Bottom: Enthalpy-entropy (Mollier) chart.

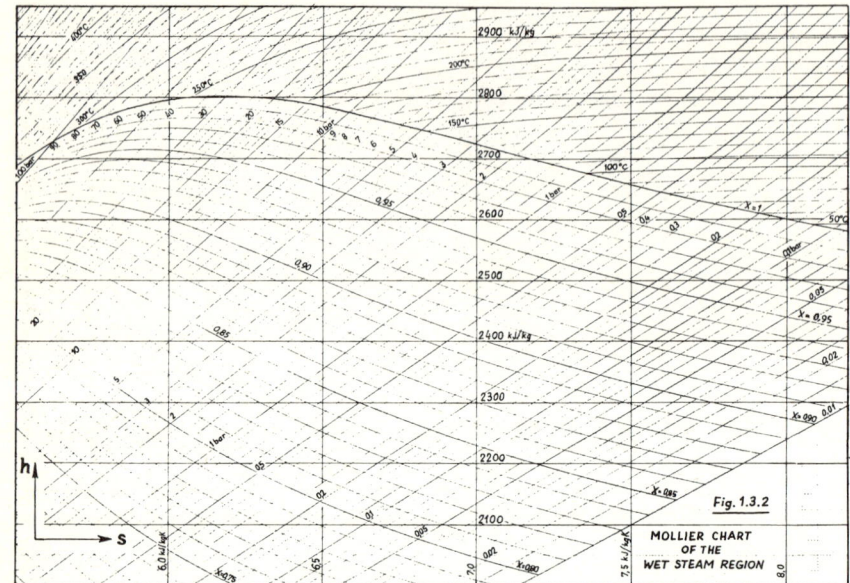

FIG. 1.3.2 Mollier chart of the wet-steam region.

A rough approximation for the vapor pressure is given by

$$\log p_s = 5.55 - \frac{2061}{T} \tag{1.3.11}$$

where the units of p_s and T are bar and degrees Kelvin, respectively. The same expression gives the saturation temperature as a function of pressure, $T_s(p)$.

A still simple but more accurate approximate description of steam is possible by using the equations of state of *ideal vapors* [1-13]. Here a real-gas factor (or "compressibility factor") Z is introduced and is assumed to be a function of entropy only. c_p is treated as variable, but the isentropic exponent n is taken to be constant. The basic equations are

$$pv_g = Z(s) \cdot RT$$

$$h_g = h_0 + \frac{nR}{n-1} ZT \tag{1.3.12}$$

$$\frac{ds_g}{Z(s)} = \frac{nR}{n-1} \frac{dh}{h-h_0} - R \frac{dp}{p}$$

The function $Z(s)$ has to be specified on the basis of empirical data. Empirical values of Z are plotted in Fig. 1.3.3 [1-13]. These show that the factor Z at

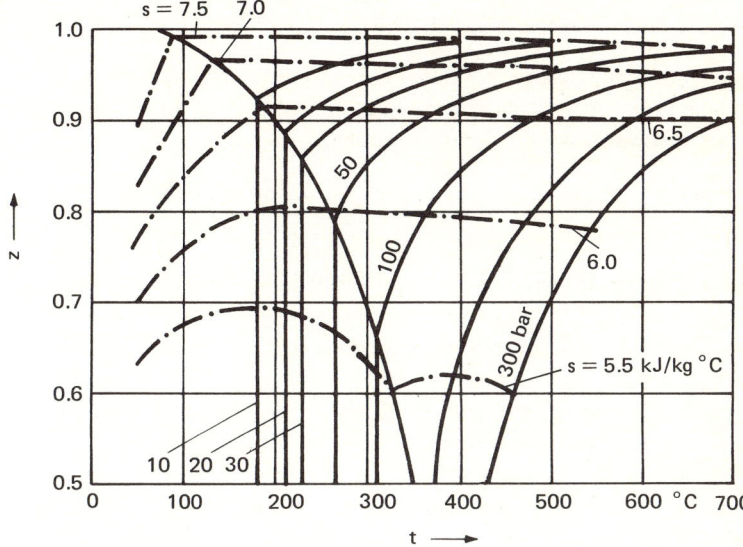

FIG. 1.3.3 Compressibility factor Z for steam [1-13].

low pressures essentially depends only on entropy. The Z values for saturated steam are plotted in Fig. 1.3.4. For numerical calculations requiring higher accuracy, comprehensive computer programs involving the complicated equations of state given in [1-12] are in standard use.

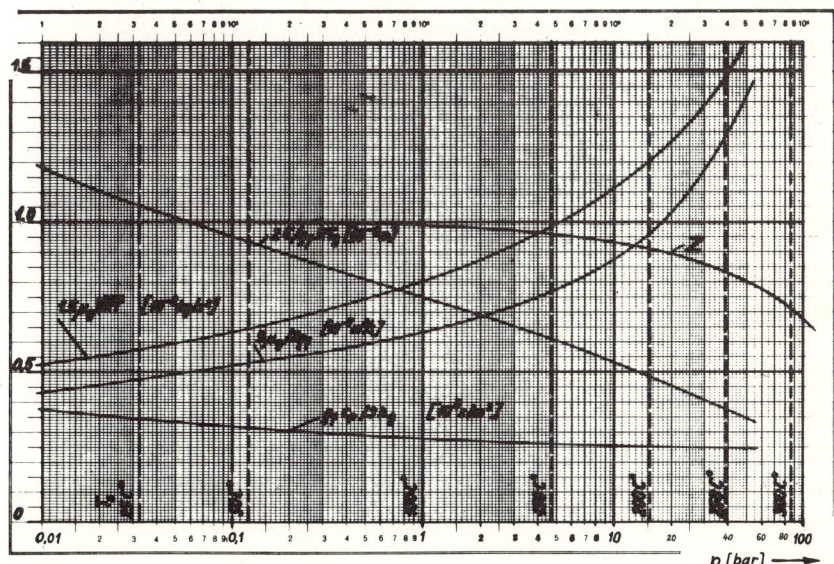

FIG. 1.3.4 Property data for saturated wet steam.

1.3.2.3 Other Property Data

The following property data are essential for the theory of condensing steam flow: specific heat c_p, dynamic viscosity μ, thermal conductivity λ, and their combination, the Prandtl number

$$\text{Pr} = \frac{\mu c_p}{\lambda} \qquad (1.3.13)$$

These data are plotted in Fig. 1.3.5 as a function of pressure, both for saturated liquid (') and for saturated vapor (''). Also included are the values of surface tension σ between water and saturated steam. All data are based on [1-12].

The isentropic exponent k of dry saturated vapor varies from $k = 1.33$ at low pressure (0.03 bar) to $k = 1.26$ at about 60 bar [1-12].

Some further data and data groups occurring in expressions presented in later sections are summarized in Figs. 1.3.4 and 1.3.6. These data are:

specific volume ratio steam/water: $\quad \dfrac{v''}{v'} \equiv \dfrac{\rho'}{\rho''}$

mean free path of molecules in (saturated) steam: $\quad \bar{l}'' \equiv \dfrac{1.5\mu''\sqrt{RT_s}}{p} \qquad (1.3.14)$

real-gas factor of saturated steam: $\quad Z \equiv \dfrac{pv''}{RT_s}$

and a few groups containing μ, ρ, σ, etc.

In wet steam of dryness fraction $X = 0.8$ or above, the liquid phase occupies only a negligible fraction, V_f, of total volume V_t. This is true up to pressures as high as 50 bar; here, from Fig. 1.3.7, $v''/v' \approx 25$ and therefore

$$\frac{V_f}{V_t} = \frac{Yv'}{Xv'' + Yv'} \approx \frac{Y}{X}\frac{v'}{v''} \leqslant \frac{0.2}{0.8}\frac{1}{25} = 0.01$$

The mean free path of steam molecules at atmospheric pressure has the value

$$\bar{l}''_{p=1 \text{ bar}} \approx 10^{-7} \text{m}$$

and varies nearly inversely with p (Fig. 1.3.5).

The optical refractive index of saturated steam is given by the expression

$$n''_{\text{opt}} = 1 + (3.06 \cdot 10^{-4} \text{ m}^3/\text{kg}) \cdot \rho'' \qquad (1.3.15)$$

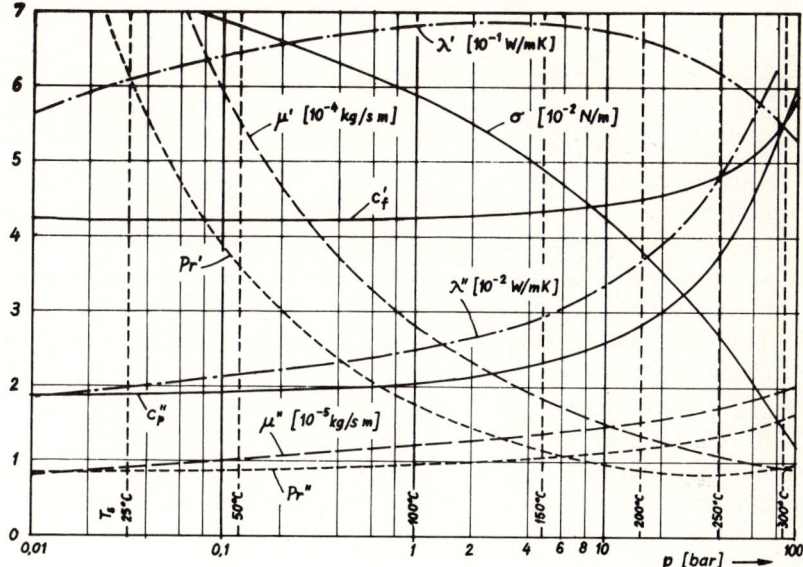

FIG. 1.3.5 Property data for saturated water and steam.

while the values for water are summarized in Table 1.3.1. The electrical conductivity of pure water is very low (about $5.5 \cdot 10^{-8} \, \Omega^{-1} \, \text{cm}^{-1}$ at 25°C [1-15]); however, even minute amounts of dissolved salts can easily cause a 10- to 100-fold increase in this value.

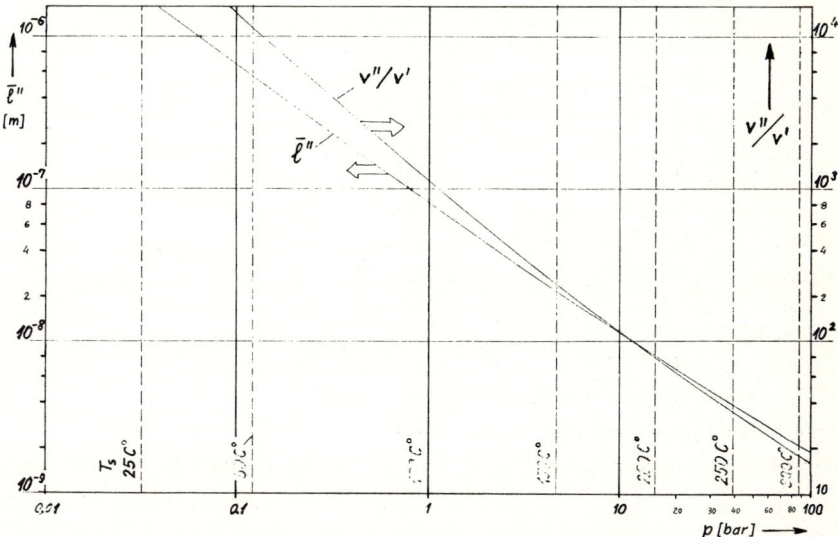

FIG. 1.3.6 Molecular mean free path and specific volume ratio.

TABLE 1.3.1 Refractive index of saturated water [1-14]

t (°C)	15	45	75	100	150	200	250
n'_{opt}	1,334	1,330	1,324	1,318	(1,304)	(1,285)	(1,255)

Typical recommended values of electrical conductivity of feed water in high-pressure, high-temperature power plants is $2 \cdot 10^{-7}$ Ω^{-1} cm^{-1} and less. The same values apply to live steam supplied to the turbines.

Conductivity values of this magnitude indicate a total content of dissolved salts of the order of $m_{salt}/m_{H_2O} = 0.1$ mg/kg $= 10^{-7}$. Estimating the mean molecular weight of the salts present to be $\bar{M}_{m,salt} \approx 80 \approx 4 M_m$, one can calculate the number of salt molecules per kilogram of water (or steam): $N_{m,salt} = N_m \cdot (m_{salt}/m_{H_2O}) M_m/\bar{M}_{m,salt} \approx 3.34 \cdot 10^{25} \cdot 10^{-7} \cdot \frac{1}{4} \approx 10^{18}$.

It will be seen in Sec. 1.3.3 and especially later in Chap. 3 that fogs produced in steam by nucleation contain between 10^{14} and 10^{18} droplets per kilogram. It is therefore conceivable that individual salt molecules do affect the nucleation in some way; however this question has not yet been thoroughly investigated.

1.3.3 Structure of Wet Steam

Real wet steam, especially in turbines, differs considerably from the idealized two-phase systems dealt with in equilibrium thermodynamics. The following facts may play a more or less crucial role:

the state parameters are subjected to fast changes (e.g., during expansion), which may cause supersaturation to occur;

the steam wetness is mostly not bulk liquid (like films), but consists of finely dispersed drops and mist (phase boundaries are not plane surfaces);

the vapor phase is not a uniform, monomolecular substance, but also contains submicroscopic-sized clusters of condensed molecules;

there is relative motion between the phases;

there is heat exchange with channel or container walls;

in the vapor phase nucleation (i.e., spontaneous formation of tiny new droplets) may occur;

existing droplets grow or evaporate or coagulate with each other or impact on walls and become absorbed in bulk liquid films.

The behavior of wet steam under given flow conditions is largely determined by the form in which the wetness is present (film or droplets, and the size or size distribution of the latter) and by the total amount condensed (value of wetness fraction).

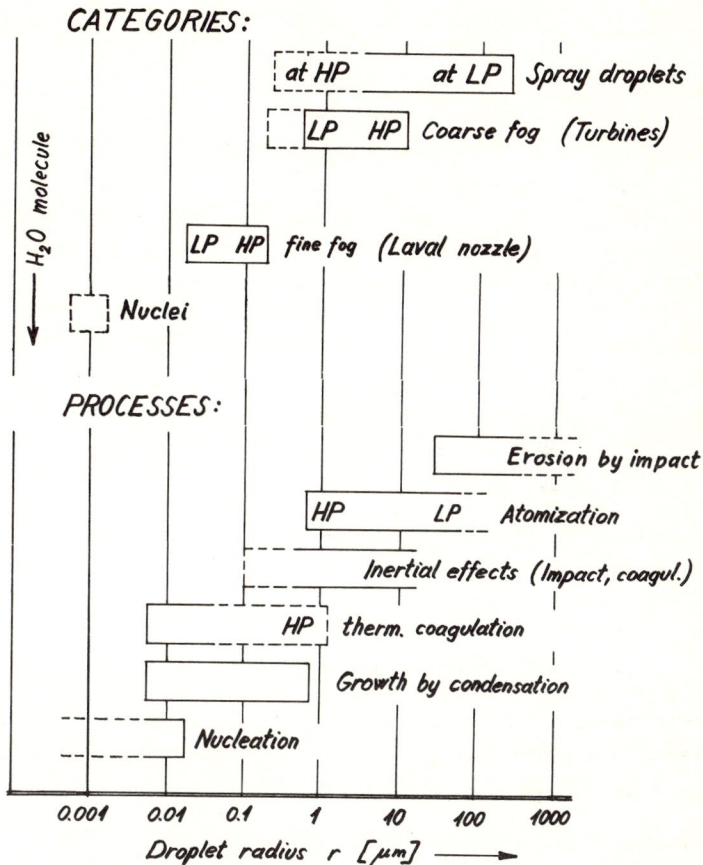

FIG. 1.3.7 Size categories of water droplets and processes they are involved in.

For practical purposes it is useful to classify droplets in steam turbines into categories shown in the upper part of Fig. 1.3.7. In the lower part, some processes are listed in which droplets of a given size preferentially participate. These processes will be discussed in detail in Sec. 1.4.

In steam turbine practice, wetness fractions of $Y = 0$ to 0.15 are encountered. In order to illustrate the spatial dispersion of wetness under various conditions, let us assume that Y is known, all droplets have the same size (radius r) and droplets are uniformly distributed in space. Then, the mass of a droplet of radius r being

$$m = \frac{4\pi}{3} \rho_f r^3 \qquad (1.3.16)$$

the specific number of droplets (number of droplets per unit mass of wet steam) is given by

$$N = \frac{Y}{m} = \frac{3Y}{4\pi\rho_f r^3} \qquad (1.3.17)$$

The volume occupied by unit mass of wet steam being $v \approx Xv''$, the number concentration of droplets (number of droplets per unit volume) is

$$C = \frac{N}{Xv''} = \frac{3Y}{4\pi\rho_f r^3 Xv''} \qquad (1.3.18)$$

Another parameter of interest is the ratio of mean interdrop distance $\bar{D}$ to drop diameter $2r$. Since the average volume occupied by a drop is $1/C$, we have

$$\frac{1}{C} = \frac{\pi}{6}\bar{D}^3 \qquad (1.3.19)$$

and the last two equations yield

$$\frac{\bar{D}}{2r} = \left(\frac{\rho_f v'' X}{Y}\right)^{1/3} \qquad (1.3.20)$$

For a number of droplet sizes and pressures, the values of N, C and $\bar{D}/2r$ have been calculated for the case of $Y = 0.10$ (i.e., $X = 0.90$). The results plotted in Fig. 1.3.8 show that

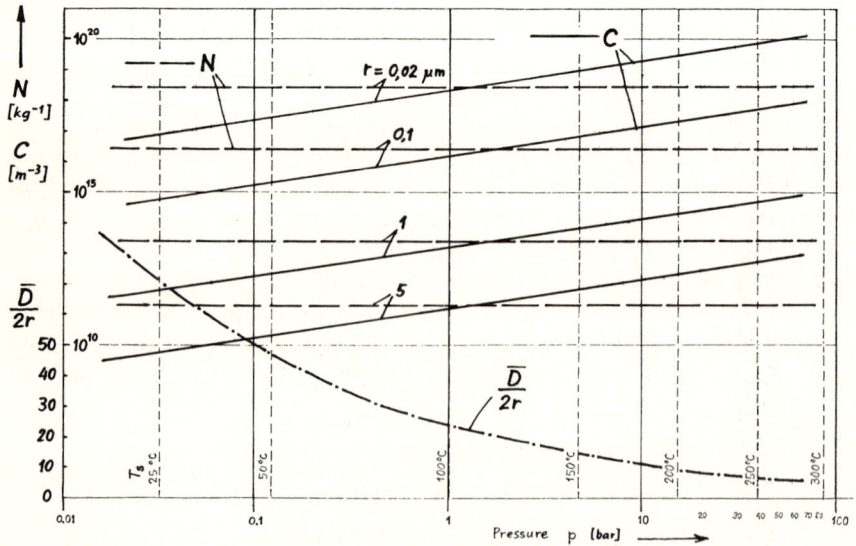

FIG. 1.3.8 Curves of specific droplet number, droplet concentration, and spacing.

TABLE 1.3.2 Number of molecules contained in droplet

r (μm)	0.001	0.002	0.005	0.01	0.1	1
i	142	1140	18,000	$1.4 \cdot 10^5$	$1.4 \cdot 10^8$	$1.4 \cdot 10^{11}$

the distance-to-diameter ratio is independent of droplet size and decreases from about 50 at 0.1 bar to 7 at 50 bars;

the specific number of droplets, which is nearly independent of pressure, has the following typical values:

for fine mists ($r = 0.02 \ldots 0.2$ μm): $N \approx 10^{18}$-10^{15} kg^{-1}
for coarse mists ($r = 0.5 \ldots 10$ μm): $N \approx 10^{14}$-10^{10} kg^{-1}

A final question of interest refers to the number of molecules i contained in a droplet of given size. This number can be obtained from m_m and from Eq. (1.3.16) as

$$i = \frac{m}{m_m} = \frac{4\pi \rho_f r^3}{3 m_m} \qquad (1.3.21)$$

Results calculated with $\rho_f = 1000$ kg/m^3 are summarized in Table 1.3.2.

1.3.4 Supersaturation and Subcooling

In many technical processes wet steam is subjected to fast changes of state (e.g., by expansion, compression, heating, or cooling). In such transient processes time is often insufficient for equilibration of phases, because evaporation and condensation involve heat transfer, the rate of which is limited by various factors. Therefore *nonequilibrium processes* occur, which are characterized by finite temperature differences between and within the phases. Equality of pressure between adjacent parts of the vapor and the liquid is usually maintained, because pressure disturbances propagate at very high (acoustic) speeds. Thus in nonequilibrium wet steam, superheated *or* saturated *or* subcooled steam, and subcooled *or* saturated *or* superheated water can coexist at pressure p. The term *superheated* means $T > T_s(p)$, and *subcooled* (also termed *supersaturated*) means $T < T_s(p)$.

Subcooled steam states and superheated water states are unstable; they occur only during dynamic processes and disappear if equilibrium is established. Superheated steam and subcooled liquid represent stable states typical of monophase systems. The deviations from the saturated equilibrium states are usually expressed by the following parameters:

Subcooling: $\qquad \Delta T = T_s(p) - T_g \qquad (1.3.22)$

Supersaturation ratio: $\qquad S = \dfrac{p}{p_s(T_g)} \qquad (1.3.23)$

Subcooled (i.e., supersaturated) dry steam is described by the same equations of state as superheated steam. This means that the Mollier chart of subcooled dry steam is obtained by extrapolating the curvilinear isobars and isotherms that apply for the superheated range (Fig. 1.3.9). The rectilinear wet-steam isobars are tangent to the dry-vapor isobars at the saturation point (e.g., point B). Due to the instability of supersaturated states, no experimental data exist with regard to the equation of state of subcooled steam. The newer high-precision equations of state, which have a thermodynamically sound structure (Formulation IFC 1967 [1-12] and especially MIT 1969 [1-12a]), are believed to yield accurate results down to enthalpies corresponding to $Y = 0.05$ and maybe more.

ΔT and S can be readily converted into each other. Using Eq. (1.3.11) to relate p_s to T_g and p to T_s, we obtain

$$\log S = \log p - \log p_s = -\frac{2061°K}{T_s} + \frac{2061°K}{T_g} = \frac{2061°K}{T_s T_g} \cdot \Delta T$$

or

$$\ln S = \frac{4746°K}{T_s T_g} \cdot \Delta T = \frac{4746°K}{T_s} \cdot \frac{\Delta T/T_s}{1 - \Delta T/T_s} \quad (1.3.24)$$

This latter relationship is plotted in the upper part of Fig. 1.3.10 for various values of T_s, i.e., for various pressures.

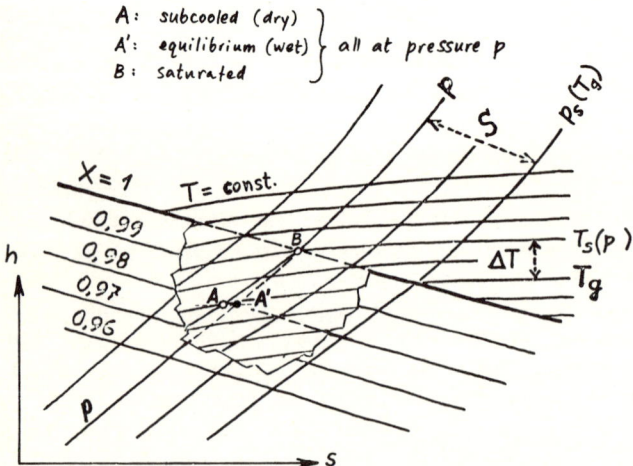

FIG. 1.3.9 Definition of subcooling and supersaturation. (The wet-equilibrium Mollier chart is "peeled off" from part of the subcooled region.) Subcooling in A: $\Delta T = T_s(p) - T_g$. Supersaturation in A: $S = p/p_s(T_g)$.

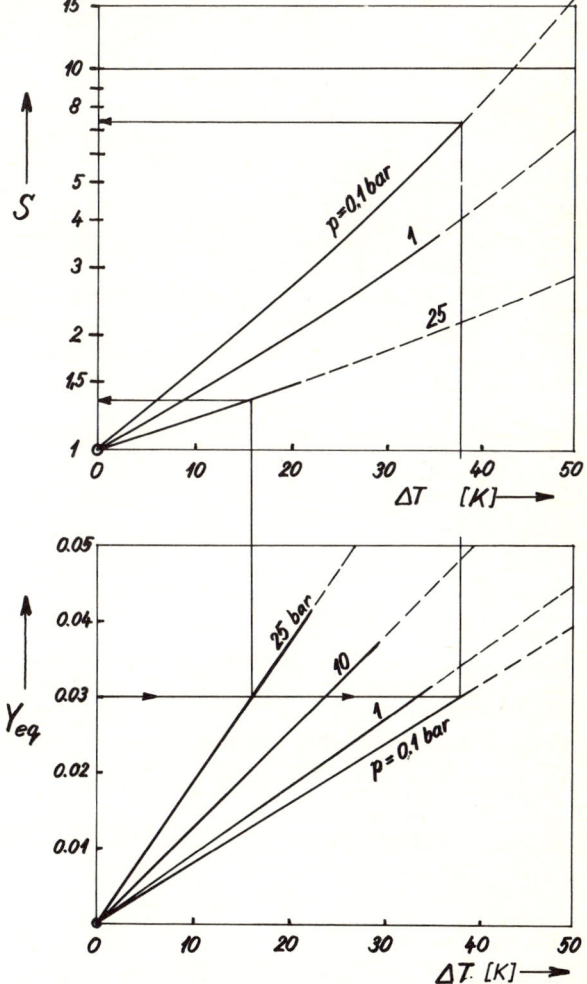

FIG. 1.3.10 The relationship between supersaturation ratio, subcooling, and equivalent wetness fraction.

If a supersaturated state is reversed into an equilibrium one, supersaturation disappears and moisture condenses. If reversion occurs under adiabatic conditions and at constant pressure, the enthalpy h remains constant (because of $dh - vdp = 0$) (points A and A' in Fig. 1.3.9), and the amount of moisture condensing can be easily calculated. For the supersaturated state (A) we have $h = h''(p) - c_p \Delta T$, and for the equilibrium state (A') we have $h = h''(p) - \Delta h_{fg} Y_{eq}$, where Δh_{fg} is the latent heat of condensation and Y_{eq} is the "equivalent" wetness fraction. By equating we get $c_p \Delta T = \Delta h_{fg} Y_{eq}$, or

$$Y_{eq} = \frac{c_p}{\Delta h_{fg}} \Delta T \qquad (1.3.25)$$

This relationship is plotted in the bottom part of Fig. 1.3.10 for various pressure levels.

Adiabatic reversion to equilibrium is accompanied by an entropy rise, because during reversion heat is exchanged between phases that have different temperatures. The magnitude of the entropy rise Δs_{rev} can be calculated for the adiabatic-isobaric reversion from a subcooling ΔT using the ideal-gas equations of state, Eqs. (1.3.10), as follows. In the subcooled state the entropy of steam is

$$s_g = s_{g0} + c_p \ln\left(\frac{T_g}{T_0}\right) - R \ln\left(\frac{p}{p_0}\right)$$

while after reversion to a wet equilibrium state having moisture Y_{eq} the entropy is

$$s_{eq} = s''(p) - \frac{\Delta h_{fg} Y_{eq}}{T_s}$$

$$= s_{g0} + c_p \ln\left(\frac{T_s}{T_0}\right) - R \ln\left(\frac{p}{p_0}\right) - \frac{\Delta h_{fg}}{T_s} Y_{eq}$$

Therefore we obtain

$$\Delta s_{rev} = s_{eq} - s_g = c_p \ln\left(\frac{T_s}{T_g}\right) - \frac{\Delta h_{fg}}{T_s} Y_{eq}$$

or by making use of Eqs. (1.3.22) and (1.3.25)

$$\frac{\Delta s_{rev}}{c_p} = -\ln\left(1 - \frac{\Delta T}{T_s}\right) - \frac{\Delta T}{T_s} \qquad (1.3.26)$$

This relationship is plotted in Fig. 1.3.11. As a typical case, let us consider reversion from $\Delta T = 40°$K at $p = 1$ bar (where $T_s = 373°$K and $c_p = 2.0$ kJ/kg · K). For $\Delta T/T_s = 40/373 = 0.107$ the plot gives $\Delta s_{rev}/c_p = 6.1 \cdot 10^{-3}$ and therefrom we have $\Delta s_{rev} = 0.0122$ kJ/kg · K.

This entropy rise means a loss of useful work of about $\Delta W = T_{ambient} \cdot \Delta s_{rev} \approx 300 \cdot 0.0122 = 3.7$ kJ/kg. The isentropic heat drop from saturation to 40°K supercooling (or $Y_{eq} = 0.035$) at one bar is, according to Fig. 1.3.2, about 110 kJ/kg. The loss corresponds to about 3.5%. This example shows that as reversion occurs in a steam turbine, the efficiency of the stage or stages involved can be appreciably affected. A loss of this type in

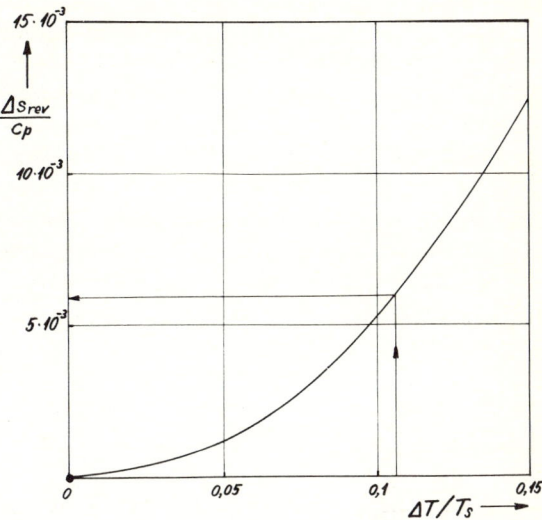

FIG. 1.3.11 Entropy rise associated with reversion.

wet-steam turbines is termed *thermodynamic loss*. In [1-17] experimental values of this loss are reported to be in good agreement with theory.

1.4 BEHAVIOR OF WET STEAM

In this section physical phenomena associated with wet steam are reviewed and discussed. The mathematical expressions describing these phenomena play an important role in calculations concerning wet-steam flow treated in the other sections.

Although wet steam comprises a multitude of droplets, most phenomena involving these droplets (except coagulation) can be examined by considering a single droplet in an infinite vapor environment. While this simplification is well justified in the low-pressure turbine, it is of doubtful validity at high pressure.

1.4.1 Thermodynamics of Steam/Droplet Mixtures

1.4.1.1 Critical Droplet Size

The vapor pressure of a convex (curved) liquid surface is larger than that of a plane one. Therefore in saturated steam, droplets tend to evaporate. The vapor pressure p_r of a droplet of temperature T_r and of radius r is given by the Kelvin-Helmholtz equation as

$$p_r = p_s(T_r) \exp\left(\frac{2\sigma}{r\rho_f R T_r}\right) \qquad (1.4.1)$$

where σ is the surface tension. It is assumed here that σ is independent[*] of r and that the vapor is an ideal gas. Since the exponent is positive, p_r is always greater than p_s. Numerical values of the group $(2\sigma/\rho_f RT)$ plotted in Fig. 1.3.4 being of the order of 10^{-9} m, the exponential factor becomes significantly different from unity only for very small droplets ($r < 10^{-8}$ m).

If the drop is surrounded by a supersaturated vapor atmosphere so that $p = p_r$ and $T_g = T_r$, it is in equilibrium. The equilibrium is unstable, however, because smaller drops (with $p_r > p$) tend to evaporate and larger ones (with $p_r < p$) tend to grow. For a given supersaturation ratio $S = p/p_s$, Eq. (1.4.1) gives the radius of the unstable droplet, called "critical droplet size," as

$$r_{\text{crit}} = \frac{2\sigma/\rho_f RT_g}{\ln S} = \frac{T_s}{4746°K} \frac{2\sigma/\rho_f RT_g}{\Delta T/T_g} \qquad (1.4.2)$$

where the second equality is based on Eq. (1.3.24) and holds only for steam. At saturation ($S = 1$), r_{crit} is infinite, as it must be for a plane surface. At significant supersaturations ($S \geqslant 2$), r_{crit} is of the order of 10^{-9} m = 0.001 μm. Table 1.3.2 shows that such critical droplets contain only about a hundred molecules. In Fig. 1.4.1, r_{crit} is plotted over the pressure for various values of the subcooling ΔT.

[*]For droplets of $r < 10^{-9}$ m this assumption may be in error. However at present the nature of the dependence $\sigma(r)$ is still a controversial issue [1-18a] and will not be further considered in the present review.

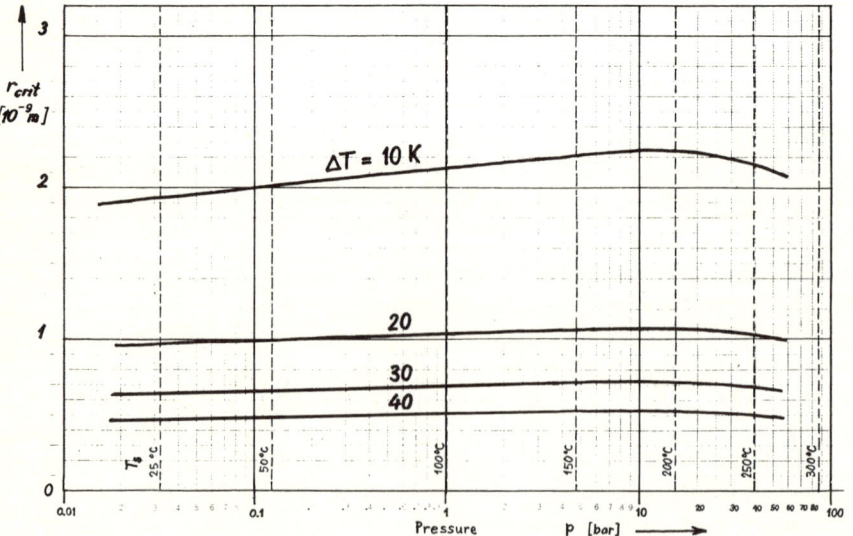

FIG. 1.4.1 Curves of r_{crit} for various ΔT values, calculated from Eq. (1.4.2). (Assumption: vapor is an ideal gas.)

A drop having a size $r \neq r_{crit}$ can be shown (cf. Sec. 3.3.2.2) to tend to maintain its surface temperature at a value

$$T_r = T_s(p) - \Delta T \frac{r_{crit}}{r} \qquad (1.4.3)$$

For the plane surface this formula yields, as required, a surface temperature equal to $T_s(p)$, and for a critical-sized droplet $T_r = T_g$. Subcritical droplets assume a temperature colder than T_g, and tend therefore to evaporate.

1.4.1.2 Molecular Clusters in the Vapor Phase

According to kinetic theory, the molecules of the vapor ("monomer" gas) are in continuous disorderly motion, and their speeds of flight are statistically distributed according to Maxwell's law. Among the many collisions that are constantly occurring among monomer molecules, in some instances low-speed molecules will meet, and intermolecular cohesion forces may cause them to stick together into a dimer, i.e., a cluster of $i = 2$ molecules. Trimers ($i = 3$), etc., also occur. In statistical thermodynamics the equilibrium number of i-mers in unit mass of gas is given by Boltzmann's law as

$$N_{i,\,eq} = N_m \exp\left(-\frac{\Delta G_i}{kT}\right) \qquad (1.4.4)$$

where $k = 1.38 \cdot 10^{-23}$ J/°K is the Boltzmann constant and where ΔG_i is the change in Gibbs free enthalpy when a number i of molecules is converted into an i-mer cluster. The magnitude of ΔG_i and its variation with i is strongly dependent on the value of supersaturation S in the vapor. Therefore, the equilibrium distribution of clusters changes rapidly with S. Figure 1.4.2, based on [1-19], shows the variation of $\Delta G_i/kT$ at $T = 273°$K and $S = 4$, together with the pertinent equilibrium distribution of i-mers. The latter is shown also for saturated conditions ($S = 1$). It is seen that in saturated steam polymers are extremely scarce. At $S = 4$, however, even clusters of rather large size become reasonably probable. The minimum of the distribution curve ($i = i_{crit}$) coincides with the critical droplet size. The distribution curves as plotted correspond to an equilibrium situation that is never truly achieved in nature; in reality, the numbers of relatively large clusters ($i > i_{crit}$) remain much smaller than the equilibrium numbers.

When initially saturated vapor becomes supersaturated, there begins an increased formation and growth of clusters, and some of the clusters fortuitously grow beyond the metastable critical size into stable sizes. This process is called "nucleation" (or spontaneous droplet formation), and will be treated in more detail in Chap. 3.

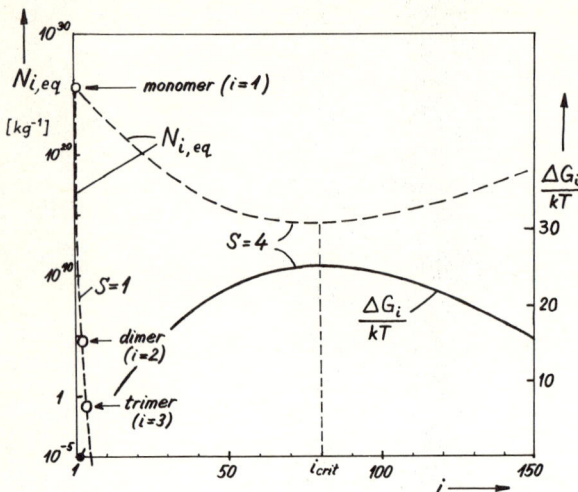

FIG. 1.4.2 Equilibrium concentration of molecular clusters as a function of cluster size.

1.4.2 Heat Transfer in Wet Steam

1.4.2.1 Temperature within Droplets; Flashing

The temperature within a droplet is determined by heat conduction to or from the surface, where temperature is influenced by the ambient vapor conditions. An appreciable difference between center and surface temperature can occur only during rapid changes of the ambient conditions. In turbine flow, such changes are usually slow enough to allow small droplets ($r < 1$ μm) to adapt their inner temperature without appreciable delay. Larger drops, however, will have inner temperatures lagging behind the surface values.

If wet steam is subjected to sudden expansion, pressure and $T_s(p)$ decrease and so do droplet surface temperatures, see Eq. (1.4.3) [here the second term remains essentially constant, as apparent from Eq. (1.4.2)]. The interior of larger droplets is still warm, but the pressure in these droplets has fallen. Cavitation boiling may occur, resulting in an explosion and fragmentation of the droplet. This phenomenon is called *flashing*.

In [1-18] an expression is derived for the maximum size r_{flash} of droplets which escape flashing when subjected to a pressure drop from p_1 to p_2 in a time interval Δt. It is assumed that droplet surface temperature decreases with time linearly, that the droplets are initially at saturation temperature $T_s(p_1)$ and that flashing occurs if the center of the drop becomes superheated by 5°K or more. Since the amount of superheat required for flashing is not exactly known, the formula gives only a rough estimate of the stable drop size. One obtains

$$r_{\text{flash}} = \sqrt{\left(\frac{\lambda_f}{\rho_f c_t}\right) \frac{\Delta t}{G_B}} \qquad (1.4.5)$$

where $\lambda_f/\rho_f c_f \approx 1.6 \cdot 10^{-7}\, \text{m}^2/\text{s}$ for water, and G_B is determined from the parameter

$$g_B = \frac{0.25}{\ln(p_1/p_2)} \qquad (1.4.6)$$

using Fig. 1.4.3. If $g_B > 1$, the pressure drop is insufficient to provoke flashing.

For example, expansion by $p_1/p_2 = 2$ in $\Delta t = 5$ ms. This gives $g_B = 0.25/\ln 2 = 0.36$, from which $G_B = 0.45$ and $r_{\text{flash}} = 1.6 \cdot 10^{-7} \cdot 5 \cdot 10^{-3}/0.45 = 4.2 \cdot 10^{-5}\, \text{m} = 42\, \mu\text{m}$. We see that flashing is limited in such expansions to rather large droplets.

1.4.2.2 Heat Transfer between Vapor and Droplets

In heat transfer the heat-carrying media are usually regarded as continua. This simplification may not be made when heat transfer to very small droplets is considered, because the molecular structure of the vapor becomes noticeable. Whether the vapor behaves with regard to a droplet as a continuum or on the other extreme as a free-molecular gas depends on the value of the Knudsen number, which is defined as

$$\text{Kn} \equiv \frac{\text{mean free path of vapor molecules}}{\text{diameter of droplet}} = \frac{\bar{l}}{2r} \qquad (1.4.7)$$

where $\bar{l}$ is given, for saturated steam ($\bar{l}''$), by Eq. (1.3.14) and its numerical values are plotted in Fig. 1.3.6. Below $\text{Kn} \approx 0.01$ there is continuum flow; above $\text{Kn} \approx 4.5$ free molecular conditions prevail; in between there is a broad transition regime. Most fog droplets studied in steam turbines fall into the transition regime, especially at low pressure.

Since small droplets have small inertia, they do not acquire large relative velocities (slip velocities) with regard to the vapor in which they are entrained,

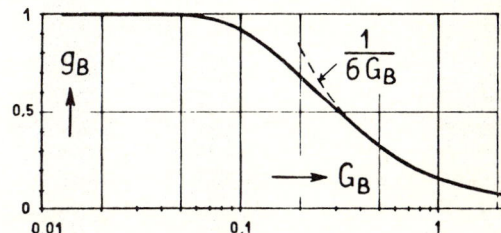

FIG. 1.4.3 The function $g_B(G_B)$ characterizing droplet flashing.

provided that no discontinuities (shock waves) are present in the flow field. Therefore in steam turbines high Kn numbers will only occur in combination with low droplet Reynolds and Mach numbers. The latter are defined with the relative velocity w_r between droplet and vapor as

$$\text{Re}_r = \frac{w_r r \rho_g}{\mu_g} \qquad (1.4.8)$$

$$\text{M}_r = \frac{w_r}{a_g} \qquad (1.4.9)$$

At 1 bar pressure, the value $\text{Re}_r = 1$ is achieved, e.g., for droplets having $r = 0.1$ μm $= 10^{-7}$ m radius, at a relative speed of 200 m/s. For small Re_r and M_r ($\text{Re}_r \leqslant 1$, $\text{M}_r \leqslant 0.1$), the heat transfer coefficient α_r of a sphere [1-18, 1-20] can be expressed as

$$\alpha_r = \frac{\lambda_g}{r} \frac{1}{1 + \dfrac{2\sqrt{8\pi}}{1.5 \, \text{Pr}_g} \dfrac{\kappa}{\kappa + 1} \dfrac{\text{Kn}}{a_{\text{th}}}} \qquad (1.4.10)$$

where λ_g is the thermal conductivity of the vapor, Pr_g is the vapor Prandtl number defined by Eq. (1.3.13) and $\kappa = c_{pg}/c_{vg}$ is the ratio of specific heats. The coefficient $a_{\text{th}} = (T_g^* - T_g)/(T_r - T_g)$ describes the thermal accommodation of vapor molecules rebounding from the droplet surface (T_g^* denotes the temperature associated with the average kinetic energy of rebounded molecules); for liquid/vapor interaction, $a_{\text{th}} \approx 1$. The above equation may be written for steam, taking $\text{Pr}_g = 1.2$, $\kappa = 1.3$, and $a_{\text{th}} = 1$, as

$$\alpha_r = \frac{\lambda_g}{r + 1.59 \, \bar{l}} \qquad (1.4.10a)$$

The heat transferred in unit time from a droplet to the surrounding vapor is expressed as

$$\dot{Q} = 4\pi r^2 \cdot \alpha_r (T_r - T_g) \qquad (1.4.11)$$

where the droplet surface temperature T_r is given by Eq. (1.4.3).

For larger droplets, which can have nonnegligible slip, only continuum situations are of interest. For spherical droplets, theoretical and experimental values [1-21, 1-22] may be approximated up to $\text{Re}_r \text{Pr}_g = 1000$ by

$$\alpha_r = 0.357 \frac{\lambda_g}{r} \sqrt{\text{Re}_r \text{Pr}_g + 7.96} \qquad (1.4.12)$$

This expression gives $\alpha_r = \lambda_g/r$ for the quiescent case, and about ten times more for $\mathrm{Re}_r \cdot \mathrm{Pr}_g = 1000$.

1.4.2.3 Heat Transfer to Walls

In supersaturated or wet-steam flows with stagnation conditions lying below the saturation curve, heat transfer to walls is greatly enhanced as compared to an equivalent noncondensing flow. The reason for this lies in the fact that walls become wetted by liquid droplets or films deposited on them by impact or condensation. These liquid layers assume a surface temperature equal to the saturation temperature $T_s(p)$ pertinent to local pressure. The complex heat transfer problems arising in the presence of liquid films have been analyzed in [1-23].

1.4.2.4 Thermal Lag Effects in Wet Steam

Time constants characterizing the response of a heterogeneous system, like wet steam, to external change of its parameters play a key role in determining deviations from equilibrium. If the time constants of the system are small with respect to the time scale of imposed change, conditions will stay near to equilibrium; in the opposite case processes will be "frozen," i.e., no phase transitions will occur.

For characterizing the thermal behavior of wet steam during expansion processes, two time constants are of particular interest:

a. Removal of subcooling (i.e., reversion to equilibrium) at constant pressure in a mixture of subcooled steam and (uniform-sized) droplets:

$$\Delta t_{\mathrm{rev}} \equiv \frac{\Delta T_{t=0}}{(-d\,\Delta T/dt)_{t=0}} \qquad (1.4.13)$$

b. Equalization of internal temperature differences within a droplet:

$$\Delta t_{\mathrm{int}} \equiv \frac{(T_c - T_r)_{t=0}}{[-d(T_c - T_r)/dt]_{t=0}} \qquad (1.4.14)$$

where T_c is the temperature at the center of the droplet.

For Δt_{rev}, an expression can be derived by considering that the heat addition rate to the steam phase at constant pressure can be described as $-(1-Y)c_p\,d\Delta T/dt = N\dot{Q}$ where N is the specific number of droplets and $\dot{Q}$ is the heat transfer rate from one droplet. By introducing, instead of N, the wetness fraction $Y = N\,m$ and using Eq. (1.4.11), Eq. (1.4.13) gives

$$\Delta t_{\mathrm{rev}} = (1-Y)\frac{c_p\,\Delta T}{N\dot{Q}} = \frac{(1-Y)c_p\,\Delta T\,m}{Y 4\pi r^2 \alpha_r (T_r - T_g)}$$

By considering only cases in which droplets are much larger than the critical size, we may set $T_r = T_s(p)$ and therefore $T_r - T_g = \Delta T$. With Eqs. (1.3.16) and (1.4.10a) we obtain

$$\Delta t_{\text{rev}} = \frac{(1-Y)4\pi r^3 \rho_f c_p}{3Y4\pi r^2 \lambda_g}(r + 1.59\bar{l})$$

or

$$\Delta t_{\text{rev}} = \frac{1-Y}{Y}\frac{\rho_f c_p}{3\lambda_g} r(r + 1.59\bar{l}) \tag{1.4.15}$$

The property data group can be read from Fig. 1.3.4. In Fig. 1.4.4, Δt_{rev} has been plotted as a function of droplet size for steam/droplet mixtures having a wetness fraction of $Y = 0.05$, at three different pressure levels.

As for Δt_{int}, nonstationary heat conduction within a droplet has to be considered. From the analytic solution of the pertinent differential equation one finds [1-24]

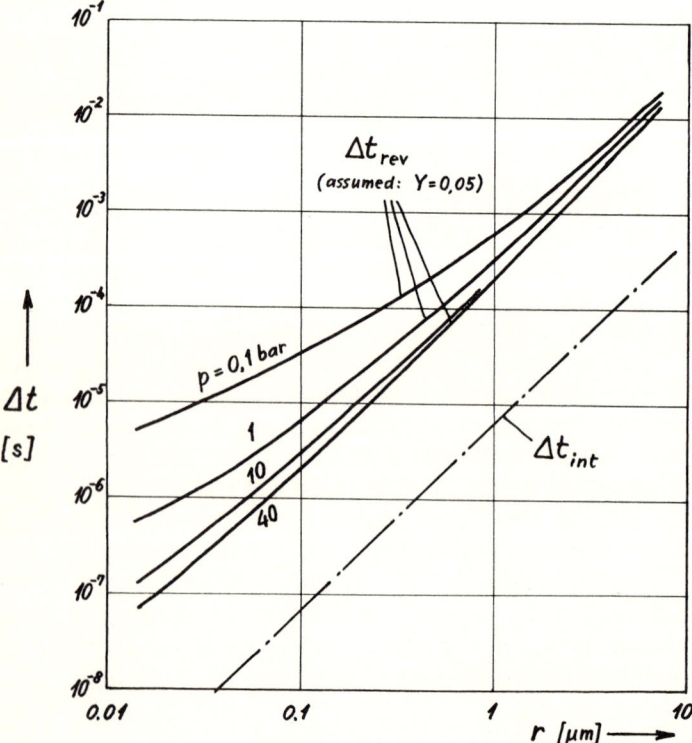

FIG. 1.4.4 Time constants of thermal effects.

Basic Notions

$$\Delta t_{int} \approx \frac{\rho_f c_f}{\lambda_f} r^2 \qquad (1.4.16)$$

The property group has the value $\rho_f c_f/\lambda_f \approx 6.2 \cdot 10^6$ s/m² for water. Also Δt_{int} is plotted in Fig. 1.4.4.

1.4.3 Mechanics of Steam/Droplet Mixtures

1.4.3.1 Drag Forces; Terminal Settling Speed

If the relative velocity w_r of a droplet with regard to the surrounding steam is not zero, a drag force given by

$$F_D = c_D(\pi r^2) \frac{1}{2} \rho_g w_r^2 \qquad (1.4.17)$$

is exerted on the droplet. Droplets are assumed to be spherical. (The limits of validity of this assumption will be discussed in Sec. 1.4.3.4.)

The drag coefficient c_D depends on the Knudsen number in a similar way as α_r (Sec. 1.4.2). The following expression [1-18] is valid for small droplets ($Re_r \leqslant 1$, $M_r \leqslant 0.1$):

$$c_D = \frac{24}{Re_r} \frac{1}{1 + 2.70\,Kn} = \frac{24\mu_g/\rho_g w_r}{r + 1.35\bar{l}} \qquad (1.4.18)^*$$

An expression for c_D, which also covers high Reynolds and Mach numbers, is derived in [1-25]; in the low-Re, low-M limit both expressions give results identical within 10%. From Eqs. (1.4.17) and (1.4.18)

$$F_D = \frac{6\pi\mu_g r^2 w_r}{r + 1.35\bar{l}} \qquad (1.4.19)$$

At higher Reynolds numbers ($0 < Re_r < 5000$) and under continuum flow conditions, c_D can be represented [1-26] as

$$c_D = 0.292 \left(\frac{9.06}{\sqrt{Re_r}} + 1\right)^2 \qquad (1.4.20)$$

A practical way to visualize the importance of drag forces in droplet motion is to determine the terminal "settling" speed w_{rt} of droplets in a gravitational or centrifugal force field of field acceleration g. At terminal speed steady motion

---

*In the reference the factor ahead of Kn is given with the incorrect value of 2.53 instead of 2.70.

being attained, $F_D = mg$, and therefore, for small droplets, Eq. (1.4.19) gives

$$\frac{6\pi\mu_g r^2 w_{rt}}{r + 1.35\bar{l}} = \frac{4\pi}{3} r^3 \rho_f g$$

from which the terminal settling speed follows as

$$w_{rt} = \frac{r(r + 1.35\bar{l})}{9\mu_g/2\rho_f} g \qquad (1.4.21)$$

The property group in the denominator is plotted in Fig. 1.3.4.

As an example let us consider turbine flow at $p = 0.1$ bar ($\bar{l} \approx 10^{-6}$ m, $9\mu_g/2\rho_f \approx 0.5 \cdot 10^{-7}$ m^2/s) at a radius $r_{\text{flow}} = 0.5$ m with tangential speed $C_\theta = 300$ m/s; we then have $g = C_\theta^2/r_{\text{flow}} = 1.8 \cdot 10^5$ m/s^2. Droplets of, e.g., $r = 0.1$ μm $= 10^{-7}$ m have a terminal speed $w_{rt} = (10^{-7} + 1.35 \cdot 10^{-6})$ $10^{-7} \cdot 1.8 \cdot 10^5/0.5 \cdot 10^{-7} = 0.52$ m/s. Despite the very strong centrifugal force field, these small droplets have a very low terminal speed and therefore cannot be centrifuged from the steam flow.

1.4.3.2 Coagulation

If droplets approaching each other in free flight touch, several possibilities exist: they may bounce off; they may coalesce, oscillate, and fall to pieces again; or they may coalesce and stay together. Apart from geometric factors (eccentricity of impact), the size and relative speed of the droplets play an important role in determining which event is going to occur [1-27, 1-28] (Fig. 1.4.5). Extrapolation of experimental data obtained with drops of $d \approx 1$ mm at $w_c \approx 1$ m/s to smaller droplet sizes and higher collision velocities w_c, according to the tentative assumption $\rho_f w_c^2 d/\sigma = $ constant (which means equal ratio of the kinetic and surface energies involved), suggests that droplets of 1 μm diameter are likely to coalesce if their collision velocity is below about 20-30 m/s. For still smaller droplets it is usually assumed [1-29] that each collision results in coalescense.

The collision of two droplets is in general a matter of coincidence (as implied by this very word), and obeys statistical laws. An exception exists in steam turbine flow in one instance: when a relatively large drop (dia. D_t) moves through steam containing many smaller droplets (of radius r). In this case the large drop scavenges out some of the small droplets in its path. The efficiency of this process is given by the "impaction (or collision) efficiency" E defined in Fig. 1.4.6a. E depends essentially on two dimensionless parameters: the "inertial impaction parameter"

$$K = \frac{2\rho_f}{9\mu_g} \frac{r^2 w_c}{D_t/2} (1 + 2.70 \text{ Kn}) \qquad (1.4.22)$$

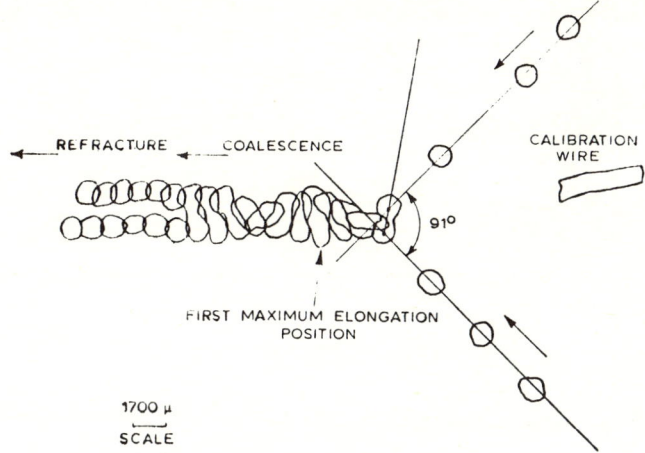

Collision velocities 115·5 cm/s and 129·5 cm/s.
Diameter of drops from each atomizer 1245 μ.

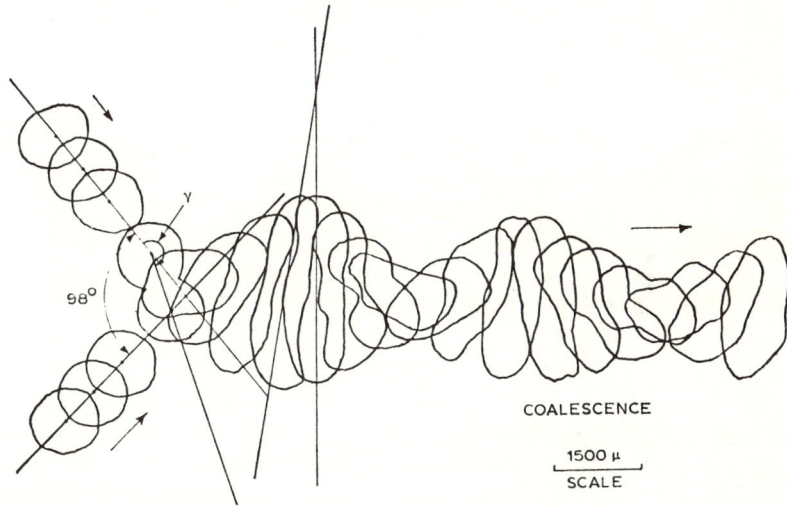

Collision velocities 83 cm/s and 77·5 cm/s.   Diameter of drops from each atomizer 1130 μ.

FIG. 1.4.5 Typical tracing of droplet collisions [1-38].

where $Kn = \bar{l}/2r$, and the modified target Reynolds number

$$\Phi = 9\frac{\rho_g}{\rho_f}\frac{w_c D_t \rho_g}{\mu_g} \qquad (1.4.23)$$

and is plotted in Fig. 1.4.6b. It is seen that above $K \approx 10$ the scavenging process is very efficient. As an example let us consider LP turbines, where large droplets ($D_t \approx 10^{-4}$ m) may fly at high speed ($w_c \approx 300$ m/s) through a

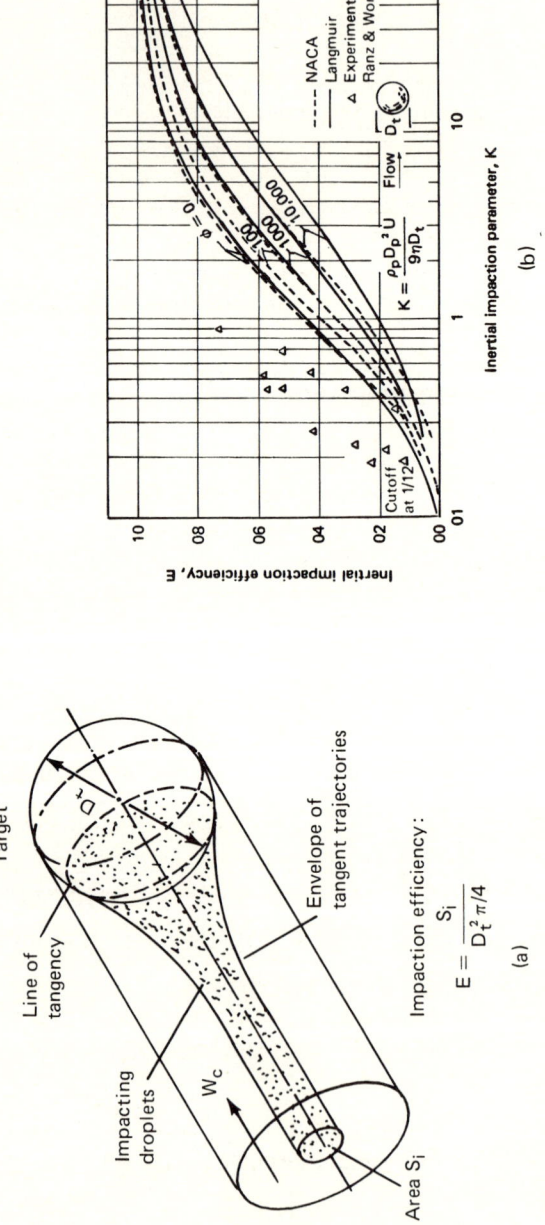

FIG. 1.4.6 Impaction efficiency between large and small drops. Reprinted with permission from Golovin and Putnam [1-40]. Copyright by the American Chemical Society.

mist containing droplets of say $r = 0.5\ \mu m = 5 \cdot 10^{-7}$ m. With $9\ \mu_g / 2\rho_f = 0.5 \cdot 10^{-7}$ m^2/s (Fig. 1.3.4), $\bar{l} = 5 \cdot 10^{-7}$ m, $\rho_g/\rho_f = v'/v'' = 10^{-4}$ (Fig. 1.3.6), and $\rho_g = 0.1$ kg/m^3, $\mu_g = 10^{-5}$ kg/s·m, we obtain $K = 80$ and $\Phi = 0.27$. The impaction efficiency for this case is seen to be virtually unity.

The importance of this scavenging process in turbines however depends decisively on the amount and motion of large droplets present in the flow. It is likely to be unimportant at low pressure but may be important in HP flows.

Collisions between droplets of comparable size during their flow through the turbine can be treated by statistical methods. In case of very small droplets, thermal (Brownian) motion is the main cause of collisions; in case of larger ones turbulence effects and changes of flow speed and direction are likely to play the main role.

Thermal (Brownian) coagulation in LP turbines has been thoroughly analyzed in [1-30]. Typically the calculations show that coagulation is present, but it can hardly change the average size of droplets by more than a few percent. In HP turbines thermal coagulation might play a more important role. The decay of a (uniform-sized) droplet population by thermal coagulation is expressed by the Smoluchowski equation [1-29]. Calculations referring to conditions typical of crossover pipes after the exit of HP turbines ($p = 10$ bar, $Y = 0.10$) are shown in Fig. 1.4.7. It is assumed that at time zero all moisture is uniformly dispersed in the form of droplets of radius r_0. As time proceeds, coagulation leads to an increase in mean droplet size $\bar{r}$. The curves for various r_0 have a bottom envelope that ascends with time. It is seen, however, that on the time scale of interest in turbines (10^{-3} to 10^{-2} s) virtually no effects on droplet size are incurred.

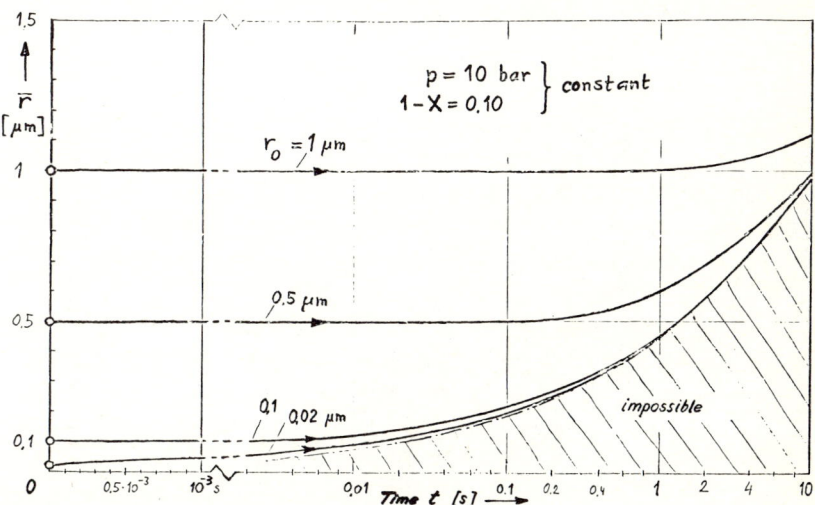

FIG. 1.4.7 Droplet size growth by agglomeration.

Coagulation due to high turbulence and to sudden deflections of flow in turbines seems not to have been analyzed in detail. Preliminary estimates made for LP turbines [1-18] indicated no appreciable effect.

1.4.3.3 Impact, Deposition, Rebounding

If a steam flow carrying droplets is suddenly deviated, some of the droplets will be centrifuged from the flow and impinge against the obstacle. In this manner turbine blades collect part of the moisture present in the flow. Laminar and turbulent diffusion of droplets may promote the process. Deposition problems will be treated in more detail in Chaps. 2 and 7.

Impingement does not necessarily result in deposition. Droplets or fractions of them may rebound [1-13]. In laboratory experiments in an air atmosphere, rebounding was found to be quite important, but the results were not considered to be fully applicable to steam atmospheres [1-32]. Figure 1.4.8, based on a different experiment [1-33], shows that droplets impinging on a dry metal surface with a normal velocity component below 6 m/s completely adhere. At high normal velocities only about half of the droplet mass is retained. If the surfaces are wet, which is the normal case in wet-steam turbines, all small droplets deposited at oblique angles are believed to be caught in the surface films.

The erosion damage caused by high-speed impact of relatively large ($> 20 \mu$m) droplets will be treated in Chaps. 5 and 6.

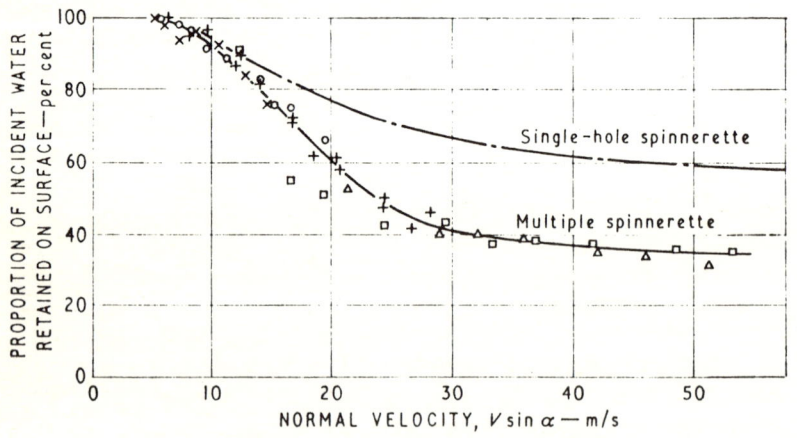

Droplet mean diameter 180 μm.
Impact angle α
× 15°
○ 30°
+ 20°
△ 60°
□ 90°

FIG. 1.4.8 Experimental data for water droplets striking a stationary surface.

1.4.3.4 Droplet Deformation and Break-Up

The stability of droplets moving through a gaseous atmosphere depends mainly on the ratio of aerodynamic pressure forces trying to deform it and surface-tension forces trying to make its shape spherical. This ratio is usually expressed by the Weber number

$$\text{We} = \frac{\rho_g w_r^2 d}{\sigma} \qquad (1.4.24)$$

At low values of We (slow relative speed w_r, or small droplet diameter d), droplets remain nearly spherical. Droplets exposed to high-We flow conditions are quickly deformed and fragmented. Droplet break-up has been the subject of numerous experimental investigations (see review in [1-34]). Several types of break-up can be distinguished. The type occurring is determined, besides We, by additional factors such as droplet viscosity, gas stream density, and the rate at which speed is increased from zero to the value w_r. In Fig. 1.4.9A the various types observed are shown schematically. Under turbine flow conditions, droplets torn off from blades usually break up in the bag mode [1-35], and the critical minimum value of We required for break-up to occur is about

$$\text{We}_{\text{crit}} \approx 20$$

Figure 1.4.9B shows stability criteria (We_{crit} values) of several authors plotted in the w_r, d diagram. The conditions assumed ($p = 0.14$ bar) are typical of LP turbines. It is seen that, according to most criteria, maximum stable droplet sizes in LP turbines (where $w_r = 200–400$ m/s) will be of the order of about $< 0.1–0.4$ mm.

If We_{crit} is sufficiently exceeded, disruption occurs within a very short time after exposure [1-34], typically of the order

$$\Delta t_{\text{rupt}} = (0.3-1) \cdot \frac{\pi}{4} \sqrt{\frac{d^3 \rho_f}{\sigma}} \qquad (1.4.25)$$

which for a 1 mm water drop has the value of 1 to 3 ms.

The size distribution of droplets resulting from break-up has been investigated in LP turbine cascades in [1-35]. The results, shown in Fig. 1.4.10 and in Table 1.4.1, reveal distributions with a well-determined droplet size range.

These deformation and break-up mechanisms are further complicated in turbines by the presence of wakes. Water detached in LP turbines from the trailing edge as a film or ligament initially finds itself in the low-velocity wake flow and is disrupted first into rather large (mm-sized) drops. As the drops are accelerated and carried into high-velocity regions of the flow, this "primary

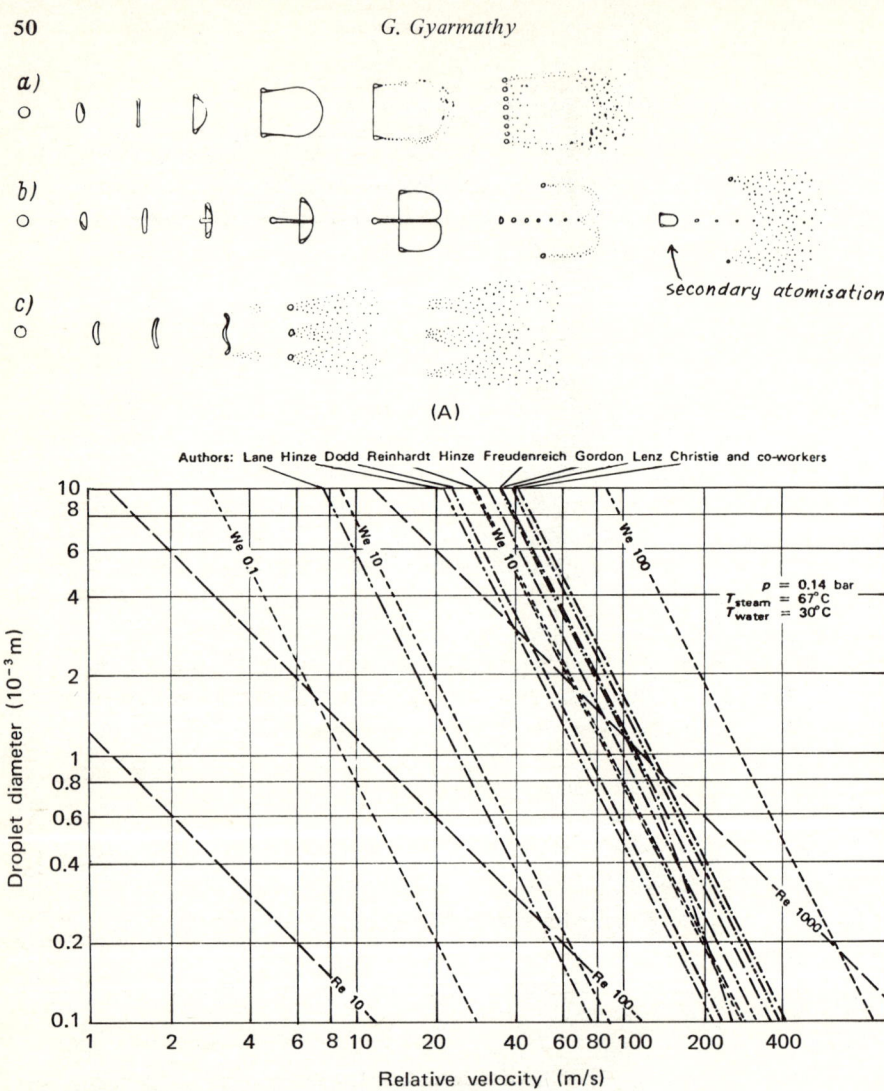

FIG. 1.4.9 (A) Types of droplet break-up: (a) bag, (b) club, (c) disc [1-34]. (B) Stability criteria for break-up in LP steam [1-34].

atomization" is soon followed by a "secondary atomization" corresponding to the mechanisms discussed above. Flow conditions and flight distances are usually such that secondary atomization occurs before the drops could reach the blades of the following rotor wheel. If, due to narrow interrow gaps, secondary atomization cannot occur, erosion danger is gravely enhanced.

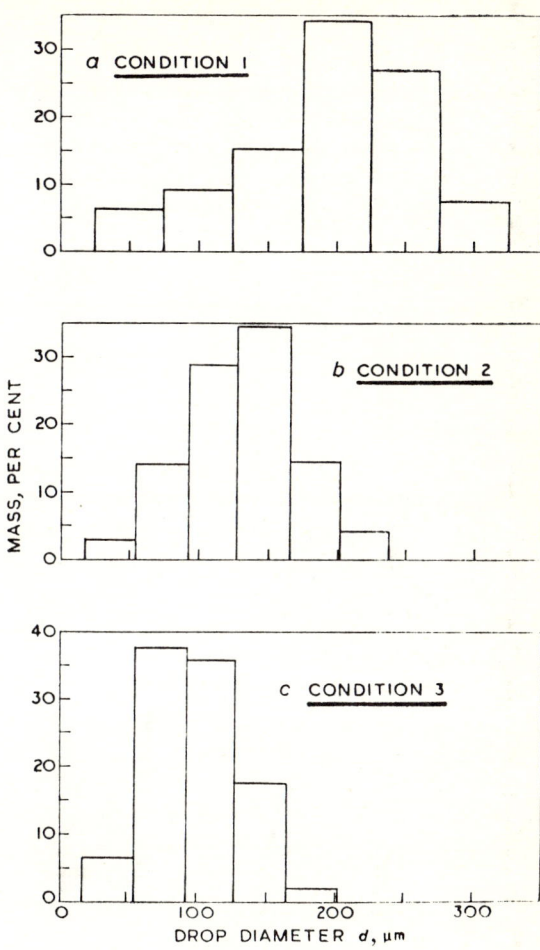

FIG. 1.4.10 Size distribution of droplets torn off from blade trailing edges. Reproduced by courtesy of the Council of the Institution of Mechanical Engineers, from [1-36].

TABLE 1.4.1 Secondary atomization: Size of drops formed and conditions at break-up [1-35]

Condition number	Calculated critical Weber number	Approximate drop velocity (m/s)	Steam velocity (m/s)	Maximum diameter (μm)	Mass mean diameter (μm)
1	18.7	15	281	310	200
2	23.4	15	340	252	135
3	21.9	15	377	200	98

1.4.4 Motion of Liquid on Channel Walls

Both liquid films and rivulets have been observed in wet-steam turbines. In regions of flow separation on stator blades, thick streams are formed [1-36], and even stationary droplets can occasionally be detected. Films, rather than rivulets, are likely to prevail in regions where moisture deposition is intense (e.g., on the hollow side of blades) or where there are intense driving forces (like shear forces of steam flow or rotor centrifugal forces) acting on the liquid layer. In such regions the films tend to become thinner. Thick films have a tendency to break up into rivulets. A theoretical criterion [1-37] given in terms of shear force, pressure gradient, surface tension, and contact angle, which specifies the maximum stable film thickness, has been found to agree within a factor of two in a (single) case investigated [1-38]. The application of such criteria to turbines is hampered by the lack of information concerning shear forces and local surface conditions (contact angle).

The influence of gravity on water motion in turbine flow channels is normally outweighed by the action of aerodynamic forces.

Actual water motion on turbine stator blades and casing walls presents a very complicated picture, as revealed by endoscopic observation [1-36] (Fig. 1.4.11). The patterns depend largely on the pressure and steam-flow fields adjacent to the surface. As a general rule, visible amounts of water on turbine walls are a sign of disorderly flow. In separation pockets and along trailing edges, deposited water can easily move in a radial direction. Such water streams tend to accumulate and detach at preferred locations, thus causing locally concentrated blade erosion.

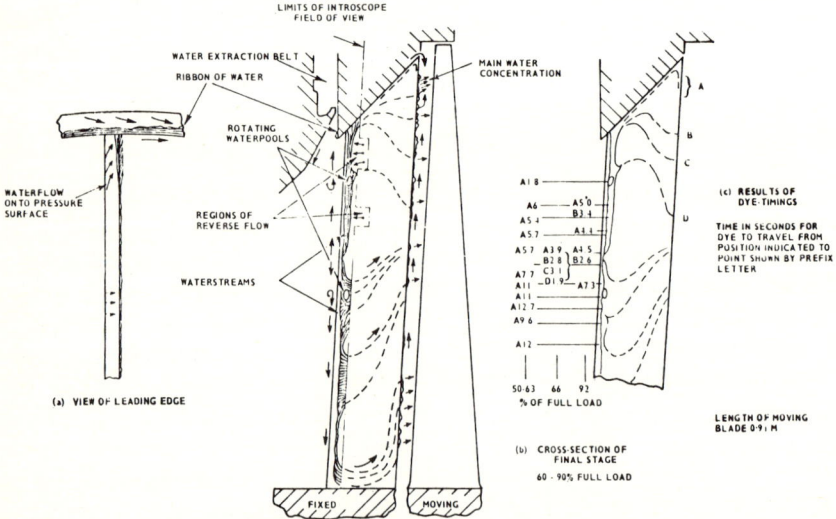

FIG. 1.4.11 Water flows on the pressure surfaces of the final-stage fixed blades. Reproduced by courtesy of the Council of the Institution of Mechanical Engineers, from [1-36].

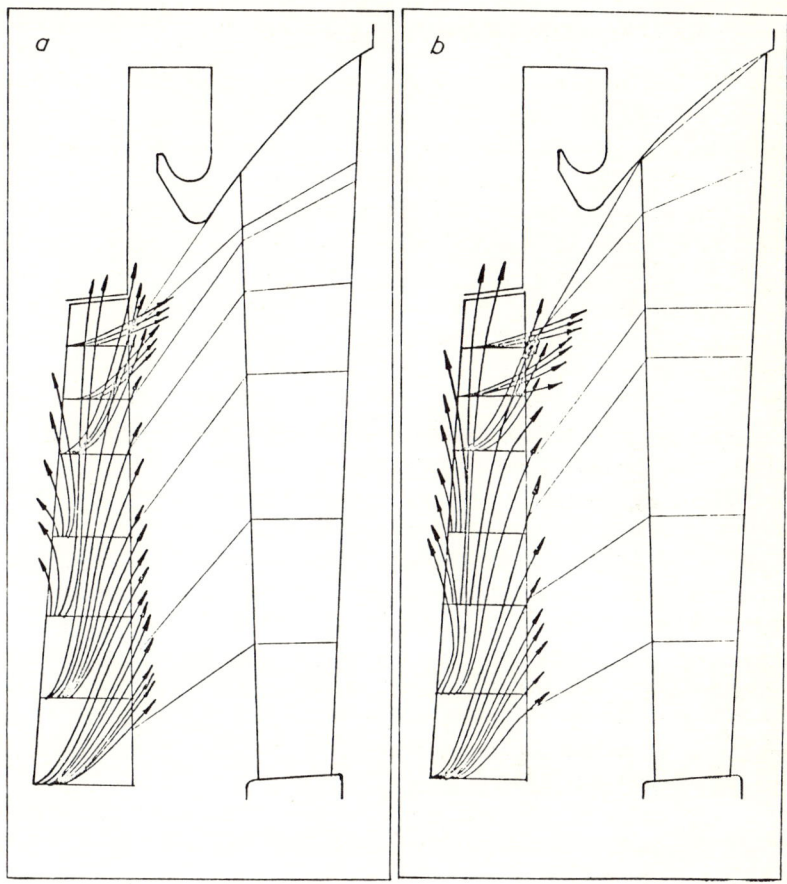

FIG. 1.4.12 Water flows on moving blades [1-39].

No water film observations on moving blades have been made up to the present. Due to the influence of the centrifugal forces, such films are believed to be thin and to move in a direction close to radial.

Theoretical treatment of water flows in thin layers is possible [1-18, 1-37], but severe idealizations are required. These are not valid for stator walls where water flow is often intermittent and rivulet formation predominates. On rotor blades, however, theoretical results may come closer to reality. For typical LP turbine conditions film thicknesses on the order of 10 μm are obtained with flow velocities on the order of several meters per second [1-18].

The direction of flow on rotor blade surfaces is determined by centrifugal and Coriolis forces. An example for flow patterns obtained is shown in Fig. 1.4.12 [1-39].

NOMENCLATURE

a	speed of sound
C	number of droplets per unit volume
c_D	drag coefficient
c_p	specific heat at constant pressure
c_V	specific heat at constant volume
C_θ	tangential speed
d, D	droplet diameter
$\bar{D}$	mean interdrop distance
E	impact efficiency (Sec. 1.4.3.)
F_D	drag force
g	field acceleration
G_B, g_B	functions characterizing droplet flashing (Sec. 1.4.2)
ΔG_i	change in Gibbs free enthalpy (Sec. 1.4.1)
h	enthalpy
i	number of molecules contained in a droplet of given size
k	index of isentropic expansion
k	Boltzmann constant (Sec. 1.4.1)
K	inertial impact parameter (Sec. 1.4.3)
Kn	Knudsen number (Sec. 1.4.2)
$\bar{l}$	molecular mean free path length
m	mass
M	Mach number (Sec. 1.4.2)
M_m	molecular weight
n	index of polytropic expansion or optical refraction index
N	number of droplets per unit mass of wet steam
$N_{i,\text{eq}}$	equilibrium number of i-mers in unit mass of gas
Pr	Prandtl number
$\dot{Q}$	heat transfer rate
r	radius of droplet
r_{crit}	critical droplet size
r_{flash}	maximum radius of droplets which escape flashing
Re	Reynolds number (Sec. 1.4.2)
R	gas constant
s	entropy
Δs_{rev}	entropy rise during reversion from a subcooling ΔT
S	supersaturation ratio
T	temperature
ΔT	subcooling
T_r	surface temperature of droplet
t_T	temperature at triple point (Sec. 1.3.2)
t_K	temperature at critical point (Sec. 1.3.2)
Δt_{rev}	time constant characterizing reversion from subcooling ΔT

Δt_{int}	time constant characterizing equalization of internal temperature differences
v	specific volume
V	volume
w_c	collision speed
w_r	relative velocity between droplet and vapor
w_{rt}	terminal settling speed
We	Weber number (Sec. 1.4.3)
X	dryness fraction
Y	wetness fraction
Y_{eq}	equivalent wetness fraction (Sec. 1.3.4)
Z	compressibility factor
α	heat transfer coefficient
η	efficiency
λ	thermal conductivity
μ	dynamic viscosity
ρ	density
τ	surface tension
ϕ	target Reynolds number (Sec. 1.4.3)

Subscripts

f	liquid phase
g	gaseous phase
fg	phase transition
r	referring to droplet
s	saturation
t	total (gaseous plus liquid phase)
0	inlet conditions

Superscripts

$^-$	mean value
$'$	saturated liquid
$''$	saturated vapor

Abbreviations

HP	high pressure
IP	intermediate pressure
LP	low pressure
CT	cooling tower
TFC	through-flow cooling
BWR	boiling-water reactor

PWR   pressurized-water reactor
RH   steam reheater
WS   water separator

REFERENCES

1-1   Stodola, A.: "Dampf- und Gasturbinen," 5th ed., Springer, Berlin, 1922.
1-2   Gardner, G. C.: Events Leading to Erosion in the Steam Turbine, *Proc. Inst. Mech. Eng.*, London, 168 (1963-1964), I, 23, 593-623.
1-3   Pink, H.-R.: Die Zukunft der Kraftwerke mit fossilen Brennstoffen, *VDI-Berichte* Nr. 208, 1974, p. 17.
1-4   Ludewig, M.: Endstufenschaufeln von Grossdampfturbinen, *Elektrizitätswirtschaft* 66 (1967), 24, 737-743.
1-5   Riollet, G.: Expérience acquise dans la construction des turbines nucléaires de grande puissance, *Paper No. 6/2, NUCLEX Technical Meeting, Basel*, Oct. 16-21, 1972.
1-6   Mühlhäuser, H.: Bau grosser Sattdampfturbinen, *Paper No. 6/1, NUCLEX Technical Meeting, Basel*, Oct. 16-21, 1972.
1-7   ———: Power Station Engineering Today, *Pamphlet by Kraftwerk-Union*, 1972.
1-8   Hegetschweiler, H.: General Electric Steam Turbine Generators for Nuclear Power Plants—A Review and Outlook, *Paper No. 6/3, NUCLEX Technical Meeting, Basel*, Oct. 16-21, 1972.
1-9   Hossli, W.: Die 1160-MW-Turbine für das Kernkraftwerk "Donald C. Cook" der AEP, *Brown Boveri Mitteilungen*, 59 (1972), 1, 4-19.
1-10 Spencer, R. C., and Miller, E. H.: Performance of Large Nuclear Turbines, *Combustion*, Aug. 1973, 24-30.
1-11 Ringeis, W. K., Strasser, W., and Peuster, K.: Kernkraftwerk Gundremmingen: Aufbau der Gesamtanlage, *Atomwirtschaft*, 10 (1965), 11, 575-588.
1-12 Schmidt, E.: "Properties of Water and Steam in SI-Units," Springer, Berlin, 1969.
1-12a Keenan, J. H., Keyes, F. G., Hill, P. G., and Moore, J. G.: "Steam Tables," Wiley, New York, 1969.
1-13 Traupel, W.: "Thermische Turbomaschinen," 1. Auflage, Band 1, Springer, Berlin, 1962.
1-14 ———: "Handbook of Chemistry and Physics," 46th ed., Chemical Rubber Co., Cleveland, 1966.
1-15 Groh, G., et al.: "Physical Chemistry" [in Hungarian], Egyet, Nyomda, Budapest, 1945.
1-16 Baehr, H. D.: "Thermodynamik," 1. Auflage, Springer, Berlin, 1962.
1-17 Dejc, M. E., and Trojanovskij, B. M.: "Untersuchung und Berechnung axialer Turbinenstufen," VEB-Verlag Technik, Berlin, 1973.
1-18 Gyarmathy, G.: "Grundlagen einer Theorie der Nassdampfturbine," Dissertation ETH, Zürich, Juris-Verlag, 1962. English Translations: C.E.G.B. (London) Rept. T-781 (1963) and USAF-FTD (Dayton, Ohio) Rept. TT-63-785.
1-18a Wegener, P. P.: "Nonequilibrium Flows," part I, Dekker, New York, 1969, especially pp. 182 ff.
1-19 Feder, J., Russell, K. C., et al.: Homogeneous Nucleation and Growth of Droplets in Vapours, *Advances in Physics (Philos. Mag. Suppl.)*, 15 (1966), 57, 111-178.
1-20 Kang, S.-W.: Analysis of Condensation Droplet Growth in Rarefied and Continuum Environments, *AIAA J.*, 5 (1967) 7, 1288-1295.
1-21 Johnstone, H. F., Pigford, R. L., and Chapin, J. H.: Heat Transfer to Clouds of Falling Particles, *Trans. Amer. Inst. Chem. Eng.*, 37 (1941), 95-133.

1-22 McAdams, W. H.: "Heat Transmission," McGraw-Hill, London, 1954.
1-23 Konorski, A., Jankowski, T., and Prokopovicz, J.: Heat Exchange and Evaporation of Water Film on Heated Guide Vanes of Steam Turbines (in Polish), *Trans. Inst. Fluid-Flow Machinery (Poland)*, 1971, no. 57, 129-159.
1-24 Carslaw, H. S., and Jaeger, J. C.: "Conduction of Heat in Solids," 2nd ed., Clarendon, Oxford, 1959.
1-25 Crowe, C. T., Babcock, W. R., and Willoughby, P. G.: Drag Coefficient for Particles in Rarified, Low-Mach-Number Flows, *Paper No. 3-3, Int. Symp. Two-Phase Flow Systems, Haifa*, Aug. 29-Sept. 2, 1971.
1-26 Abraham, F. F.: Functional Dependence of Drag Coefficient of a Sphere on Reynolds number, *Phys. Fluids*, 13 (1970), 8, 2194-2195.
1-27 Ryley, D. J., and Wood, M. R.: The Collision, in Free Flight, of Water Droplets in Atmospheres of Air and Steam, *Proc. Inst. Mech. Engs.*, 180 (1965-1966). Pt. 30, 73-87.
1-28 Ryley, D. J., Ralph, W. J., and Tubman, K. A.: The Collision Behaviour of Water Drops Within a Low-pressure Steam Turbine, *Int. J. Mech. Sci.*, 12 (1970), 589-596.
1-29 Fuchs, N. A.: "Mechanics of Aerosols," Pergamon, Oxford, 1964.
1-30 Ryley, D. J.: Condensation Fogs in Low Pressure Steam Turbines, *Int. J. Mech. Sci.*, 9 (1967), 729-741.
1-31 Krzeczkowski, S.: Motion of Drops in Flow Ahead of a Cylindrical Obstacle (in Polish), *Trans. Inst. Fluid-Flow Machinery (Poland)*, 40 (1968), 3-22.
1-32 Parker, G. J.: The Collision of Drops with Dry and Wet Surfaces in an Air Atmosphere, *Proc. Inst. Mech. Eng. (London)*, 184 (1969-1970), Pt. 3G (III), 57-63.
1-33 McAllister, D. H.: Discussion Contribution to [1-32], *ibid.*, pp. 78-79.
1-34 Hässler, G.: "Untersuchungen zur Verformung und Auflösung von Wassertropfen...," dissertation, University of Karlsruhe, Karlsruhe, 1971.
1-35 Moore, M. J., Langford, R. W., and Tipping, J. C.: Research at C.E.R.L. on Turbine Blade Erosion, *Inst. Mech. Eng. Conf. Bristol, Publ.* 2, "Wet Steam," 1968, pp. 1-8.
1-36 Moore, M. J., and Sculpher, P.: Conditions Producing Concentrated Erosion in Large Steam Turbines, *Proc. Inst. Mech. Eng., (London)*, 184 (1969-1970), Pt. 3G (III), 45-56.
1-37 Gardner, G. C.: Some Aspects of the Flow of Liquids as Films and Drops, *Central Electricity Generating Board (London)*, Rept. RD/P/M10.
1-38 Ryley, D. J., and Small, J.: Re-entrainment of Deposited Liquid from Simulated Steam Turbine Fixed Blades, *Inst. Mech. Eng. Conf., Warwick, Publ.* 3, 1973, *Paper* C 21/73, pp. 9-18.
1-39 Kirillov, I. I., and Yablonik, R. M.: The Flow of Wet Steam in a Turbine, *Inst. Mech. Eng. Conf., Bristol, Publ.* 2, "Wet Steam," 1968, pp. 78-85.
1-40 Golovin, M. N., and Putnam, A. A.: Inertial Impaction on Single Elements, *Ind. Eng. Chem. Fund.* 1 (1962), 1, 264-273.

CHAPTER 2

Gas Dynamics of Wet Steam and Energy Losses in Wet-Steam Turbines

M. J. Moore

The aim of this chapter is to provide a theoretical background to the study of wet-steam flows in turbines. Although the problems of two-phase flow have attracted interest over many years, the theoretical representation of the various phenomena has lagged considerably behind the experimental development of wet-steam turbines. Thus at the present time a definitive theoretical design method for a wet-steam turbine is not available. However recent advances in theoretical methods and instrumentation should redress the balance, and in the following pages an attempt has been made to describe the latest theories and their relevance to turbine design.

2.1 EQUATIONS OF STATE FOR STEAM

Steam in the superheated and partially condensed state in a turbine departs appreciably from ideal gas laws. The equations of state relating the properties of the medium are complex, and design calculations are therefore usually carried out by using tabulated properties [2-1]. However, in wet-steam flow calculations it is often more convenient to relate steam properties using simple functions, in particular where steam becomes supersaturated and tabulated values are not available. In the following paragraphs some of the equations of state for steam given in the literature are described.

2.1.1 Equations for Superheated and Supersaturated Steam

For accuracy over a wide range of pressure and temperature in the superheated range the Steam Table formulation [2-1] is, of course, to be preferred. For many calculations, the use of these complex equations may involve appreciable computer storage and processing time. Simpler functions have been proposed by several authors, e.g., [2-2 to 2-5], the basis of references [2-3, 2-4] being as follows:

Horlock [2-3] points out that we can make the approximation

$$R = R(S)$$
$$c_p = c_p(S)$$

over a useful range of steam pressure and temperature as shown in Fig. 2.1.1. On this basis he develops the following equations for gas "constant" over small ranges of pressure and temperature

1 Effective gas constant (J/kg · K)

$$R = \frac{pv}{T} = 455.3 - 23.14 \left[\frac{p}{27.586(10^5)}\right]^{1.35} \left(\frac{588.9}{T}\right)^{5.85} \qquad (2.1.1)$$

where units of p, v, T are N/m², m³/kg, °K, respectively.

2 Specific heat at constant pressure (kJ/kg · K)

$$c_p = 11542.0 - 21.026R \qquad (2.1.2)$$

3 Index of isentropic expansion k for

$$pv^k = \text{constant}$$
$$k = \text{constant} = 1.30 \qquad (2.1.3)$$

and therefore enthalpy change Δh_s for small temperature change ΔT is given by

$$\Delta h \simeq \left(\frac{k}{k-1}\right) R \Delta T \qquad (2.1.4)$$

4 Polytropic expansion index n for constant isentropic efficiency $\eta (= dh/dh_s)$

$$pv^n = \text{constant}$$
$$n = \frac{k}{1 - kR(\eta - 1)/c_p} \qquad (2.1.5)$$

This equation is applicable to an expansion process but *not* a compression process because of their differing definitions of efficiency.

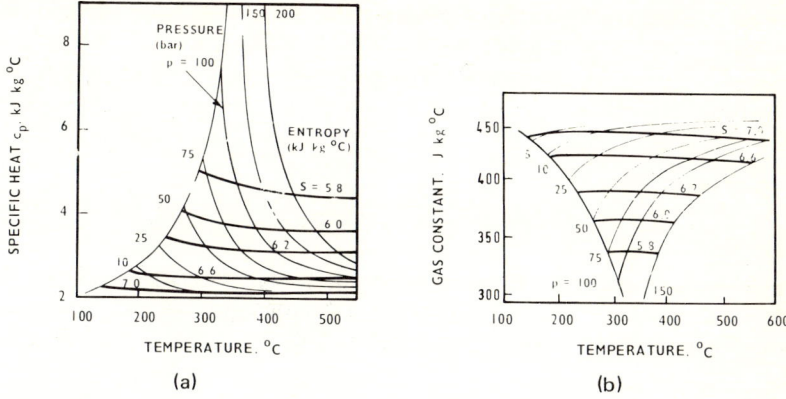

FIG. 2.1.1 Variation of (a) specific heat and (b) gas constant.

A somewhat similar approach was presented by Dzung [2-4], who classifies nonideal gases in terms of the parameters

$$Z = \frac{pv}{RT} \tag{2.1.6}$$

$$\begin{aligned} k &= -\frac{v}{p}\left(\frac{\partial p}{\partial v}\right)_s \\ m &= -\frac{v}{p}\left(\frac{\partial p}{\partial v}\right)_h \end{aligned} \tag{2.1.7}$$

Superheated or supersaturated steam is termed a polytropic vapor where

$$Z = Z(S)$$

$$m = 1.0$$

$$k = \text{constant}$$

Dzung shows that equivalent functions of state θ, σ, k may then be used to replace T, S, χ in the thermodynamic equations where

$$\text{temperature } \theta = ZT$$

$$\text{entropy } \sigma = \int Z^{-1} \, dS$$

A polytropic change is then described by the equations

$$pv^n = \text{constant}$$

$$\frac{p}{\theta^{n/(n-1)}} = \text{constant}$$

$$\Delta h_{1-2} = \frac{k}{k-1} R\theta_1 \left[\left(\frac{p_2}{p_1}\right)^{(n-1)/n} - 1 \right]$$

A formulation by Vukalovich et al. [2-5] provides probably the most accurate method of calculation of gas constant but at the expense of complexity. The relevant equation given below can be seen to be appropriate for computer solution only.

$$\frac{pv}{RT} = 1 + F_1(T)p + F_2(T)p^2 + F_3(T)p^3 \qquad (2.1.8)$$

where $F_1(T) = (b_0 + b_1\phi + \ldots b_5\phi^5) \times 10^{-9}$
$F_2(T) = (c_0 + c_1\phi + \ldots c_8\phi^8) \times 10^{-16}$
$F_3(T) = (d_0 + d_1\phi + \ldots d_8\phi^8) \times 10^{-23}$
$\phi = 10^3/T$
and the constants of the series are

$b_0 = -5.01140$	$c_0 = -29.133164$	$d_0 = -34.551360$
$b_1 = +19.6657$	$c_1 = +129.65709$	$d_1 = +230.69622$
$b_2 = -20.9137$	$c_2 = -181.85576$	$d_2 = -657.21885$
$b_3 = +2.32488$	$c_3 = +0.704026$	$d_3 = +1036.1870$
$b_4 = +2.67376$	$c_4 = +247.96718$	$d_4 = -977.45125$
$b_5 = -1.62302$	$c_5 = -264.05235$	$d_5 = +555.88940$
	$c_6 = +117.60724$	$d_6 = -182.09871$
	$c_7 = -21.276671$	$d_7 = +30.554171$
	$c_8 = +0.5248023$	$d_8 = -1.9917134$

2.1.2 Equations for Wet Steam

In Dzung's classification, wet steam in equilibrium may be considered a semipolytropic vapor where

$$Z = Z(S)$$

$$m \cong \text{constant}$$

$$k \cong \text{constant}$$

In general, therefore, properties cannot be expressed by simple equations.

The equation of state now becomes the relationship between temperature T_{sat} and pressure p_{sat} at saturation (i.e., independent of specific volume), which in [2-1] is given as

1   for $T_{sat} < 373°K$

$$\log_{10}(p_{sat}) = a_0 + a_1 \log_{10}\phi + a_2 \phi + \frac{a_3}{\phi}$$

where $\phi = T_{sat}$
$a_0 = 28.59051$
$a_1 = -8.20$
$a_2 = 2.4804 \times 10^{-3}$
$a_3 = -3142.31$

2   for $T_{sat} > 373°K$

$$\log_{10}(p_{sat}) = b_0 + \frac{b_1}{\phi} + \frac{b_2(\phi^2 - b_3)(10^{b_4(\phi^2-b_3)^2} - 1)}{\phi}$$
$$+ b_5 10^{b_6(647-T_{sat})^{1.25}}$$

where $\phi = T_{sat}$
$b_0 = 5.432368$
$b_1 = 2.0057 \times 10^3$
$b_2 = 1.3869 \times 10^{-4}$
$b_3 = 2.9370 \times 10^5$
$b_4 = 1.1965 \times 10^{-11}$
$b_5 = -4.40 \times 10^{-3}$
$b_6 = -5.7148 \times 10^{-3}$

Of particular interest for calculation of wet-steam flows is the isentropic index k for expansion of wet steam in equilibrium. Several formulas have been suggested, a recent example [2-6] giving

$$k = 0.603 + a_0 \left[1 - \left(\frac{p}{p_*}\right)^{3/2}\right]^{a_1} \quad (2.1.9)$$

where
$a_0 = 0.5220 - 0.1418 Y/(1-Y)$
$a_1 = 1.34565 - 0.76825(1-Y)$
critical pressure $p_* = 221.3 \times 10^5$ N/m²

As long ago as 1859, Rankine [2-7] noted "it has been deduced by trial that for such pressure as usually occur in the working of steam engines

$$pv^{1.11} = \text{constant''} \quad (2.1.10)$$

and for isentropic wet-steam expansions, Zeuner [2-8] gives

$$n = 1.035 + 0.10X_1 \qquad (2.1.11)$$

where X_1 is the initial dryness fraction.

However these values pertain to equilibrium conditions only and we shall now consider the characteristics of a nonequilibrium or relaxing medium.

2.2 RELAXATION

2.2.1 General System

When some property ϕ of a medium is displaced from its equilibrium value ϕ_0 and restoration to equilibrium occurs at a finite rate, the substance is termed a relaxing medium. The simplest system is defined by the proportionality:

$$\frac{d\phi}{dt} \propto (\phi - \phi_0)$$

The system can be characterized by the constant of proportionality and may be written

$$\frac{d\phi}{dt} = -\frac{1}{\tau}(\phi - \phi_0) \qquad (2.2.1)$$

where τ is known as the relaxation time (or in some cases, the time constant).

This equation can be integrated simply to give the familiar exponential decay of ϕ to equilibrium

$$\frac{\phi - \phi_0}{\phi_1 - \phi_0} = e^{-t/\tau}$$

where ϕ_1 is the initial displaced value.

The physical significance of relaxation time may be seen from Fig. 2.2.1a. It represents the time from release at ϕ_1 in which the displacement amplitude reduces to $1/e \,(= 0.3679)$ of the original value.

Alternatively, from Eq. (2.2.1) we may deduce

$$\tau = (\phi_1 - \phi_0)\left[-\frac{d(\phi - \phi_0)}{dt}\right]^{-1}_{t=0}$$

that is, the time at which equilibrium would be restored if decay continued at the initial rate.

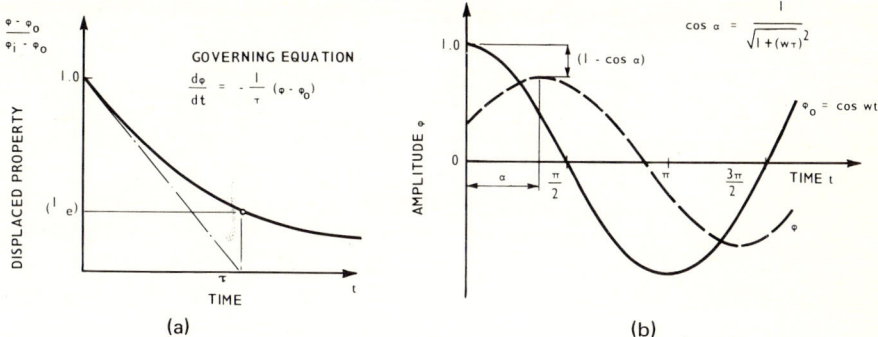

FIG. 2.2.1 (a) Response of simple system displaced from fixed equilibrium; (b) system response ϕ to sinusoidal oscillation of equilibrium value ϕ_0.

Some further aspects of relaxation may be seen from the response of property ϕ to a time-varying equilibrium ϕ_0. For example if ϕ_0 executes a simple harmonic oscillation at frequency ω

$$\phi_0 = A \cos \omega t$$

then

$$\frac{d\phi}{dt} = -\frac{\phi}{\tau} + \frac{A}{\tau} \cos \omega t$$

This is a standard form of linear differential equation with solution

$$\phi = Be^{-t/\tau} + \frac{A}{\sqrt{1 + \omega^2 \tau^2}} \cos(\omega t - \alpha) \qquad (2.2.2)$$

where phase angle $\alpha = \tan^{-1}(\omega\tau)$ and B is an arbitrary constant.

In Eq. (2.2.2) we see a transient decay with relaxation time τ. The steady response displays a lag and amplitude reduction which are again simple functions of τ. Phase angle α has a maximum value of $\pi/2$ at $\omega\tau = 0$. That is, the maximum lag occurs either at a very high frequency ω, or where the relaxation time τ of the system is very large.

The deviation or slip from equilibrium, defined as the amplitude difference $\Delta\phi(= \phi_0 - \phi)$, may be written for the steady solution

$$\Delta\phi = A \cos \omega t - \frac{A}{\sqrt{1 + \omega^2 \tau^2}} \cos(\omega t - \alpha)$$

It can be seen that relaxation introduces a reactive or out-of-phase component

$$\phi' = \frac{A \sin \alpha}{\sqrt{1 + \omega^2 \tau^2}} \sin \omega t$$

and a power factor for the in-phase component of

$$\cos \alpha = \frac{1}{\sqrt{1 + \omega^2 \tau^2}}$$

Finally, if we assume that the rate of dissipation of energy in the system is proportional to the deviation $|\Delta\phi|$, we can then integrate the energy E dissipated during one cycle

$$E \propto \int_0^{2\pi/\omega} \left| \cos \omega t - \frac{1}{\sqrt{1 + \omega^2 \tau^2}} \cos (\omega t - \alpha) \right| dt$$

which can be shown to give

$$E \propto \frac{4\omega\tau^2}{1 + \omega^2 \tau^2}$$

An important aspect of relaxation time may be now seen from Fig. 2.2.2, where energy dissipation E/τ is plotted versus dimensionless frequency $\omega\tau$. Energy dissipation per cycle is a maximum at a frequency

$$\omega = \frac{1}{\tau}$$

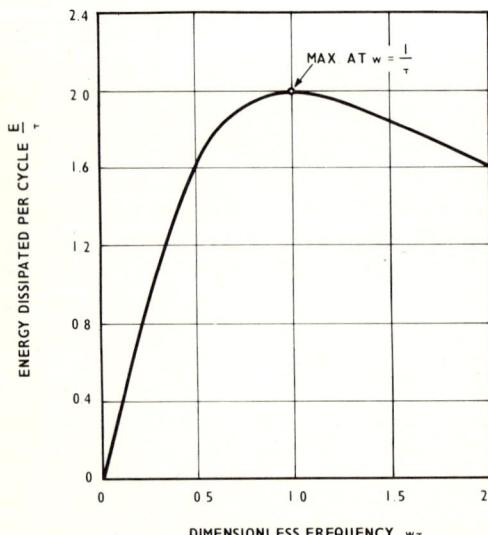

FIG. 2.2.2 The effect of frequency ω on the energy absorbed during the oscillation of a relaxing medium.

2.2.2 Thermal Relaxation of Wet Steam

A mixture of water and steam is displaced from thermodynamic equilibrium if the temperature T_g of the gaseous phase differs from that of the water. The liquid phase may be assumed to remain at the saturation temperature T_{sat} corresponding to the prevailing pressure. Equilibrium is restored by a change of phase, the latent heat transferred bringing the temperature of the gaseous phase toward the saturation value.

We will assume an initial water mass fraction Y in the form of a monodispersion of droplets of diameter d. The number n_m drops per unit mass of wet steam is given by

$$n_m = \frac{Y}{(\pi/6)\,d^3\,\rho_f} \tag{2.2.3}$$

If the droplets are immersed in supersaturated steam at temperature T_g, where $T_g < T_{sat}$, the rate of return to equilibrium is determined by the rate of transfer to the steam of the latent heat released by condensation of some of the vapor phase of the surface of the drops. Assuming the process occurs at constant pressure, the rate of enthalpy increase of the steam may be written

$$c_{pg}(1 - Y)\frac{dT_g}{dt} \cong n_m \pi d^2 \alpha (T_{sat} - T_g) \tag{2.2.4}$$

where it is assumed that the rate of enthalpy reduction due to the decrease in the proportion of vapor is small (this being so in all practical cases).

It is shown in [2-9] that the heat transfer coefficient α for small drops in steam can be expressed as

$$\alpha = \frac{2\lambda_g}{d}\left(\frac{1}{1 + 3.18\,\mathrm{Kn}}\right) \tag{2.2.5}$$

Knudsen number Kn is defined in Eq. (2.3.16).

Combining Eqs. (2.2.3), (2.2.4), and (2.2.5) we may obtain the rate equation

$$\frac{dT}{dt} = -\left(\frac{Y}{1-Y}\right)\frac{12\lambda_g}{d^2 c_{pg}\rho_f(1 + 3.18\,\mathrm{Kn})}(T_g - T_{sat})$$

which can be seen to be linear and similar to Eq. (2.2.1).

The thermal relaxation-time τ_T is therefore given by

$$\tau_T = \left(\frac{1-Y}{Y}\right)\left[\frac{d^2 c_{pg}\rho_f(1 + 3.18\,\mathrm{Kn})}{12\lambda_g}\right] \tag{2.2.6}$$

We may also consider a general form of the rate equation in terms of dimensionless variables

$$T' = \frac{c_{pg}T}{C_0^2}$$

$$t' = \frac{tC_0}{L}$$

where C_0 and L are reference flow velocity and length respectively. The equation becomes

$$\frac{dT'_g}{dt'} = -P_T(T'_g - T'_{sat})$$

and provides similar solutions of $T'_g(t)$ for particular values of the dimensionless thermal parameter P_T where

$$P_T = \left(\frac{Y}{1-Y}\right)\frac{12\lambda_g L}{d^2 c_{pg}\rho_f C_0(1 + 3.18\,\text{Kn})} \tag{2.2.7}$$

2.2.3 Inertial Relaxation of Water Droplets

An alternative form of inequilibrium occurs in a wet-steam flow when the velocities of gaseous and liquid phases are not equal. Drag forces of the steam on water drops in the flow will tend to reduce the velocity of the steam relative to the drops according to the equation of motion

$$\left(\frac{\pi}{6}d^3\rho_f\right)\frac{du}{dt} = D \tag{2.2.8}$$

Drag force D may be expressed in terms of a drag coefficient C_D

$$D = \frac{\pi d^2}{4}\frac{\rho_g(C_0 - u)^2}{2}C_D \tag{2.2.9}$$

where it has been assumed that the drop velocity u is less than steam velocity C_0. It is also assumed that C_0 is constant and, from [2-9], for small $|C_0 - u|$

$$C_D = \frac{24\mu}{|C_0 - u|\rho_g d}\left(\frac{1}{1 + 2.70\,\text{Kn}}\right) \tag{2.2.10}$$

Combining Eqs. (2.2.8), (2.2.9), and (2.2.10) we may again produce a linear rate equation

$$\frac{du}{dt} = -\frac{1}{\tau_I}(C_0 - u)$$

where the inertial relaxation time τ_I is defined as

$$\tau_I = \frac{d^2 \rho_f (1 + 2.70\,\mathrm{Kn})}{18\mu} \qquad (2.2.11)$$

The equations can be made dimensionless by using parameters $u'(=u/C_0)$ and $t'(=tC_0/L)$, the solution being determined by the dimensionless inertia parameter P_I

$$P_I = \frac{18\mu L}{d^2 C_0 \rho_f (1 + 2.70\,\mathrm{Kn})} \qquad (2.2.12)$$

2.2.4 Typical Values of Relaxation Time

From Eqs. (2.2.6) and (2.2.11) we can see that relaxation times for wet steam are not particularly sensitive to the pressure and temperature of the steam over the normal range of conditions found in turbines.

For low-pressure steam ($p < 1$ bar) we may insert the following values into the equations to obtain typical values of τ

$$c_{pg} = 1.9 \times 10^3 \text{ J/kg} \cdot \text{C}$$
$$\rho_f \cong 1.0 \times 10^3 \text{ kg/m}^3$$
$$\lambda_g \cong 0.024 \text{ J/ms} \cdot \text{C}$$
$$\mu \cong 1.0 \times 10^{-5} \text{ N} \cdot \text{s/m}^2$$

Then for a wet-steam mixture of

$$Y = 0.10$$
$$d = 1.0 \text{ μm}$$
$$\tau_T \cong 60 \text{ μs}$$
$$\tau_I \cong 5 \text{ μs}$$

We may now consider typical turbine flows based on the dimensions given in Fig. 2.2.3. Times for the flow to pass through the turbine will be approximately

	HP	LP
Through the turbine	8 ms	5 ms
Through one blade row	300 μs	200 μs

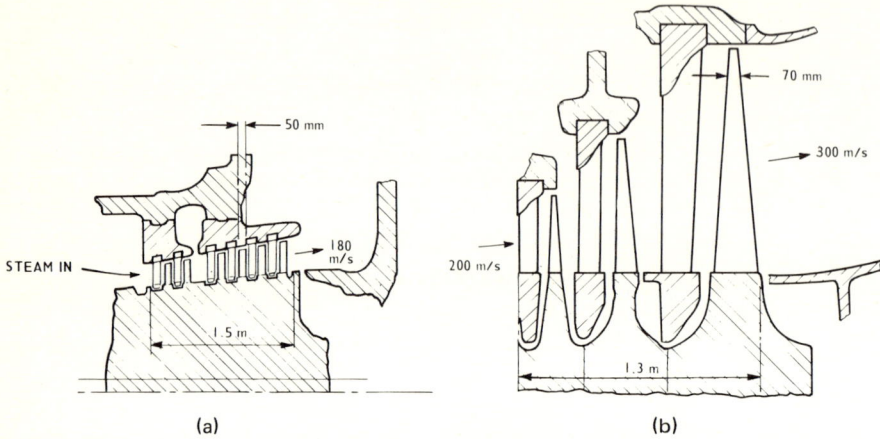

FIG. 2.2.3 Section through (*a*) typical HP turbine and (*b*) wet-steam stages of a typical LP turbine.

Significant relaxation is therefore unlikely over the complete turbine but within the blade rows thermal relaxation appears quite probable, particularly where the fog droplets are greater than 1 μm in diameter.

It will also be noticed that inertial relaxation time is considerably less than the thermal value. From Eqs. (2.2.6) and (2.2.11) we can write the ratio

$$\frac{\tau_T}{\tau_I} = \frac{1-Y}{Y} 1.5 \frac{c_{pg}\mu}{\lambda_g}$$

The dimensionless group on the right-hand side, the Prandtl number for the gaseous phase, is approximately 0.85 and is substantially independent of pressure and temperature. The above ratio is therefore dominated by the phase ratio P_Y where

$$P_Y = \frac{Y}{1-Y} \qquad (2.2.13)$$

2.3 ONE-DIMENSIONAL EQUATIONS FOR WET-STEAM FLOW

To calculate relaxation effects in a turbine we require the equations of conservation of mass, momentum, and energy for a wet-steam mixture. These equations are developed below for unsteady and steady one-dimensional flow through a stream tube of arbitrary cross section. Limiting assumptions at this stage are that

1. the liquid phase is present as a monodispersion
2. the total number of droplets in the flow remains constant (i.e., no spontaneous condensation occurs)

3 the drops are small ($<5\ \mu m$ dia) such that their velocity relative to the steam phase is also small
4 the drops are not very small ($>0.05\ \mu m$ dia) so that the pressure within the drop is substantially equal to the pressure of the surrounding steam

The influence of spontaneous condensation on the flow equations is considered in Sec. 2.4. The adherence to one-dimensional flow is considered adequate to describe the influence of relaxation and, with the advance of streamline-curvature techniques, is not limiting for the calculation of blade-to-blade or axisymmetric flows in wet-steam turbines.

2.3.1 Continuity Equation

The equations are developed for a fixed control volume dV in the flow, as shown in Fig. 2.3.1. As we assume no formation or complete evaporation of waterdrops we can write a droplet number conservation equation

$$\frac{\partial}{\partial t}\int_V n\,dV + \oint_A n u_n\,dA = 0$$

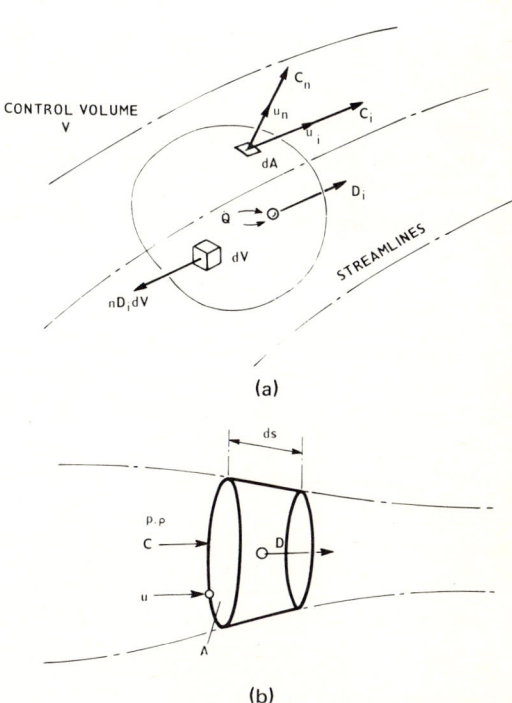

FIG. 2.3.1 Control volume: (a) general and (b) stream tube.

where n is the number of drops/unit volume and u_n is the component of droplet velocity normal to element dA of the control volume surface area for outgoing droplets.

For calculation of flow along a stream tube we may perform the integrations for cross-sectional area $A = A(s)$ and specify that conditions are uniform at each cross section. Further, for small $dV(= A\,ds)$ conditions are uniform within the control volume. The one-dimensional stream tube equation then becomes

$$A\frac{\partial n}{\partial t} + \frac{\partial}{\partial s} nuA = 0$$

Finally, expressing droplet number as $N(= nA)$, the number of droplets per unit length of stream tube, the number conservation equation becomes

$$\frac{\partial N}{\partial t} + \frac{\partial}{\partial s} Nu = 0$$

The continuity equation for the liquid phase may now be developed in a similar manner. For the control volume dV, the net rate of increase in mass of the liquid phase equals the rate of condensation from the gaseous phase, thus

$$\frac{\partial}{\partial t}\int_v nm\,dV + \int_A nmu_n\,dA = \int_v \dot{Q}n\,dV$$

where m is the mass of a droplet and $\dot{Q}$ the rate of mass increase per droplet by condensation. For a stream tube we again specify uniformity over any cross-sectional area, the continuity equation becoming

$$A\frac{\partial}{\partial t} nm + \frac{\partial}{\partial s} nmuA = \dot{Q}nA$$

which reduces to the expected result

$$\frac{\partial m}{\partial t} + u\frac{\partial m}{\partial s} = \dot{Q} \tag{2.3.1}$$

Continuity for the gaseous phase follows in a similar manner

$$\frac{\partial}{\partial t} A\rho_g + \frac{\partial}{\partial s} A\rho_g C = -N\dot{Q} \tag{2.3.2}$$

For steady flow, partial derivatives with respect to time t become zero, and the set of continuity equations may be written

$Nu = $ constant

$$u \frac{dm}{ds} = \dot{Q}$$

$$\frac{d}{ds} A\rho_g C + N\dot{Q} = 0 \qquad (2.3.3)$$

2.3.2 Momentum Equation for the Liquid Phase

The vector relationship given by Newton's second law—force equals rate of change of momentum—may be applied in direction i to a fixed control volume as in Fig. 2.3.1, to give

$$\Sigma F_i = \frac{D}{Dt} mnu_i \, dV - n\dot{Q}C_i \, dV$$

The summation sign indicates the total external forces acting on the control volume dV containing the water drops and in this case equals the component of total aerodynamic drag of the steam on the drops. It is assumed that the drag D per droplet acts in the direction of the relative velocity $(C-u)$. The droplets themselves exert no pressure on the surface of the control volume, and it is assumed also that forces on the drops due to gradients of pressure in the gaseous phase are negligible. It can be seen that the mass increase of the liquid phase due to condensation is assumed to have the velocity component C_i of the gaseous phase. If evaporation occurs, this term would be modified to contain the droplet velocity u_i.

Expanding the basic equation

$$\int_v nD_i \, dV = \frac{\partial}{\partial t} \int_v nmu_i \, dV + \int_A nu_n mu_i \, dA - \int_v \dot{Q}nC_i \, dV$$

and making the appropriate simplifications for flow through a stream tube

$$ND = \frac{\partial}{\partial t} Num + \frac{\partial}{\partial s} Nu^2 m - N\dot{Q}A$$

Expanding terms and using the continuity relationship (2.3.1), the liquid phase momentum equation becomes

$$D = m\left(\frac{\partial u}{\partial t} + u \frac{\partial u}{\partial s}\right) - (C - u)\dot{Q} \qquad (2.3.4)$$

The second term on the right-hand side of the equation can now be seen to be the impulsive force produced by condensing particles undergoing sudden deceleration from gas velocity C to droplet velocity u. For evaporation this term does not appear in the equation.

For steady flow we omit the partial differential $\partial u/\partial t$ and obtain

$$D = mu\frac{du}{ds} - (C - u)\dot{Q} \qquad (2.3.5)$$

2.3.3 Momentum Equation for the Gaseous Phase

The momentum equation for the gaseous phase in control volume dV can be derived in a similar manner

$$\Sigma F_i = \frac{D}{Dt}(\rho_g C_i\,dV) + n\dot{Q}C_i\,dV$$

which can be expanded as follows

$$\int_A (p\,dA)_i - \int_v nD_i\,dV = \frac{\partial}{\partial t}\int_v \rho_g C_i\,dV + \int_A \rho_g C_i C_n\,dA + \int_v Qn C_i\,dV$$

For a stream tube it can be readily shown that this equation together with the continuity equation gives the momentum equation for the gaseous phase

$$-A\frac{\partial p}{\partial s} - ND = \rho_g A\left(\frac{\partial C}{\partial t} + C\frac{\partial C}{\partial s}\right) \qquad (2.3.6)$$

which for steady flow becomes

$$A\frac{dp}{ds} + ND + \rho_g AC\frac{dC}{ds} = 0 \qquad (2.3.7)$$

2.3.4 Energy Equation

Applying the First Law of Thermodynamics to the control volume dV, making the common assumption that no heat is conducted or radiated across the control volume surface, we write

$$\frac{dE}{dt} + \frac{dW}{dt} = 0$$

where E is the total energy and W is the work done by the system. For a two-phase mixture we can expand this expression

$$\frac{\partial}{\partial t}\int_v (E_g\rho_g + E_f nm)\,dV + \int_A (E_g\rho_g C_n + E_f nmu_n)\,dA$$
$$+ \int_A \left(\frac{p}{\rho_g}\right)\rho_g C_n\,dA + P = 0$$

The third and fourth terms represent the work done on the boundaries of volume dV and the work extracted mechanically by some moving surface immersed in the control volume respectively. The energy exchange on change of phase is totally within volume dV and therefore does not appear in the equation for the mixture.

For a stream tube we assume zero mechanical work extracted ($P=0$), and in terms of total enthalpy H the above equation can be rewritten

$$\frac{\partial}{\partial t}\left(A\rho_g c_{vg} T_g + A\rho_g \frac{C^2}{2}\right) + \frac{\partial}{\partial t}\left(Nmc_{vf}T_f + Nm\frac{u^2}{2}\right) + \frac{\partial}{\partial s}(H_g A\rho_g C)$$
$$+ \frac{\partial}{\partial s}(H_f Nmu) = 0 \quad (2.3.8)$$

For steady flow we omit time derivatives and expand to give

$$A\rho_g C^2 \frac{dC}{ds} + (Nu)mu\frac{du}{ds} + A\rho_g C\frac{dh_g}{ds} + (Nu)m\frac{dh_f}{ds}$$
$$- Nu\left(h_{fg} + \frac{C^2 - u^2}{2}\right)\frac{dm}{ds} = 0 \quad (2.3.9)$$

2.3.5 Equations of State

From Sec. 2.1 we can insert an appropriate expression for gas constant $R\,[=R(p,T)]$ into the usual equation

$$\frac{p}{\rho_g} = RT_g \quad (2.3.10)$$

which in differential form becomes

$$\frac{dp}{p}\left[1 - \frac{p}{R}\left(\frac{\partial R}{\partial p}\right)_{T_g}\right] - \frac{d\rho_g}{\rho_g} = \frac{dT_g}{T_g}\left[1 + \frac{T_g}{R}\left(\frac{\partial R}{\partial T_g}\right)_p\right] \quad (2.3.11)$$

Similarly, for a nonideal gas, enthalpy h_g is also a function of pressure and temperature so that

$$dh_g = \left(\frac{\partial h_g}{\partial T_g}\right)_p dT_g + \left(\frac{\partial h_g}{\partial p}\right)_{T_g} dp$$

From the Maxwell equations for basic thermodynamic relationships [2-13], it can be shown that

$$\left(\frac{\partial h_g}{\partial p}\right)_{T_g} = \frac{1}{\rho_g} + \frac{T_g}{\rho_g^2}\left(\frac{\partial \rho_g}{\partial T_g}\right)_p$$

Combining the above equations and using a differential form of Eq. (2.3.10), we can derive the following general relationship

$$dh_g = c_{pg}\, dT_g - \frac{T_g}{\rho_g R}\left(\frac{\partial R}{\partial T}\right)_p dp \qquad (2.3.12)$$

Enthalpy change for the liquid phase can be written

$$dh_f = c_{vf}\, dT_f$$

and from Clapeyron's equation [2-13] a change in saturation temperature can be related to pressure change

$$\frac{dT_f}{dp} = \frac{T_f v_{fg}}{h_{fg}}$$

For the liquid phase we may therefore write

$$dh_f = \left(\frac{c_{vf} T_f v_{fg}}{h_{fg}}\right) dp \qquad (2.3.13)$$

2.3.6 Supplementary Equations

It is assumed that droplet temperature corresponds at any time to the saturation value for the prevailing pressure. The latent heat released to the droplets by condensing vapor is therefore returned by conduction/convection to the vapor phase, or

$$h_{fg} \dot{Q} = \pi d^2 \alpha (T_f - T_g)$$

Inserting the expression for heat transfer coefficient α from Eq. (2.2.5) gives

$$\dot{Q} = m(T_f - T_g)\left[\frac{12\lambda_g}{d^2 \rho_f h_{fg}(1 + 3.18\, \mathrm{Kn})}\right] \qquad (2.3.14)$$

Drag force D may also be obtained from Eqs. (2.2.9) and (2.2.10), where for small relative velocity or slip

$$D = \frac{3\pi\mu\, d(C-u)}{1 + 2.70\, \mathrm{Kn}} \qquad (2.3.15)$$

In the above equations, allowance has been made for deviation from continuum flow at low pressure by the introduction of Knudsen number Kn where from [2-9]

$$\mathrm{Kn} = \frac{l}{d} = \frac{1.5\mu}{d\rho_g}\left(\frac{1}{RT_g}\right)^{1/2} \qquad (2.3.16)$$

2.3.7 Dimensionless Equations for Steady Flow

Equations (2.3.3), (2.3.5), (2.3.7), and (2.3.9) for steady flow will now be considered in dimensionless form with certain simplifying assumptions. For brevity we will consider the equations for the gaseous phase as an ideal gas, a reasonable approximation for low-pressure steam. Similarly in practice the change in enthalpy of the liquid phase is small. We therefore assume

$$R = \text{constant}$$
$$dh_g = c_{pg}\, dT_g$$
$$dh_f = 0$$

We may also safely neglect the impulse term in Eq. (2.3.5).

Introducing reference values A_0, C_0, d_0, ρ_0, L we now define dimensionless variables as follows

$$\bar{A} \equiv \frac{A}{A_0} \qquad \bar{C} \equiv \frac{C}{C_0} \qquad \bar{u} \equiv \frac{u}{C_0}$$

$$\bar{d} \equiv \frac{d}{d_0} \qquad \bar{s} \equiv \frac{s}{L} \qquad \bar{\rho}_g \equiv \frac{\rho_g}{\rho_0}$$

$$\bar{h} \equiv \frac{h}{C_0^2} \qquad \bar{p} \equiv \frac{p}{\rho_0 C_0^2} \qquad \bar{D} \equiv \frac{NDL}{A_0 \rho_0 C_0^2}$$

The equations for steady flow may then be written as follows (for clarity bars have been omitted and all variables are dimensionless)

$$\frac{du}{ds} = P_I \left(\frac{C-u}{ud^2}\right)$$

$$\frac{dd}{ds} = \frac{P_T}{P_I}\left[\frac{c_{pg}}{c_{vf}}\,h_f - h_g\right](udh_{fg})^{-1}$$

$$D = P_Y P_I \frac{d(C-u)}{u}$$

$$\frac{1}{C}\frac{dC}{ds} = (G_2 - G_1 - G_3)\left(\frac{\rho_g C^2}{p} - \frac{C^2}{c_{pg}T_g} - 1\right)^{-1}$$

$$\frac{1}{\rho}\frac{d\rho}{ds} = G_1 - \frac{1}{C}\frac{dC}{ds}$$

$$\frac{1}{p}\frac{dp}{ds} = G_2 - \frac{\rho_g C}{p}\frac{dC}{ds}$$

$$\frac{1}{h_g}\frac{dh_g}{ds} = G_3 - \frac{C}{h_{fg}}\frac{dC}{ds}$$

where

$$G_1 = P_Y \frac{3d^2}{A\rho_g C}\frac{dd}{ds} - \frac{1}{A}\frac{dA}{ds}$$

$$G_2 = -\frac{D}{pA}$$

$$G_3 = -\frac{Du}{h_g A\rho_g C} + P_Y 3d^2 \frac{dd}{ds}\left(\frac{h_{fg}}{h_g} + \frac{C^2 - u^2}{2h_g}\right)$$

The solution for the dimensionless variables can be seen to depend only upon the values of the parameters P_T, P_I, P_Y, defined in Eqs. (2.2.7), (2.2.12), and (2.2.13), respectively.

2.3.8 Application of the Flow Equations

The calculation of flow conditions in an arbitrary stream tube is an initial value problem for which several numerical methods are available [2-14]. For example a fourth-order Runge-Kutta method has been used in the calculations shown later and is generally stable if the interval ds is small. However before applying the equations, the following definitions should be noted.

2.3.8.1 Wetness Fraction Y and Specific Volume v_m

Where the phase velocities are not equal, differing values of Y and v for the flow may be defined, based on volume and flow rate respectively. In terms of unit volume of stream tube, we may write

$$Y = \frac{1}{1 + A\rho_g/Nm}$$

$$v_m = \frac{v_g}{1 + Nm/A\rho_g}$$

whereas on a mass and volume flow-rate basis

$$Y = \frac{1}{1 + CA\rho_g/Num}$$

$$v_m = \frac{v_g}{1 + Num/A\rho_g C}$$

The latter definition is used in the general flow equations developed in the previous section.

2.3.8.2 Expansion Index

To indicate the degree of relaxation occurring in the flow it is convenient to define an effective expansion index for the gaseous phase. Assuming the expansion corresponds to

$$pT_g^{n/(1-n)} = \text{constant}$$

then

$$n = \frac{1}{1 + (T_g/p)(dp/dT_g)} \tag{2.3.17}$$

From Sec. 2.1, for nonequilibrium expansions $n \to 1.30$ and for equilibrium isentropic flow $n \to 1.12$. Equilibrium flow with frictional losses may produce values of $n < 1.12$.

2.3.8.3 Polydisperse Liquid Phase

In practice the dispersed phase will not consist of droplets of equal size. Typical size spectra are shown diagrammatically in Fig. 2.3.2 and various formulae are available in the literature to correlate the probability-size distribution. That of Mugele and Evans [2-15] is particularly useful, being based on a realistic upper limit in drop diameter d_{max}.

It is often convenient to represent a spectrum by an equivalent monodispersion of drops of a mean diameter, denoted d_{ab}, where

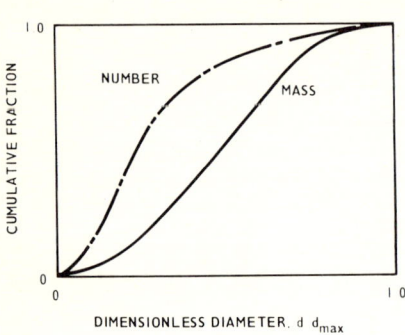
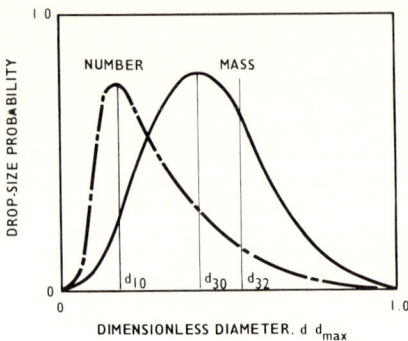

FIG. 2.3.2 Droplet size spectra.

$$d_{ab}{}^{a-b} = d_{\max}^{a-b} \frac{\int_0^1 \left(\frac{d}{d_{\max}}\right)^a \frac{dN^*}{d(d/d_{\max})} \, d(d/d_{\max})}{\int_0^1 \left(\frac{d}{d_{\max}}\right)^b \frac{dN^*}{d(d/d_{\max})} \, d(d/d_{\max})}$$

and N^* is the cumulative number fraction of droplets below diameter d in the spectrum. For example, the most probable diameter (number mean) is d_{10}; the mass mean is d_{30}.

However, for use in flow equations, the effective mean diameter is determined by the type of interaction between the phases that is controlling the process. For example, where condensation is occurring onto existing drops from subcooled vapor, the equivalent monodispersion must produce the same rate of increase in wetness fraction as the actual spectrum. Thus equating the heat transfer rate to the rate of latent heat release

$$n_{\text{tot}} \int_0^{d_{\max}} \pi d^2 \alpha (T_f - T_g) \frac{\partial N^*}{\partial d} \, dd = h_{fg} \dot{Q}_{\text{tot}}$$

where n_{tot} is the total number of droplets in the spectrum per unit mass of mixture. Introducing Eq. (2.2.5) for heat transfer coefficient α and equating to the heat transferred from the equivalent monodispersion gives

$$n_{\text{tot}} \int_0^{d_{\max}} \frac{d}{1 + 3.18 \, \text{Kn}} \frac{\partial N^*}{\partial d} \, dd = \frac{(n_{\text{tot}})_e \, d_e}{1 + 3.18 \, \text{Kn}_e}$$

The droplet number $(n_{\text{tot}})_e$ for the monodispersion may be defined to give the same total wetness fraction as the spectrum, thus

$$(n_{\text{tot}})_e = \left(\frac{1}{d_e^3}\right) \int d^3 n_{\text{tot}} \frac{\partial N^*}{\partial d} \, dd = \frac{Y}{(\pi/6) d_e^3 \rho_f}$$

and substituting gives the effective drop diameter

$$d_e^2 (1 + 3.18 \, \text{Kn}_e) = \frac{\int d^3 (\partial N^*/\partial d) \, dd}{\int d/(1 + 3.18 \, \text{Kn})(\partial N^*/\partial d) \, dd} \quad (2.3.18)$$

For practical cases this expression may be simplified to give, at high pressure, where $\text{Kn} \to 0$

$$d_e = d_{31}$$

and at low pressure, when $3.18 \text{Kn} \gg 1$

$$d_e = d_{32}$$

The use of an equivalent monodispersion in computations of wet-steam flow depends implicitly upon the spectrum shape remaining unchanged, i.e.

$$\frac{d}{dt}\left[\frac{\partial N^*}{\partial (d/d_{\max})}\right] = 0$$

which is approximately the case at low pressure, but less so at high pressure.

2.3.9 Examples of Flow Calculations

To demonstrate the extent of relaxation in typical wet-steam flows, some examples have been calculated as shown in Figs. 2.3.3–2.3.5.

In Fig. 2.3.3, expansion in a steam engine cylinder has been simulated for an assumed monodispersion of 3 μm dia droplets at two values of expansion rate, $(1/V_1)(dV/dt)$. At the lower rate of 50 s^{-1}, corresponding in practice to an engine speed of 300 rpm, it can be seen that a near-equilibrium expansion occurs at an effective index near the value given by Rankine. Thermal relaxation time for the wet steam is approximately 0.6 ms, compared with a stroke time of 40 ms. At the much higher engine speed, 3000 rpm, the expansion rate of 500 s^{-1} reduces the stroke time to 4 ms, and it can be seen that the expansion is now decidedly not in equilibrium. In practice of course steam engines do not operate at such a speed.

The second example concerns inertial relaxation only, at the mouth of a Pitot tube. The calculation is described in more detail in [2-16]. Figure 2.3.4 shows the vapor streamlines and drop trajectories for two values of inertia parameter. The decrease in P_I from 5.71 to 0.186 corresponds to an increase

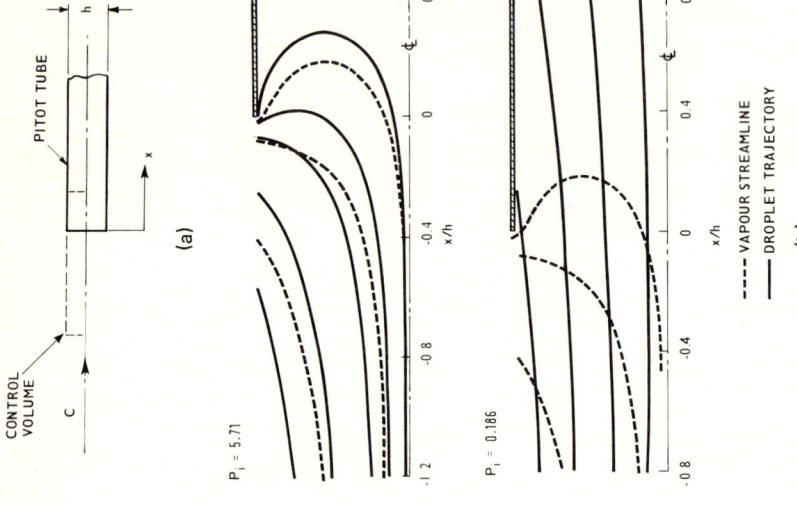

FIG. 2.3.4 (a) Pitot-tube nomenclature; (b) droplet trajectories at the mouth of a Pitot tube.

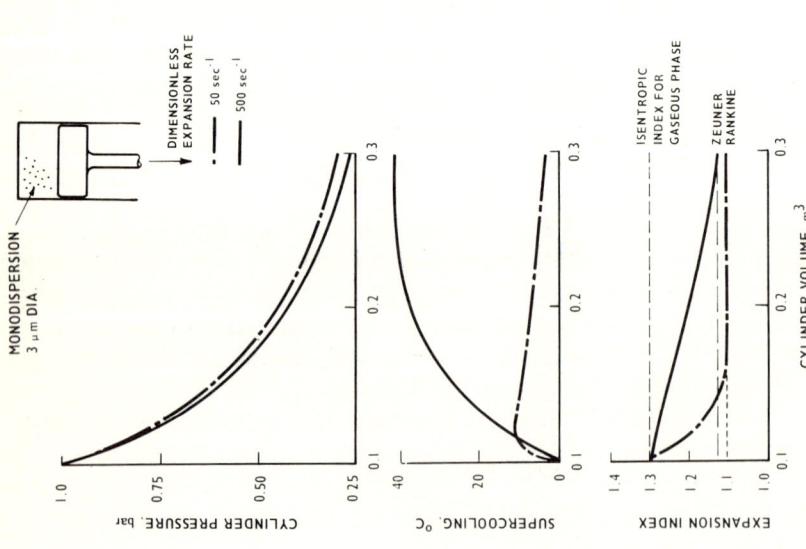

FIG. 2.3.3 Nonequilibrium expansion in a steam engine.

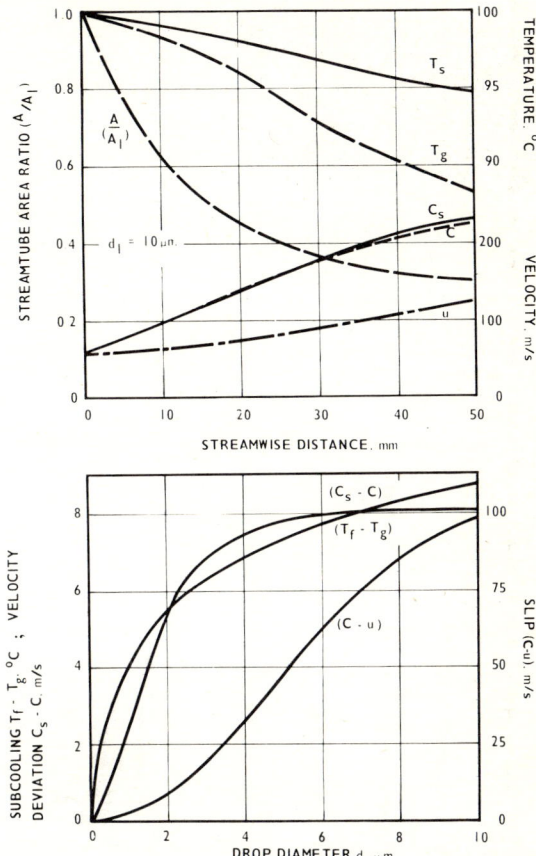

FIG. 2.3.5 One-dimensional wet-steam expansion.

of ~ 5 times in droplet diameter and the resulting nonequilibrium flow can be clearly seen.

Finally, Fig. 2.3.5 shows the calculated wet-steam flow through a simple convergent nozzle of the proportions of a turbine blade row. The various nonequilibrium effects are again evident, but it can be seen that slip is small for droplets of less than 1 μm dia.

2.4 ONE-DIMENSIONAL FLOW WITH SPONTANEOUS CONDENSATION

2.4.1 Flow Equations

To complete the mathematical description of two-phase flows the effect of spontaneous condensation (i.e., the formation of new droplets) on the flow must be included. The details of the nucleation process will be described in

Chap. 3; the present section considers the solution of the flow equations when spontaneous condensation occurs. We shall assume for simplicity that the steam and droplets in the flow move at the same velocity; as shown in the previous section this is reasonable for droplets of less than 1 μm dia.

The continuity equation can now be expressed conveniently in terms of wetness fraction Y as follows

$$\frac{1}{A}\frac{dA}{ds} + \frac{1}{C}\frac{dC}{ds} + \frac{1}{1-Y}\frac{dY}{ds} - \frac{1}{v_g}\frac{dv_g}{ds} = 0$$

The component dY/ds is composed of the increase of the liquid phase, by condensation onto existing drops and by formation of new drops. That is

$$\frac{dY}{ds} = \int_{d_*}^{d\,\max} \frac{\alpha \pi d^2 (T_{f,d} - T_g) n_m}{C h_{fg}} \left(\frac{\partial N^*}{\partial d}\right) dd + J\frac{\pi d_*^3}{6v_f C} = \frac{\dot{Q}_Y}{C}$$

where J is the rate of formation of new drops per unit mass of wet steam. Formulae for J will be presented in Chap. 3 that show nucleation rate is critically dependent on the degree of supersaturation of the steam. New drops nucleate at a critical drop size d_* where

$$d_* = \frac{4\sigma v_f}{h_{fg} \ln(T_g/T_{\text{sat}})}$$

The temperature T_f of small droplets deviates from the flat-surface saturation value T_{sat} as the result of surface tension effects, and can be written

$$T_{f,d} = T_{\text{sat}} - (T_{\text{sat}} - T_g)\left(\frac{d_*}{d}\right)$$

Further details of these relationships will be given in Chap. 3.

The equations of momentum and energy conservation may be written for zero slip as

$$v_g(1-Y)\frac{dp}{ds} + C\frac{dC}{ds} = 0$$

$$(1-Y)c_{pg}\frac{dT_g}{ds} - h_{fg}\frac{dY}{ds} + C\frac{dC}{ds} + Yc_{vf}\frac{dT_f}{ds} = 0$$

respectively, treating the gaseous phase as a perfect gas and neglecting small changes in the enthalpy of the condensing vapor.

The equations of state and of heat transfer from droplets to steam are presented in Sec. 2.3.

2.4.2 Transonic Solution

The closed set of simultaneous differential equations describing the flow can be expressed in matrix vector form

$$[B] \frac{d\mathbf{w}}{ds} = \mathbf{e}$$

where **w** and **e** are column vectors with transposes

$$\mathbf{w}^T = C, v_g, T_g, p, Y$$

$$\mathbf{e}^T = -Cv_g(1-Y)\frac{dA}{ds}, 0, -Yc_{pg}\frac{dT_f}{ds}, 0, \dot{Q}_Y$$

and [B] is the matrix

$$\begin{bmatrix} v_g(1-Y)A & -CA(1-Y) & 0 & 0 & CAv_g \\ C & 0 & 0 & v_g(1-Y) & 0 \\ C & 0 & (1-Y)c_{pg} & 0 & -h_{fg} \\ 0 & pT_g & -pv_g & v_gT_g & 0 \\ 0 & 0 & 0 & 0 & C \end{bmatrix}$$

Matrix B is singular and the equations therefore have no solution where the determinant B is zero, i.e., where

$$C^2 = \frac{pv_g T_g c_{pg}(1-Y)}{c_{pg}T_g - pv_g} = a^2 \qquad (2.4.1)$$

the local sonic velocity through the wet-steam mixture.

Numerical integration becomes difficult near this sonic point, particularly where a choked or sonic flow is required, as can be seen from the equation for velocity C

$$\frac{1}{C}\frac{dC}{ds} = \frac{(1/A)(dA/ds) + [1/(1-Y)](\dot{Q}_Y/C)}{1 - M^2} \qquad (2.4.2)$$

where $M(= C/a)$ is the flow Mach number. Calculation of transonic flow in nozzles is a familiar problem for a single phase flow [2-13] where Eq. (2.4.2) shows that the sonic condition occurs at the minimum area position, or throat. In such cases the solution is found by systematic modification of the inlet conditions until the sonic throat condition is obtained. For wet steam the position of the sonic point is not known a priori, and the iteration on inlet conditions must be based on the accurate location of a singularity, a sensitive and potentially unstable process. The integration of ordinary differential equations through singularities has been studied previously [2-17 to 2-19] and the method of Davidson [2-17] is described here.

The procedure is basically the introduction of a new variable x such that

$$\frac{ds}{dx} = 1 - M^2 \qquad (2.4.3)$$

Multiplying each of the flow equations by ds/dx then removes the singularity; for example Eq. (2.4.2) becomes

$$\frac{1}{C}\frac{dC}{dx} = -\frac{1}{A}\frac{dA}{ds} + \frac{1}{1-Y}\frac{\dot{Q}_Y}{C}$$

the gradient dC/dx being zero at the sonic point. Equation (2.4.3) is added to the new set of simultaneous equations, which may now be integrated using standard techniques. Note that in order to make x continuous, the sign of s must reverse at the sonic point, as in Eq. (2.4.3).

The iteration to determine the critical mass flow is based on the simultaneous satisfaction of the equations

$$\frac{ds}{dx} = 0 \qquad \frac{dC}{dx} = 0$$

As shown diagrammatically in Fig. 2.4.1, the calculation begins at (C_1, s) and proceeds towards the sonic point. In successive solutions $C_1 \to C_{1c}$, the critical value, on the basis

$$\frac{ds}{dx} \to 0 \qquad C_1 \text{ is decreased}$$

$$\frac{dC}{dx} \to 0 \qquad C_1 \text{ is increased}$$

until a sufficiently accurate value C_1^* is obtained. Using this inlet condition, integration proceeds to an optimum point C_{opt}, where the gradient $(dC/ds)_{\text{opt}}$ is close to the slope at the sonic point. The optimum position corresponds to the minimum value of

$$\left| \frac{dC/dx}{d^2C/dx^2} - \frac{ds/dx}{d^2s/dx^2} \right|$$

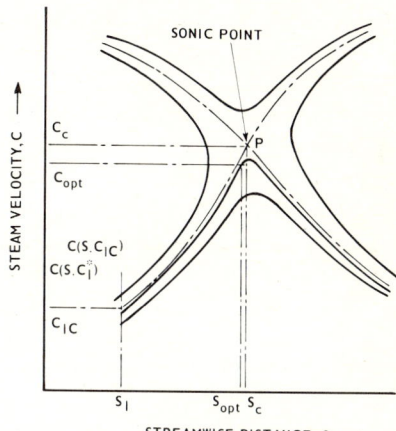

FIG. 2.4.1 Diagram of the saddle-point singularity at sonic velocity in a nozzle.

A property of saddle-point singularities is that the slope of the critical solution is linear in the region of the critical point. The integration proceeds from C_{opt}, therefore, by assuming a simple linear extrapolation to a position s_2 downstream of the critical value s_c

$$C_2 = C_{opt} + \left(\frac{dC}{ds}\right)_{opt} (s_2 - s_{opt})$$

Integration can then continue in the supersonic region in the normal way. Some examples of calculations of this type for convergent-divergent nozzles are shown in Fig. 2.4.2a-d and Fig. 2.4.3 from [2-23].

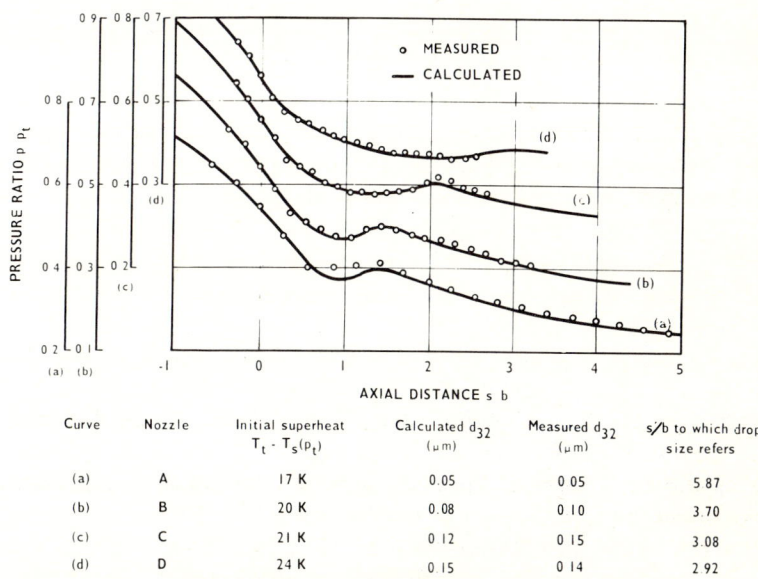

Curve	Nozzle	Initial superheat $T_t - T_s(p_t)$	Calculated d_{32} (μm)	Measured d_{32} (μm)	s/b to which drop size refers
(a)	A	17 K	0.05	0.05	5.87
(b)	B	20 K	0.08	0.10	3.70
(c)	C	21 K	0.12	0.15	3.08
(d)	D	24 K	0.15	0.14	2.92

FIG. 2.4.2 Examples of nozzle flow with initially superheated steam.

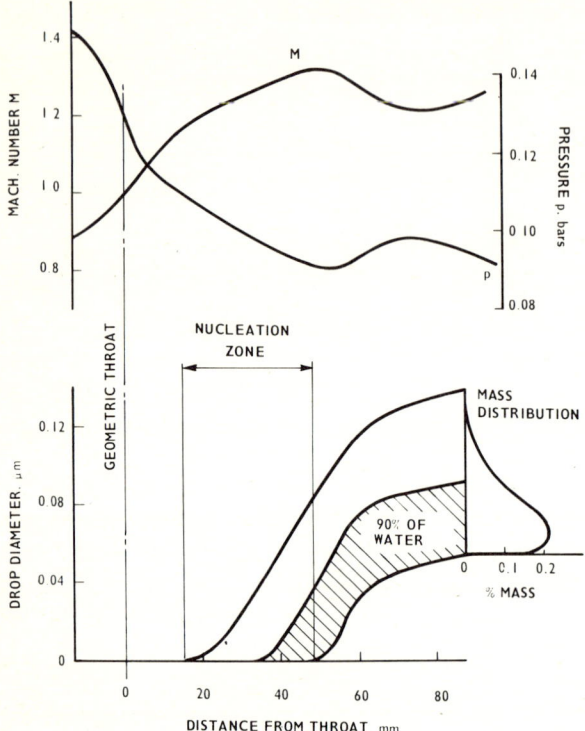

FIG. 2.4.3 Calculated details of spontaneous condensation in a nozzle.

2.4.3 Sudden Heat Release on Condensation

The examples shown in Figs. 2.4.2 and 2.4.3 clearly show the rapid deceleration of the flow and pressure rise following the release of latent heat by spontaneous condensation. The effect of heat addition on subsonic and supersonic flows has been described, for example in [2-12]. In a similar manner the influence coefficients for wet steam are shown in Table 2.4.1.

Inserting appropriate values for c_p, h_{fg}, etc., it can be shown that the effects are qualitatively similar to single phase flows, i.e.

$$dY > 0 \quad \begin{array}{lll} M < 1 & dC > 0 & dp < 0 \\ M > 1 & dC < 0 & dp > 0 \end{array}$$

Spontaneous condensation therefore has the effect of moving flow Mach number toward unity, the effect produced when $M = 1$ being described by Pouring [2-20] as "thermal choking." Heat addition in excess of this quantity is termed supercritical and produces particular phenomena that have been

TABLE 2.4.1 Influence coefficients for wet steam (zero slip, perfect gas)

	$\dfrac{dA}{A}$	$\dfrac{dY}{1-Y}$
$\dfrac{dC}{C}$	$-\dfrac{1}{1-M^2}$	$\dfrac{1}{1-M^2}\dfrac{h_{fg}-c_pT}{c_pT}$
$\dfrac{(1-Y)dp}{\rho C^2}$	$\dfrac{1}{1-M^2}$	$-\dfrac{1}{1-M^2}\dfrac{h_{fg}-c_pT}{c_pT}$

analyzed in detail by Barschdorff [2-21]. Depending on the position of heat release, the ensuing flow may be steady or unsteady as shown diagrammatically in Fig. 2.4.4. Heat release (i.e., spontaneous condensation) in the region of the throat produces pressure changes at the throat that can propagate upstream and hence modify the nozzle inlet conditions. Oscillatory flows are therefore produced.

Condensation downstream of the throat can produce supercritical heat addition in a stable flow over a limited range, as indicated in Fig. 2.4.4a. Here heat release initially produces a strong deceleration of the flow to subsonic conditions, but continued condensation reestablishes supersonic conditions.

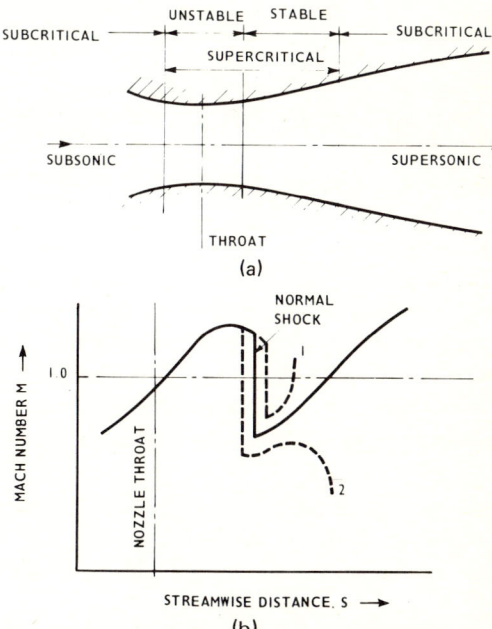

FIG. 2.4.4 (a) Zones of sub- and supercritical heat addition during condensation in a nozzle; (b) typical stable configuration of supercritical heat addition.

The stable configuration results in a reinforcement of pressure waves to form an "aerodynamic" normal shock as shown in Fig. 2.4.4b. The calculation procedure in such cases is therefore to select a position a in the subsonic part of the flow and to apply the normal Rankine-Hugoniot equations for the gaseous phase as given for example in [2-22] as

$$M_b^2 = \frac{1 + [(\chi - 1)/(\chi + 1)](M_a^2 - 1)}{1 + [2\chi/(\chi + 1)](M_a^2 - 1)}$$

$$\frac{p_b}{p_a} = 1 + \frac{2\chi}{\chi + 1}(M_a^2 - 1)$$

$$\frac{\rho_a}{\rho_b} = 1 - \frac{2}{\chi + 1}\left(1 - \frac{1}{M_a^2}\right)$$

If the continuing integration of the flow from position b downstream of the shock does not pass through a sonic point, a new shock position is selected. For example, Fig. 2.4.4b shows diagrammatically the effect of

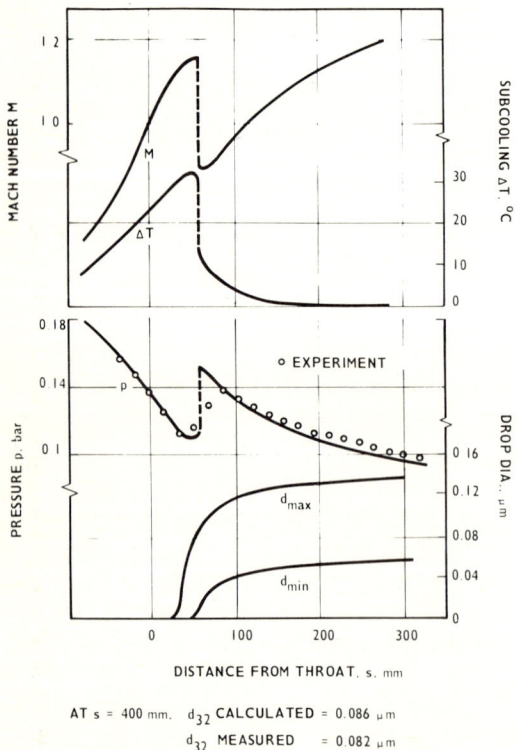

AT s = 400 mm. d_{32} CALCULATED = 0.086 μm
d_{32} MEASURED = 0.082 μm

FIG. 2.4.5 Example of condensation in a nozzle with supercritical heat addition.

selecting shock positions too far downstream (1) and too far upstream (2). The iteration for the correct position is analogous to the selection of the critical mass flow for the duct, and the method of Davidson previously described is appropriate. An actual example of supercritical heat addition is shown in Fig. 2.4.5.

2.5 CALCULATED EXAMPLES OF WET-STEAM FLOWS

2.5.1 Condensation in a Multistage HP Turbine

An attempt has been made to calculate the condensation in a typical HP turbine for a nuclear application. Flows relative to fixed and moving blades are considered as a series of one-dimensional convergent nozzles, with appropriate matching of absolute and relative flow angles by notional velocity triangles at each blade row outlet. The inlet conditions selected were

$$p = 60 \text{ bar (dry-saturated)}$$
$$C = 80 \text{ m/s (axial)}$$

The calculation was carried out for the first seven stages of an HP turbine assuming for each stage

Degree of reaction	50%
Fixed and moving blade exit angles (from circumferential direction)	25°
Moving blade velocity	166 m/s
Fixed and moving blade axial chord	50 mm
Isentropic efficiency	0.85

The equation of state (2.1.8) was used in the calculation.

Some results of the calculation are given in Fig. 2.5.1 and show nucleation occurring in the stage 2 moving blades. The range of drop sizes formed is small compared for example with supersonic condensation in a nozzle (Fig. 2.4.5). No secondary nucleation occurs within the remaining five stages, the low value of supercooling indicating a near-equilibrium expansion after stage 3. The first three stages however operate with supersaturated steam and correspondingly higher blade efflux velocities. Fog drop size is predicted to be relatively large at $\sim 2 \ \mu$m dia, and deposition of water on the blading would be significant. However, no allowance is made for deposition and re-entrainment in this example.

2.5.2 Transonic LP Blading

Flow through the tip sections of an LP turbine final-stage rotor is complex, the performance of blading being usually determined by cascade testing using

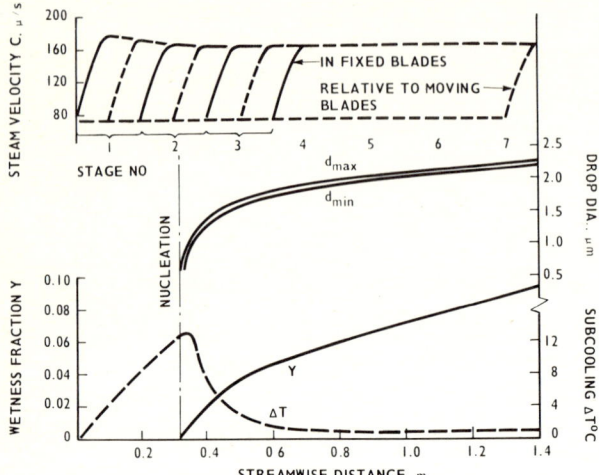

FIG. 2.5.1 Calculated expansion of wet steam through a seven-stage HP turbine.

a single-phase flow. In order to indicate the effect of wet steam on flow through blading, typical bounding stream tubes have been constructed for a tip section cascade as shown in Fig. 2.5.2. The area variation along the stream tubes was based on approximate calculations for equilibrium wet steam. The assumed inlet conditions to the cascade were

$$p = 0.30 \text{ bar (saturated, equilibrium)}$$

$$Y = 0.03$$

$$d_{32} = 0.50 \text{ μm}$$

The calculated flow in Fig. 2.5.2 shows that the droplet dispersion entering the blade was not sufficiently fine to prevent the steam from becoming highly supersaturated. Secondary nucleation is predicted for both stream tubes, stream tube A producing considerably smaller drops than stream tube B as a result of the extremely rapid Prandtl-Meyer expansion around the trailing edge of the pressure surface. However, the drop sizes predicted are approaching molecular dimensions, leading to the suspicion that we have exceeded the meaningful range of the nucleation equations used.

Although clearly an oversimplication of the flow, the calculations show that secondary nucleation with its attendant losses is likely to occur in the rotor tip sections of large LP turbines.

2.5.3 Condensation in a Two-dimensional Flow Field

We have so far considered wet-steam flows in a single dimension to demonstrate the main effects without unnecessary complexity. However,

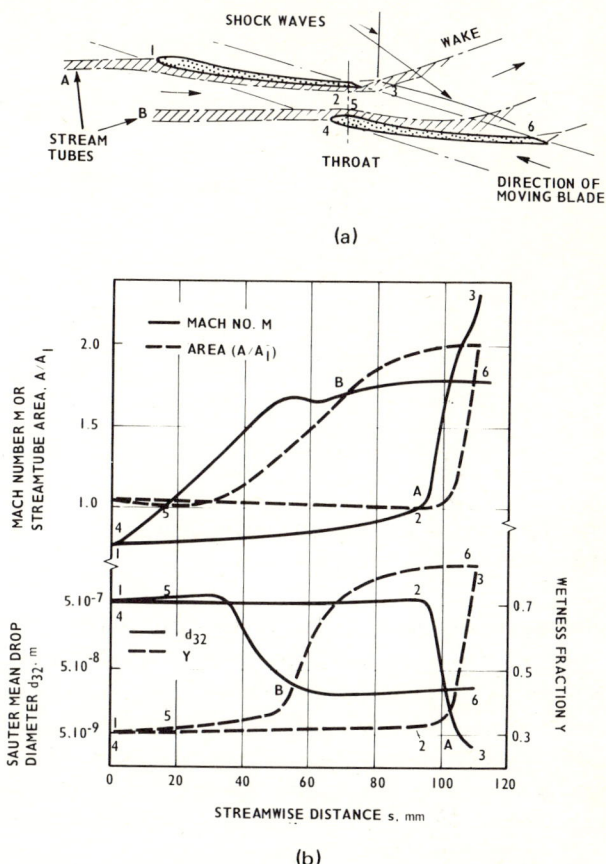

FIG. 2.5.2 (*a*) Example of tip section of a final stage LP rotor blade; (*b*) calculated condensation in stream tubes.

within turbine blade rows, particularly for cascades of high deflection angle, cross-channel gradients of steam properties may be appreciable. To determine the effects on spontaneous condensation of gradients of supersaturation normal to streamlines, the equations of Sec. 2.4 have been incorporated in a two-dimensional calculation procedure.

The theoretical method is based on the streamline curvature approach, in which the conservation equations along a stream tube are satisfied simultaneously with the momentum equation normal to the flow. An initial estimate of the position of the streamlines is obtained from a single-phase solution of the flow. Nucleation equations are then introduced and modified streamline positions determined by systematic iteration in the usual way. In the region of nucleation and initial droplet growth, fluid properties change rapidly and calculation step size must be reduced accordingly. To facilitate local control of calculation step size, it was found desirable to work in natural coordinates

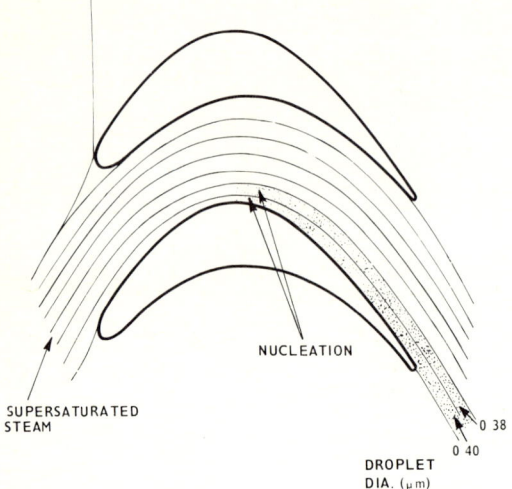

FIG. 2.5.3 Example of calculated two-dimensional flow through an impulse cascade.

(s, n—streamwise and normal) rather than to use the more usual blade-to-blade calculation procedure of specifying quasiorthogonals in the circumferential direction [2-24].

Results for an impulse blade section in Fig. 2.5.3 show that nucleation can occur at one side of the flow passage only. The water quantity formed is small, but the drop sizes are relatively large compared with those nucleated in the transonic blade of the previous example. The relevance of this phenomenon to the general problem of control of fog-drop size in turbines is the subject of continuing research.

2.6 ACOUSTIC VELOCITY IN WET STEAM

Steam flow in turbines is generally in the compressible regime, with some regions of transonic and supersonic flow. Mach number is therefore an important design parameter, requiring a knowledge of the acoustic velocity in wet steam. The subject has received considerable attention both theoretically and by experiment [2-11, 2-25 to 2-29] the most recent theory, by Petr [2-29], being summarized below.

The unsteady flow equations developed in Sec. 2.3 may be adapted to describe the propagation of a plane pressure wave of very small amplitude through stationary wet steam by replacing the main variables with the following perturbed values

$$\rho_g = \rho_{go} + \rho'_g$$
$$C = C'$$

$$u = u'$$
$$Y = Y_0 + Y'_e + Y'$$
$$T_g = T_{g0} + T'_{\text{sat}} + T'_g$$
$$E_g = E_{g0} + E'_g$$
$$p = p_0 + p'$$

Subscript zero denotes stationary value, the remaining components being small perturbations. Y'_e, T'_{sat} are perturbations of the equilibrium conditions resulting from a change in pressure; E is the internal energy of the gaseous phase. For plane flow, area A is taken as constant and equal to unity.

On this basis we may form the linearized small perturbation (or acoustic) equations, omitting subscripts on stationary terms.

1 For continuity

$$\frac{1}{\rho_g}\frac{\partial \rho'_g}{\partial t} + \frac{\partial C'}{\partial s} = \frac{1}{1-Y}\frac{Y' + Y'_e}{\tau_T}$$

$$\frac{dY}{dt} = \frac{dY_e}{dt} + \frac{dY'}{dt} = -\frac{Y' + Y'_e}{\tau_T}$$

2 From momentum conservation

$$\frac{\partial C'}{\partial t} + \frac{1}{\rho_g}\frac{\partial p'}{\partial s} = -\frac{Y}{1-Y}\frac{C' - u'}{\tau_I}$$

$$\frac{\partial u'}{\partial t} = \frac{C' - u'}{\tau_I}$$

3 The energy equation becomes

$$\frac{\partial E'_g}{\partial t} + \frac{p}{\rho_g}\frac{\partial C'}{\partial s} = -\frac{E_g - E_f}{1-Y}\frac{Y' + Y'_e}{\tau_T}$$

4 The equation of state provides

$$\frac{\partial \rho'_g}{\partial t} = \rho_g\left(\beta_g - \alpha_g\frac{RT^2_{\text{sat}}}{h_{fg}p}\right)\frac{\partial p'}{\partial t} - \rho_g\alpha_g\frac{\partial T'_g}{\partial t}$$

$$\frac{\partial E'_g}{\partial t} = c_{vg}\left(\frac{\partial T'_g}{\partial t} + \frac{RT^2_{\text{sat}}}{h_{fg}p}\frac{\partial p'}{\partial t}\right)$$

The perturbation of wetness due to change in saturation pressure is determined using the Clapeyron equation, giving

$$Y'_e = (1-Y)\left(\frac{c_{pg}}{h_{fg}} - \frac{1}{T_{sat}}\right)\frac{RT_{sat}^2}{h_{fg}p}p'$$

Terms α_g and β_g are the thermal coefficient of expansion and the isothermal compressibility respectively.

The sound wave is assumed to be a harmonic attenuating wave such that each variable V' has a solution

$$V'_{t,s} = e^{j\omega t} + e^{-(\gamma + j\omega/a)s}$$

We may therefore substitute in the acoustic equations for

$$\frac{\partial}{\partial t} \to j\omega \quad \frac{\partial}{\partial s} \to \gamma + j\frac{\omega}{a}$$

Making this substitution and eliminating all parameters except p', Petr shows that the real and complex parts of the resulting equation provide

$$\frac{a}{a_f} = \left(\frac{(1/X + \omega^2\tau_I^2)^2 + (Y/X)^2\omega^2\tau_I^2}{1/X^2 + \omega^2\tau_I^2}\right)\left\{\left(\frac{1}{X} + \omega^2\tau_I^2\right)\right.$$
$$\left. + \frac{1/X + \omega^2\tau_I^2 - (Y/X)\omega^2\tau_I\tau_T}{1 + \omega^2\tau_T^2}\left[X\left(\frac{a_f}{a_e}\right)^2 - 1\right]\right\}^{-1}\right)^{1/2} \quad (2.6.1)$$

and

$$\gamma_\lambda = \pi\left(\frac{a}{a_f}\right)^2\left\{\frac{Y}{X}\omega\tau_I + \frac{(Y/X)\omega\tau_I + \omega\tau(1/X + \omega^2\tau_I^2)}{1 + \omega^2\tau_T^2}\left[X\left(\frac{a_f}{a_e}\right)^2 - 1\right]\right\}$$
$$\cdot \left[\frac{(1/X + \omega^2\tau_I^2)^2 + (Y/X)^2\omega^2\tau_I^2}{1/X^2 + \omega^2\tau_I^2}\right]^{-1} \quad (2.6.2)$$

where a and γ_λ are the acoustic velocity and the absorption coefficient per wavelength respectively. The latter is defined from the ratio of successive amplitudes of the variables

$$\frac{V_s}{V_{s+2\pi a/\omega}} = e^{-(2\pi a/\omega)\gamma_\lambda}$$

The equations are expressed in terms of the "frozen" and "equilibrium" speeds of sound a_f, a_e. The "frozen" sound speed is defined as

$$a_f^2 = \frac{kp}{\rho_g} \quad (2.6.3)$$

that is, the gaseous phase speed of sound inferring that all interaction with the liquid phase is "frozen." The "equilibrium" speed of sound is given by

$$a_e^2 = \frac{k_m p(1 - Y)}{\rho_g} \qquad (2.6.4)$$

where k_m is the isentropic expansion index for the wet-steam mixture, implying that during the passage of the sound wave the phases remain in equilibrium with zero slip. Values of k_m are given in Sec. 2.1.

Petr has calculated the effect of the frequency ω of the sound wave from Eqs. (2.6.1) and (2.6.2). As shown in Fig. 2.6.1a–c, acoustic velocity increases from $a_e \to a_f$ as frequency ω is increased. As may be anticipated for dispersions of small droplet size, the acoustic velocity tends toward the equilibrium value a_e. Similarly for reduced wetness fraction and increased density of the gaseous phase (increased pressure), the mixture tends towards a single-phase medium with acoustic speed a_f. Figure 2.6.2a clearly shows the peak energy absorption at frequencies equal to $1/\tau_I$ and $1/\tau_T$, as shown in Sec. 2.2.

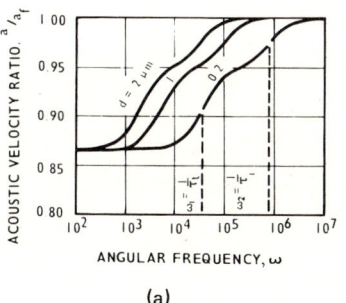

(a)

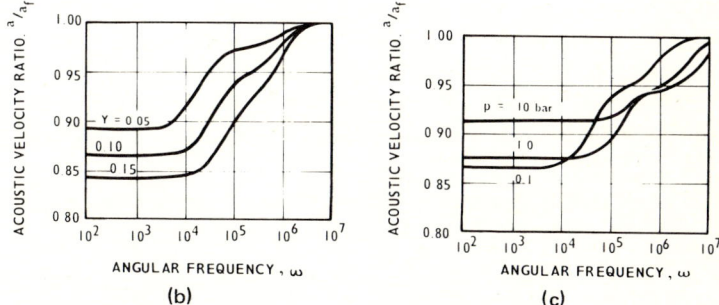

(b)   (c)

FIG. 2.6.1 Variation of acoustic velocity (*a*) with drop size and sound wave frequency, (*b*) with wetness fraction and frequency, and (*c*) with pressure level and frequency ($Y = 0.10$, $p = 0.1$ bar, and $d = 0.2$ μm unless shown otherwise).

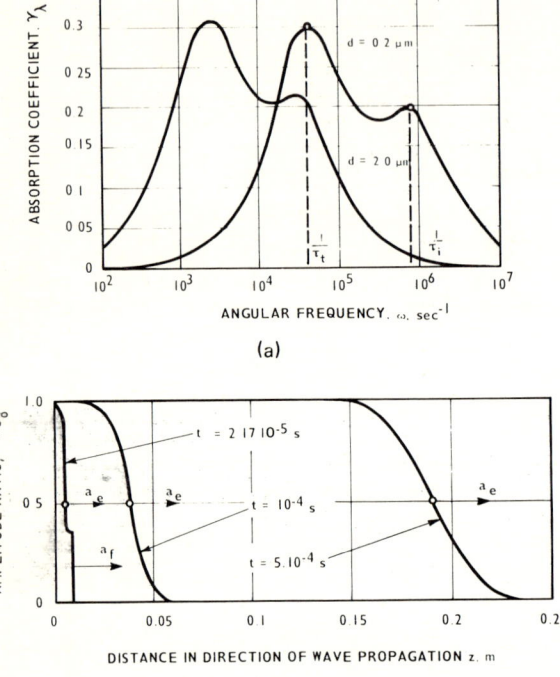

FIG. 2.6.2 (a) Variation of absorption coefficient with sound wave frequency; (b) progression of a nonperiodic small disturbance ($p = 0.1$ bar, $Y = 0.10$, $d = 0.20$ μm).

The dependence of acoustic velocity on frequency has been shown in other theoretical analyses [2-11, 2-25], but this result does not indicate the appropriate sound speed to employ in duct or blading flow. However, Petr considers also the propagation of a nonperiodic disturbance representing the random waves present in practice in the flow field. The wave is represented by a step change in C' at $s = 0$, $t = 0$ from zero to C'_0. That is

$$C'(s, 0) = 0 \quad \text{for} \quad s > 0$$
$$C'(0, t) = 0 \quad \text{for} \quad t < 0$$
$$C'(0, t) = C'_0 \quad \text{for} \quad t \geq 0$$

The corresponding Fourier series for the disturbance can be shown to be

$$C'(s, t) = \frac{C'_0}{2\pi j} \int_{-\alpha}^{\alpha} \frac{e^{j(\omega t - Ks)}}{\omega} \, d\omega$$

and the complex wave vector **K** is obtained from the acoustic equations making the substitutions

$$\frac{\partial}{\partial t} \to j\omega \qquad \frac{\partial}{\partial s} \to -j\mathbf{K}$$

The analysis of the Fourier integral then gives the results shown diagrammatically in Fig. 2.6.2b. The leading edge of the disturbance propagates at the frozen speed a_f with an exponentially decaying amplitude. There follows a dispersed wave with mean velocity a_e.

The frozen speed a_f is therefore the maximum velocity at which "information" is transmitted through a wet-steam mixture and will define Mach number–dependent phenomena, such as choking flow or the formation of shock waves. The acoustic speed a defined in Eq. (2.4.1) can be seen to be an intermediate value between a_f and a_e and arises as a result of the simplifying assumption of zero slip in the analysis of Sec. 2.4.

2.7 SHOCK WAVES IN WET STEAM

From the preceding section we may now consider the formation of a shock wave in wet steam when both thermal and inertial relaxation are present. For simplicity we shall consider a two-dimensional flow field (this represents also the most usual application for blading flows).

The accepted procedure for perfect gases has been to develop theoretical relationships for normal shock waves, and to combine these results with an undisturbed flow parallel to the shock wave, to obtain oblique shock relationships [2-22]. However, this procedure cannot be adopted for wet steam with relaxation, as the governing equations include absolute values of local conditions and are nonlinear. We therefore consider the general case of a weak oblique shock, of which the normal shock is then a special case.

The model of a shock wave in wet steam is based on the following assumptions:

1. The flow is initially in equilibrium and at supersonic Mach number M_1.
2. The initial discontinuity is an "aerodynamic" shock in the gaseous phase, at a position determined by the flow deflection through the shock, and the upstream "frozen" Mach number.
3. The droplets in the flow pass through the shock without evaporating and do not affect the shock strength or position.
4. Downstream of the shock, evaporation and slip will occur until equilibrium is restored. As shown in Fig. 2.7.1, it is anticipated that the equilibrium position will occur a significant distance beyond the shock.

The conditions immediately downstream of the shock may be obtained from the usual single-phase relationships for oblique shocks, in terms of

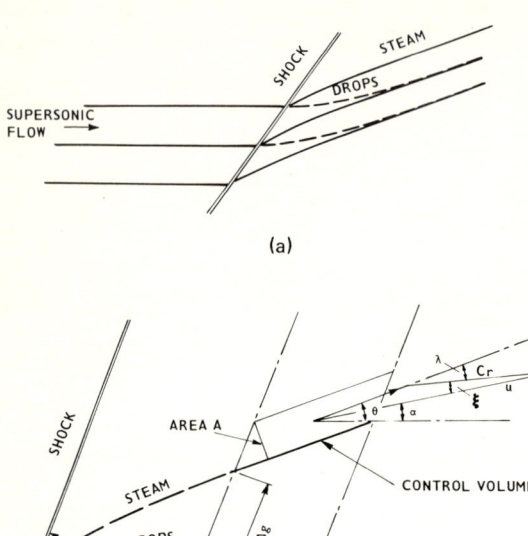

FIG. 2.7.1 (a) Oblique shock in wet steam; (b) nomenclature for oblique shock theory.

upstream values and deflection angle θ. The nomenclature employed is shown in Fig. 2.7.1b. From [2-22] we can then write

$$\tan \theta = 2 \cot \beta \left[\frac{M_1^2 \sin^2 \beta - 1}{M_1^2 (k + \cos 2\beta) + 2} \right]$$

$$p_2 = p_1 \left[1 + \frac{2k}{k+1} (M_1^2 \sin^2 \beta - 1) \right]$$

$$\rho_{g2} = \rho_{g1} \left[\frac{(k+1) M_1^2 \sin^2 \beta}{(k-1) M_1^2 \sin^2 \beta + 2} \right]$$

$$C_2^2 = \left(\frac{k p_2}{\rho_{g2}} \right) \operatorname{cosec}^2 (\beta - \theta) \left[\frac{(k-1) M_1^2 \sin^2 \beta + 2}{(k+1) M_1^2 \sin^2 \beta} \right]$$

where $M_1 = C_1 / a_{f1}$.

The shock wave represents an extremely large local pressure gradient, in which we may expect the droplets to be decelerated and deflected. Typical shock thickness δ, defined in Fig. 2.7.2a, is given by Shapiro [2-30] as

$$\delta = 4 \left[\frac{16}{5\pi} \frac{\mu}{p} \left(\frac{\pi R T_g}{2} \right) \right]^{1/2}$$

where the term in brackets is the molecular mean free path length. For low-pressure steam ($p < 0.25$ bar) we may expect values of $\delta > 1$ μm. Droplets of <0.5 μm dia will therefore be for a short time under the influence of the pressure gradient in the shock which produces a force F acting in a direction normal to the shock front where

$$F = \frac{\pi d^2}{3} \left[p_0 + \left(\frac{dp}{ds_n} \right) \frac{d}{2} \right]$$

On this basis we can calculate the approximate deflection and deceleration of a drop during its passage through the wave as

$$\Delta \alpha = \frac{\delta \cos \beta}{d \rho_f u_1^2} (p_1 + p_2)$$

$$\Delta u = \frac{\delta \sin \beta}{d \rho_f u_1} (p_1 + p_2)$$

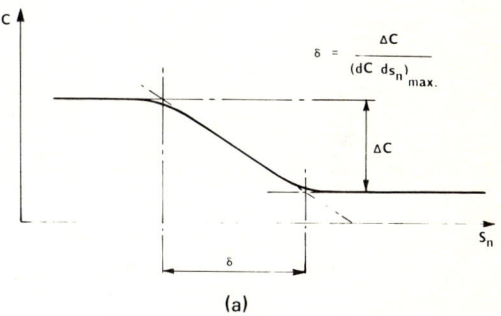

(a)

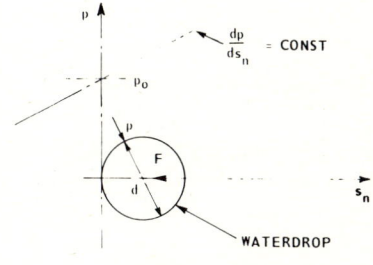

(b)

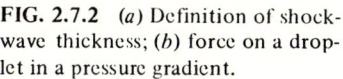

FIG. 2.7.2 (a) Definition of shock-wave thickness; (b) force on a droplet in a pressure gradient.

Again inserting typical values for low-pressure steam, we find that for a droplet of 0.1 μm dia in a flow at

$$p_1 = 0.35 \text{ bar}$$
$$u_1 = 700 \text{ m/s}$$

then

$$\Delta\alpha = 0.03°$$
$$\Delta n = 0.50 \text{ m/s}$$

Therefore except for the smallest drops we may neglect the effect of the shock wave on droplet motion.

Downstream of the shock we can write the steady-flow equations as in Sec. 2.3 with some added complexity because of the two-dimensional flow field.

1 For continuity

$$\frac{d}{ds} A\rho_g C + \frac{d}{ds} Num \cos\alpha = 0$$

2 From momentum conservation

$$A\frac{dp}{ds} + A\rho_g C \frac{dC}{ds} - ND\cos\lambda = 0$$

$$-A\rho_g C \frac{d\theta}{ds} = ND\sin\lambda$$

$$mu\frac{du}{ds} = -D\cos\xi$$

$$mu^2 \frac{d\alpha}{ds} = D\sin\xi$$

3 The energy equation becomes

$$Nu\cos\alpha \left[m\frac{d}{ds}\left(\frac{u^2}{2}\right) - \left(h_{fg} + \frac{C^2 - u^2}{2}\right)\frac{dm}{ds}\right] + A\rho_g C \frac{dH_g}{ds} = 0$$

4 Geometric relationships required are

$$A = \sin(\beta - \theta)\,\text{cosec}\,\beta$$

$$\frac{dA}{ds} = -\text{cosec}\,\beta\cos(\beta - \theta)\frac{d\theta}{ds}$$

$$\lambda = \sin^{-1}\left[\frac{u}{C_{rel}} \sin(\theta - \alpha)\right]$$

$$\xi = \lambda - \theta + \alpha$$

$$C_{rel} = [C^2 + u^2 - 2Cu \cos(\theta - \alpha)]^{1/2}$$

The supplementary equations for drag and heat transfer from droplets are as given in Sec. 2.4, with the exception of the expression for drag coefficient C_D. We can no longer assume slip is small, and a correlation for a wider Reynolds number range is required. A useful expression is given in [2-31] for $0 < \text{Re} < 3000$ as follows

$$C_D = \frac{24}{\text{Re}}\left[1 + \left(\frac{\text{Re}}{60}\right)^{5/9}\right]^{9/5}$$

where Re is defined as $C_{rel}d(\rho_g/\mu)$.

The closed set of equations can be integrated as before by any suitable numerical method such as Runge-Kutta, and typical results are shown in Fig. 2.7.3.

All the examples shown are for an approach flow at

$$M_1 = 1.8$$
$$p_1 = 0.042 \text{ bar}$$
$$Y_1 = 0.10$$

to an oblique shock of angle $\beta = 45°$, corresponding to a gaseous phase deflection of $\sim 11°$. In Fig. 2.7.3a the equalization of droplet and steam velocities can be seen to occur long before droplet evaporation can reestablish saturation. Pressure level downstream of the shock is determined mainly by the droplet momentum dissipation, the deceleration of the drops producing (somewhat surprisingly) a pressure decrease.

Distance δ_I, δ_T downstream of the shock at which inertial and thermodynamic equilibrium occur have been plotted in Fig. 2.7.3b versus the diameter of the fog droplets in the approach flow. The equilibria are arbitrarily defined as when residual slip and superheat is less than 0.5 m/s and 2°C respectively. Distances are measured in the direction of the approach flow.

The equilibrium "thickness" δ_T can be seen to be a constant factor of ~ 10 greater than δ_I, reflecting approximately the ratio of relaxation times τ_I, τ_T for droplets (Sec. 2.2). It may also be noted that even for 0.1 μm drops, thickness δ_I is ~ 2 mm and $\delta_T > 20$ mm. The disturbance of an aerodynamic shock therefore produces a significant zone of relaxation and, depending upon

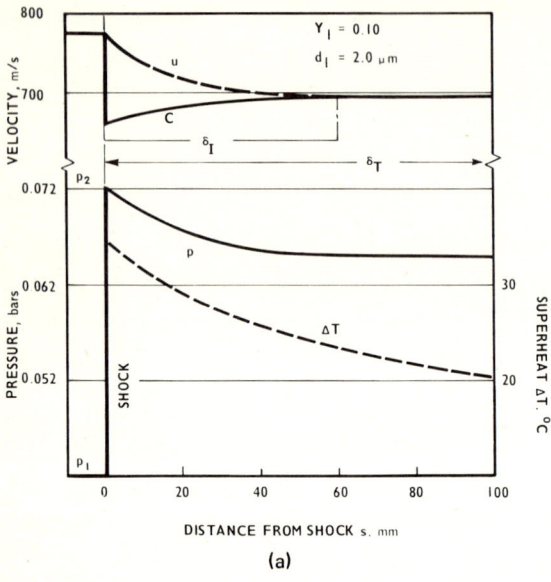

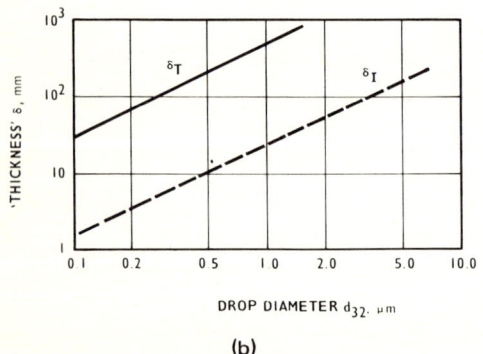

FIG. 2.7.3 (*a*) Relaxation behind an oblique shock; (*b*) effect of drop size on relaxation zones.

fog-drop size, complete equilibrium is not reestablished for some considerable distance downstream.

2.8 WETNESS LOSSES IN TURBINES

2.8.1 Empirical Approach

It is well known that the presence of water in a turbine produces a reduction in efficiency, although the mechanisms by which the extra losses are produced cannot yet be accurately quantified. One of the earliest empirical corrections to turbine efficiency to account for wetness loss was that of Baumann [2-32]

$$\eta = \eta_{\text{dry}}(1 - 0.5\alpha Y) \qquad (2.8.1)$$

relating the turbine efficiency to the predicted value η_{dry} for a machine operating with superheated steam. Y is the exhaust wetness and the constant α is known as the Baumann factor. From turbine tests Baumann deduced a value of α near unity, the above relationship being equivalent to a reduction of 1% in efficiency for each 1% of mean wetness. However, as may be expected, results from a large number of turbine tests give a fairly wide variation in value of α, mean values listed in [2-10] being reproduced in Table 2.8.1. This correction is still widely used [2-33], together with the assumption that the expanding steam condenses in equilibrium immediately on attaining saturated conditions.

The wide use of wet-steam turbines for nuclear application has encouraged a closer examination of the occurrence of wetness loss. For example the data of [2-34], shown in Fig. 2.8.1, indicate that supersaturation produces a significant part of the total loss. However, although convenient to apply, the empirical approach can give very little insight into the physical processes, and if these significant wetness losses are to be minimized, a more fundamental knowledge of their origin is required.

2.8.2 Theoretical Approach

The occurrence of supersaturation and spontaneous condensation in an expanding steam flow has been appreciated for many years [2-35]. Similarly, the basic processes producing erosion of turbine moving blades by water-drop impact are qualitatively described in [2-36]. One of the first attempts to combine the various phenomena into a theory for the wet-steam turbine was made by Gyarmathy [2-9].

TABLE 2.8.1 Mean values of Baumann factor α

Year	Author	Turbine type	α
1912	K. Baumann	Reaction	1.0
1925	W. Blowney and H. Warren	Impulse	1.15
1926	H. Guy	Impulse	1.03
1927	Ivon Freudenrich	Reaction	1.40
1938	D. Smith	–	1.0
1939	F. Flatt	Impulse	0.4–2.0
1954	N. Beldecos and K. Smith	Reaction	1.20
1958	W. Traupel	Impulse	1.15
1958	W. Traupel	Reaction	1.40

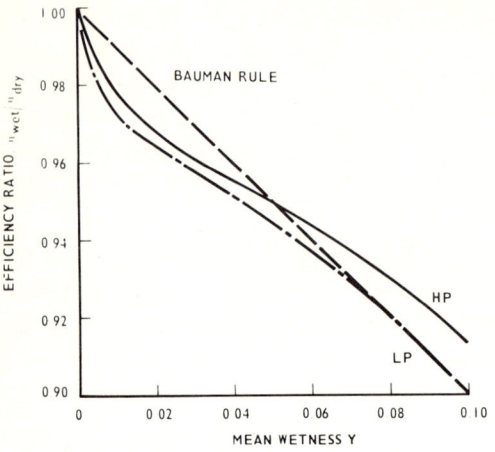

FIG. 2.8.1 Wetness loss data from tests on laboratory turbines [2-34].

It was recognized that water could be present in several forms in the turbine, as shown diagrammatically in Fig. 2.8.2. A large proportion of the water would be in the form of a fine fog, the diameter of the drops, from condensation theory, being on the order of 1 μm. The small proportion of fog deposited on the blading is re-entrained into the flow as a spray of much larger drops (typically 20-200 μm in an LP turbine), which will impinge on subsequent blading. The component of water not in the form of fog is termed "coarse" water.

Gyarmathy [2-9] derived simplified expressions for the losses incurred by the frictional drag of water drops on the steam, the limited rate of heat transfer from water drops to steam, and more directly by the impact and flow of water on the turbine blading. A similar treatment is also given by Kirillov and Yablonik [2-10]. The theoretical expressions for the various loss components are relatively simple to derive; the difficulty lies in the selection of a two-phase flow model that adequately represents the complex situation in a multistage turbine. At present, insufficient data are available to enable accurate estimates to be made of the fog and coarse-water quantities, mainly as a result of the difficulties of measurement in operating turbines (Chap. 4). However, more information is now being obtained, and detailed theories of condensation and deposition are being developed. A more general method of wetness-loss calculation is therefore described in Sec. 2.8.3, which can incorporate the results of theories or experiments directly.

2.8.3 Isentropic Wet-Steam Stage

Before considering the inefficiencies of wet-steam expansion, it is necessary to define the ideal process with reference to a typical stage consisting of fixed and moving blade rows. The usual method of quantifying losses in a "dry" turbine is to define an overall efficiency η_0 as [2-37]

$$\eta_0 = \frac{H_i - H_e}{H_i - h_{es}}$$

where i and e denote inlet and exhaust values. The energy extracted (numerator) is expressed as a fraction of the theoretical isentropic maximum output (denominator) obtainable between specified inlet total and exit static pressures.

For a multistage machine, the overall efficiency can be expressed in terms of individual stage efficiencies as

$$\eta_0 = \frac{\Sigma(H_1 - H_{3ss})\eta}{H_i - h_{es}}$$

where stage efficiency is defined on a total-to-total basis

$$\eta = \frac{H_1 - H_3}{H_1 - H_{3ss}} \tag{2.8.2}$$

The nomenclature for a stage is shown on the Mollier diagram (Fig. 2.8.3a). For conventional turbines operating with gaseous phase only, the stage

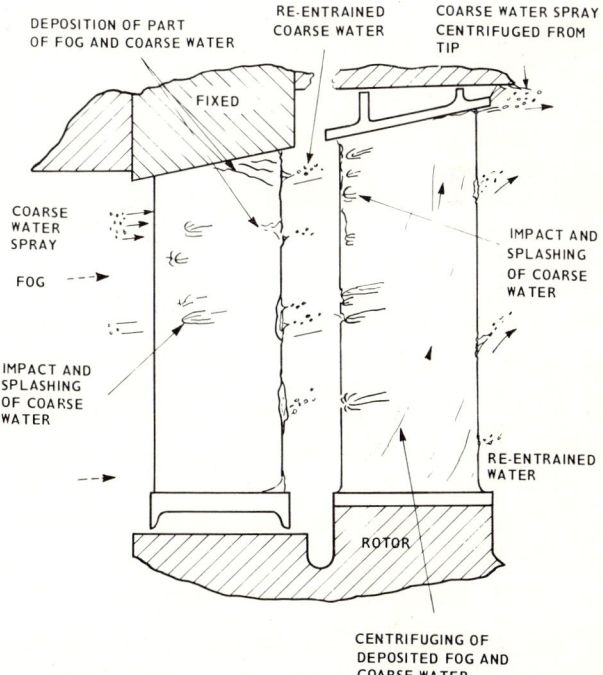

FIG. 2.8.2 Diagram of the forms of water in a wet-steam turbine stage.

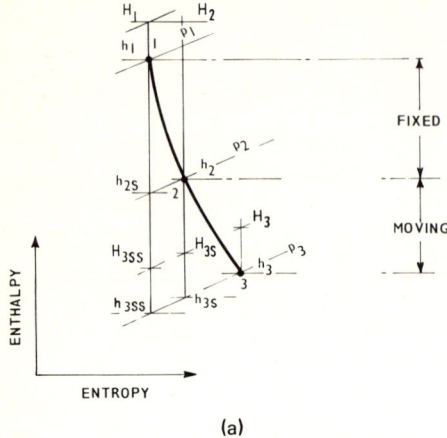

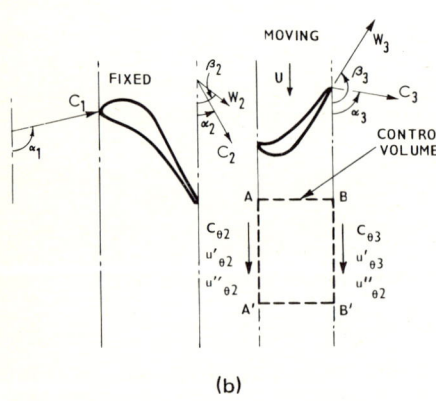

FIG. 2.8.3 (a) Single-stage condition line; (b) control volume for definition of work-loss coefficient.

efficiency can be obtained from the enthalpy loss coefficients ξ_n and ξ_m of the nozzles and moving blades

$$\eta = \left[1 + \frac{\xi_n C_2^2 + \xi_m W_3^2}{2(H_1 - H_3)}\right]^{-1}$$

where

$$\xi_n = \frac{h_2 - h_{2s}}{(1/2)C_{2s}^2}$$

$$\xi_m = \frac{h_3 - h_{3s}}{(1/2)W_{3s}^2}$$

However, in a wet-steam stage the losses are not confined to the enthalpy losses of the vapor phase in flowing through fixed and moving blading. It is therefore necessary to define carefully the ideal stage upon which the wet-steam loss estimation can be based.

The use of a total-to-total enthalpy criterion is appropriate for an intermediate turbine stage, but introduces some arbitrary aspects to the definition of the ideal stage. Obviously the maximum possible output from a stage is obtained when the exit velocity of all phases is zero. However, this is not a realistic criterion for turbine stages other than the final stage of the machine. Alternative criteria involve some aspect of the design of the actual turbine, and the following definition has therefore been devised.

The isentropic wet-steam stage is assumed to operate such that

1. the wet steam is in thermodynamic equilibrium
2. the liquid phase is uniformly distributed throughout the vapor phase
3. slip between phases is zero
4. viscous losses between phases and solid boundaries is zero
5. no water is deposited on blading surfaces
6. inlet total enthalpy is based on the assumption that the phases are in equilibrium at the inlet static pressure and velocity of the vapor phase
7. the stage operates between the same inlet, intermediate, and exit static pressure as the real stage
8. exit total enthalpy is defined by specifying that the vapor-phase velocity is in the axial direction, and equals the axial component of velocity at exit to the real turbine stage

Therefore, for specified inlet and exit conditions for the real stage, the specific output of the isentropic stage is obtained from

$$H_1 - H_{3ss} = H_1 - h_{3ss} - \frac{1}{2}(C_3 \sin \alpha_3)^2 \qquad (2.8.3)$$

and the flow angles and velocities within the isentropic stage are

$$\tan \beta_{3s} = \frac{C_3 \sin \alpha_3}{U_3}$$

$$\cos \alpha_{2s} = \frac{H_1 - H_{3ss}}{U_2 C_{2s}} \qquad (2.8.4)$$

$$W_{3s} = U_3 \sec \beta_{3s}$$

$$C_{2s} = [2(H_1 - h_{2s})]^{1/2}$$

2.8.4 Stage Work-Loss Coefficient

For a wet-steam stage it is convenient to derive expressions for the energy losses due to the various flow components by using the Euler turbine equation. The output from a stage can be obtained from the change in angular momentum flux across the control volume $AA'BB'$ (Fig. 2.8.3b). The control volume is bounded by the surfaces of a stream tube passing through the rotor, and surfaces of AA' and BB' are drawn immediately upstream and downstream of the moving-blade leading and trailing edges. Thus for the isentropic stage

$$H_1 - H_{3ss} = U_2 C_{\theta 2s} - U_3 C_{\theta 3s}$$

and since C_3 is in the axial direction, the second term on the right-hand side of the equation is zero.

For the real stage the output will be

$$H_1 - H_3 = U_2[(1 - Y_2)C_{\theta 2} + Y_2' u_{\theta 2}' + \Sigma Y'' u_{\theta 3}''] \\ - U_3[(1 - Y_3)C_{\theta 3} + Y_3' u_{\theta 3}' + \Sigma Y'' u_{\theta 3}'']$$

where Y, Y', and Y'' are the wetness fractions of the total, fog, and coarse-water components respectively (i.e., $Y = Y' + \Sigma Y''$), and u_θ' and u_θ'' are the circumferential velocity components of the fog and coarse water. The summation of coarse-water components implies that several forms will be present at differing mass-flow rates and velocities.

From Eq. (2.8.2), we can define a stage work-loss coefficient ζ such that

$$\zeta = 1 - \eta = \frac{(H_1 - H_{3ss}) - (H_1 - H_3)}{H_1 - H_{3ss}} \qquad (2.8.5)$$

Substituting for the actual and isentropic stage output quantities from the previous equations and rearranging terms, the loss coefficient can also be written

$$\zeta = (H_1 - H_{3ss})^{-1} \{U_2 Y_2'(C_{\theta 2s} - u_{\theta 2}') - U_3 Y_3'(C_{\theta 3s} - u_{\theta 3}') \\ + \Sigma [U_2 Y_2''(C_{\theta 2s} - u_{\theta 2}'') - U_3 Y_3''(C_{\theta 3s} - u_3'')] \\ + U_2(1 - Y_2)(C_{\theta 2s} - C_{\theta 2}) - U_3(1 - Y_3)(C_{\theta 3s} - C_{\theta 3})\} \quad (2.8.6)$$

The overall loss coefficient can now be seen to be composed of a sum of component loss coefficients

$$\zeta = \zeta_{(i)} + \zeta_{(ii)} + \zeta_{(iii)} + \cdots \qquad (2.8.7)$$

where, for example,

$$\zeta_{(i)} = \frac{U_2}{H_1 - H_{3ss}} [Y'_2(C_{\theta 2s} - u'_{\theta 2})]$$

$$\zeta_{(ii)} = -\frac{U_3}{H_1 - H_{3ss}} [Y'_3(C_{\theta 3s} - u'_{\theta 3})]$$

Expressions for each loss coefficient can now be formulated, and the complete stage efficiency then obtained.

2.8.5 Components of Stage-Work Loss

2.8.5.1 Fog Droplets

The flow of the fog-droplet suspension through a blade row can be calculated approximately, using the one-dimensional stream-tube theory developed in Sec. 2.3. The application of the one-dimensional method is limited to the flow of small droplets, as may be seen by considering the blade row as a stream tube of constant radius of curvature r_m. For small slip it is assumed that droplet velocities u and u_r in the streamwise and normal directions are constant. The equations of motion for a droplet are therefore

$$\frac{mu^2}{r_m} = D = \frac{mu_r}{\tau_I}$$

For typical values of r_m and u of 0.05 m, and 200 m/s, respectively, the normal velocity u_r will be less than 1% of the streamwise value if

$$\tau_I < 2 \times 10^{-6} \text{s}$$

For low-pressure steam, this corresponds to fog droplets of <2 μm dia. For droplets smaller than 2 μm, we may therefore use the results of one-dimensional flow calculation to predict the streamwise velocity u' of fog drops at the exit to blade rows.

Calculation of fog-drop velocity u'_2 at exit to the fixed blade row can be obtained directly; flow through the rotor can also be calculated relative to the moving blading. It can be shown that, for a stream tube passing through a moving blade row, the so-called rothalpy I is constant, where

$$I = H - U[(1 - Y)C_\theta - \Sigma Y u_\theta]$$

$$= (1 - Y)\left(h_g + \frac{1}{2}W^2\right) + \Sigma Y\left(h_f + \frac{1}{2}u_{\text{rel}}^2\right) - \frac{1}{2}U^2$$

$$= H_{\text{rel}} - \frac{1}{2}U^2$$

The adiabatic one-dimensional stream-tube calculation can therefore be applied, using relative values of enthalpy, velocity, etc., and incorporating the blade velocity term if the stream tube is not at constant radius from the turbine axis. The calculated relative velocities W and u_{rel} can be introduced into the loss coefficient expressions, using

$$C_\theta = U + W_\theta$$

$$u_\theta = U + u_{\theta\,\text{rel}}$$

and the work-loss coefficients for fog droplets are

$$\zeta_{(i)} = \left(\frac{U_2^2}{H_1 - H_{3ss}}\right)\left(\frac{C_{2s}\cos\alpha_{2s}}{U_2}\right) Y_2' \left(1 - \frac{u_2'\cos\alpha_2}{C_{2s}\cos\alpha_{2s}}\right)$$

$$\zeta_{(ii)} = \left(\frac{U_3^2}{H_1 - H_{3ss}}\right)\left(-\frac{W_{3s}\cos\beta_{3s}}{U_3}\right) Y_3' \left(1 - \frac{u_{\text{rel}3}'\cos\beta_3}{W_{3s}\cos\beta_{3s}}\right)$$

The terms, including blade velocity U, depend mainly upon the degree of reaction of the stage; to demonstrate the magnitudes of the various loss components, the following simplifying assumptions are made. The stage is assumed to be at constant radius and at approximately 50% reaction such that

$$U_2 = U_3$$

$$\alpha_2 = \alpha_{2s} \qquad \beta_3 = \beta_{3s}$$

$$\frac{U_2^2}{H_1 - H_{3ss}} = 0.5 \tag{2.8.8}$$

$$\frac{C_{\theta 2s}}{U_2} = -\frac{W_{\theta 3s}}{U_3} = 1.0$$

The expressions for work-loss coefficient now simplify to

$$\zeta_{(i)} = \frac{1}{2} Y_2' \left[1 - \frac{u_2'}{C_2}\left(\frac{C_2}{C_{2s}}\right)\right]$$

$$\zeta_{(ii)} = \frac{1}{2} Y_3' \left[1 - \frac{u_{\text{rel}3}'}{W_3}\left(\frac{W_3}{W_{3s}}\right)\right] \tag{2.8.9}$$

2.8.5.2 Losses Due to Coarse Water

In order to derive approximate expressions for coarse-water losses, certain simplifying assumptions are necessary to describe the motion of the liquid phase in the stage. These assumptions are based where possible on experimental data.

Deposition of fog droplets It is assumed that the steam enters the blade rows at design incidence, and that the fog droplets which contact the blade surfaces adhere and join films or rivulets running to the trailing edge. Special cases such as off-design incidence or separated flows are excluded, although these effects are known to increase fog-drop disposition [2-38].

Coarse water re-entrained from fixed blades Water detaches from the blade trailing edge, to be atomized and accelerated by the steam in the direction of the main flow.

Impact of coarse water on the moving blades A proportion of the coarse water adheres to the blade, joining the quantity deposited as fog. The remainder of the coarse water rebounds as a spray of smaller drops that are accelerated by the steam flow, leaving the rotor control volume in the direction of the main steam flow.

Re-entrainment of coarse water from moving blades The initial velocity of the drops relative to the moving blades will be low, their absolute velocity being approximately equal to the blade velocity U_3. The large drops leaving the blades therefore move at high speed into the slowly moving main steam flow and are atomized by the decelerating drag forces. The coarse-water spray produced is transported by the steam to the following stage. The proportion of water remaining on the moving blade surfaces will be centrifuged to the outer end of the blade as shown diagrammatically in Fig. 2.8.2.

Coarse water entering the fixed blades It will be in the form of a spray of droplets from the previous stage. The water will impinge on the fixed blade surfaces and a proportion will adhere, adding to the water deposited as fog. The remainder of the coarse water will rebound as a spray of smaller drops that are accelerated by the steam in the direction of the main steam flow, at the upstream surface AA' of the rotor control volume.

From these assumptions we may define the following constants for the calculation

a   fraction of fog mass-flow rate that deposits on the fixed blade surfaces
b   fraction of coarse water adhering to the fixed blades
c   fraction of fog mass-flow rate deposited on moving blade surfaces
f   fraction of coarse water that adheres to the moving blade
g   fraction of water deposited on the moving blade surface that is centrifuged to the blade tip.

Constants *a* and *c* are particularly important and are the subject of calculations by Crane [2-38] for inertial and turbulent diffusion deposition mechanisms. Some typical results for HP and LP turbine blades are shown in Fig. 2.8.4. Very little experimental data are available to verify the calculations, but the theories show clearly that deposition is probably a strong function of fog droplet size.

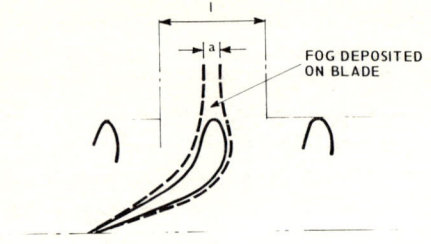

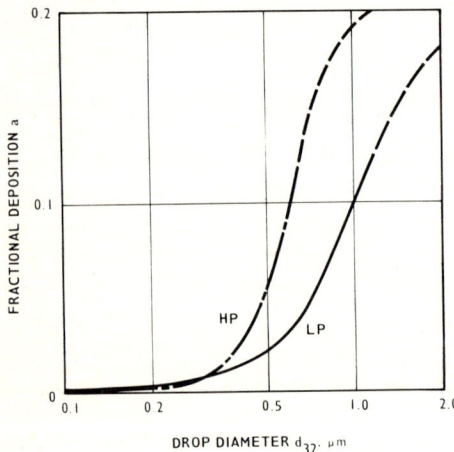

FIG. 2.8.4 Approximate fractional deposition of fog drops on blading.

Experiments have been carried out [2-39, 2-40] on the re-entrainment of water from blading. Moore et al. [2-39] show that large drops detaching from the trailing edge of fixed blades are broken up by the steam flow into a spray, the diameter of the largest stable droplets corresponding to a limiting Weber number We of 22

$$d_{max} \cong \frac{22\sigma}{C_2^2 \rho_g}$$

The mass spectrum of droplet size can be correlated by the upper limit function of [2-15], as shown in Fig. 2.8.5a, where

$$\frac{dm}{dd} = \frac{0.56 d_{max}}{d(d_{max} - d)} e^{-\omega^2}$$

Here m is the cumulative mass fraction of droplets with diameter $<d$, and the distributed quantity

$$\omega = \ln\left(\frac{d}{d_{max} - d}\right)$$

Following atomization the spray is accelerated by the steam, the smaller drops in the spectrum attaining higher velocities than the larger drops, as shown in Fig. 2.8.5b. It was found that the velocity of the largest drops could be predicted if a drag coefficient for highly distorted drops were used. From data by Laws [2-41], an approximate expression for drag coefficient can be extracted

$$C_D = 0.0306 \text{ We} + 0.500$$

for the range of rapid droplet acceleration. The equation of motion for a large drop may then be written

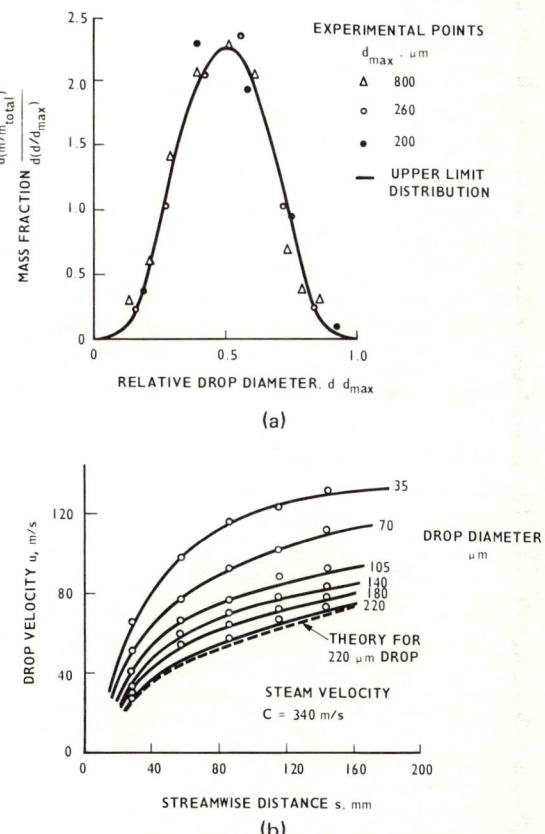

FIG. 2.8.5 (a) Normalized spectrum for re-entrained drops; (b) example of acceleration of a re-entrained drop spray.

$$\frac{\pi}{6}d^3\rho_f u \frac{du}{ds} = \frac{\rho_g(C_2-u)^2}{2}\frac{\pi d^2}{4}\frac{0.0306\rho_g(C_2-u)^2 d}{\sigma} + 0.05$$

For constant steam velocity C_2, the equation can be integrated to give the dimensionless distance $\bar{s}(\bar{u})$ downstream of the trailing edge, at which the droplet attains dimensionless velocity $\bar{u}$. The expression is also plotted in Fig. 2.8.6.

$$\bar{s} = 2.667\left\{0.5\ln\left[\frac{1.67(1-\bar{u})^2}{(1-\bar{u})^2 + 0.67}\right] + \frac{\bar{u}}{1+\bar{u}} + 1.22\left(\tan^{-1}\frac{1-\bar{u}}{0.82} - 0.88\right)\right\}$$

where $\bar{s} = (\rho_g/\rho_f)s/d$
$\bar{u} = u/C_2$

It is found in practice that the smaller drops in the spray are shielded by the larger drops and therefore do not attain their theoretical "isolated-drop" velocity. From Fig. 2.8.5, it can be seen that little error is introduced if it is assumed that the bulk of the coarse water accelerates to u''_{21}, the theoretical velocity for drops of maximum diameter.

We may now compose the work-loss coefficients for surfaces AA' and BB' of the rotor control volume on the basis of the assumption made for the passage of coarse water through the stage. The fractional flow rate of water re-entrained from fixed blade trailing edge is $a\,Y'_1 + b\,Y''_1$; the value for water rebounding from the fixed blades is $(1-b)\,Y''_1$. The work-loss coefficient $\zeta_{(iii)}$ for the coarse water at AA' is therefore

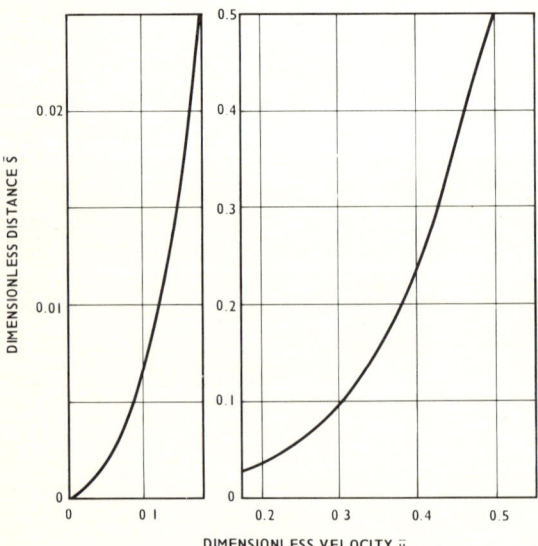

FIG. 2.8.6 Dimensionless velocity of largest drop in a re-entrained spray.

$$\zeta_{(iii)} = \frac{U_2(aY_1' + bY_1'')\cos\alpha_2(C_{2s} - u_{21}'')}{H_1 - H_{3ss}} + \frac{U_2(1-b)Y_1''\cos\alpha\,(C_{2s} - u_{22}'')}{H_1 - H_{3ss}}$$

where u_{22}'' is the velocity of the rebounding water drops at control surface AA'. Substituting from Eq. (2.8.8)

$$\zeta_{(iii)} = 0.50(aY_1' + bY_1'')\left[1 - \frac{u_{21}''}{C_2}\left(\frac{C_2}{C_{2s}}\right)\right] + 0.5(1-b)Y_1''\left[1 - \frac{u_{22}''}{C_2}\left(\frac{C_2}{C_{2s}}\right)\right] \quad (2.8.10)$$

For the rotor, the water rebounding from the rotor surface produces a loss coefficient, but no term appears for re-entrained water. Control volume surface BB' is drawn immediately downstream of the rotor trailing edge, and therefore re-entrained coarse water will cross the surface with blade velocity U_3, for both actual and isentropic cases. Hence the term for this component is zero. The coarse-water loss for the rotor $\zeta_{(iv)}$ is therefore

$$\zeta_{(iv)} = -\frac{U_3(1-f)Y_2''(C_{\theta 3s} - u_{3\theta}'')}{H_1 - H_{3ss}}$$

where u_3'' is the velocity of rebounding drops at B'. In terms of relative velocities and from Eq. (2.8.8)

$$\zeta_{(iv)} = 0.5(1-f)Y_2''\left[1 - \frac{u_{3\text{rel}}''}{W_3}\left(\frac{W_3}{W_{3s}}\right)\right] \quad (2.8.11)$$

Finally, the energy E per unit mass-flow rate of wet steam absorbed in centrifuging water to the tip of the moving blade may be estimated from the Euler equation

$$E = g(cY_2' + fY_2'')(U_{\text{tip}}^2 - U^2)$$

$$\therefore \zeta_{(v)} = g(cY_2' + fY_2'')\left(\frac{U_{\text{tip}}^2}{H_1 - H_{3ss}}\right)\left[1 - \left(\frac{r}{r_{\text{tip}}}\right)^2\right]$$

For a typical LP turbine we assign a value of $r/r_{\text{tip}} = 0.70$, and the centrifuging loss coefficient becomes

$$\zeta_{(v)} = 0.25g(cY_2' + fY_2'') \quad (2.8.12)$$

2.8.5.3 Losses of the Gaseous Phase

The angular momentum of the gaseous phase deviates from the isentropic value as the result of nonequilibrium flow, enthalpy losses during the acceleration of water drops, and the frictional losses of the blade boundary layers and wakes. In addition to these "two-dimensional" losses, there will in practice be significant losses associated with casing wall boundary layers, tip leakage, and lacing wires. We can assume that these losses will be similar in conventional and wet-steam stages so we shall exclude them from the present analysis.

Losses due to fog-droplet transport will occur in the main flow and produce an overall reduction in gas velocity. The additional losses due to acceleration of coarse water and blading surface friction will occur locally, and may therefore be combined in the form of a dimensionless momentum thickness δ_M at the surfaces of the control volume. Thus the momentum of the vapor phase is assumed to be further reduced by a factor $(1-\delta_M)$, as a result of coarse-water and blade-surface friction losses. For example, the work-loss coefficient for the control surface AA' may be written

$$\zeta_{(vi)} = \left(\frac{U_2}{H_1 - H_{3ss}}\right)(1 - Y_2' - \Sigma Y_2'')[C_{\theta 2s} - C_{\theta 2}(1 - \delta_M)]$$

Vapor velocity and wetness at the control surface is obtained from one-dimensional calculations of flow through the blade rows. Momentum thickness is obtained conveniently from the corresponding energy thickness δ_E. The relationship between these parameters is relatively insensitive to the detailed form of the circumferential distribution of C_2, and we can assume

$$\delta_M = \frac{1}{2}\delta_E$$

The energy deficit is composed of the work done by the vapor phase in accelerating the coarse-water and the friction enthalpy loss. For a constant-velocity vapor flow, the work to accelerate droplets can be shown [2-9] to be independent of the droplet frictional resistance, and is determined by the maximum velocity attained by the drops. For unit mass of drops, the energy loss δ_{E2} at control surface AA' for example is

$$\delta_{E2} = C_2 u_{21}'' - \frac{1}{2}u_{21}''^2$$

Blade surface frictional loss is expressed conventionally as an enthalpy loss coefficient ξ. No data are available to indicate whether this surface loss is increased or reduced by the presence of water on the blade surfaces, and it is necessary to assume that the normal "dry" loss correlations apply.

Introducing the losses and mass-flow components, the work-loss coefficients for surfaces AA' and BB' are respectively

$$\zeta_{(vi)} = 0.5 X_2 \left[1 - \frac{C_2}{C_{2s}} (1 - \delta_{M2}) \right]$$

$$\zeta_{(vii)} = 0.5 X_3 \left[1 - \frac{W_3}{W_{3s}} (1 - \delta_{M3}) \right] \qquad (2.8.13)$$

where

$$\delta_{M2} = \frac{0.5}{X_2} \left[(aY_1' + bY_1'') \frac{u_{21}''}{C_2} \left(1 - \frac{1}{2}\frac{u_{21}''}{C_2}\right) \right.$$

$$\left. + (1-b)Y_1'' \frac{u_{22}''}{C_2}\left(1 - \frac{1}{2}\frac{u_{22}''}{C_2}\right) - Y_1'' \frac{u_1''}{C_1}\left(1 - \frac{1}{2}\frac{u_1''}{C_1}\right) \right] + 0.5\xi_n$$

$$\delta_{M3} = \frac{0.5}{X_3} \left[(1-f) Y_2'' \frac{u_{3\mathrm{rel}}''}{W_3}\left(1 - \frac{1}{2}\frac{u_{3\mathrm{rel}}''}{W_3}\right) \right] + 0.5\xi_m$$

$$X = 1 - Y' - \Sigma Y'' \qquad (X \text{ is the dryness fraction})$$

In Eq. 2.8.13, the momentum input of coarse water entering the fixed blades has also been included, the velocity of this component at inlet to the stage being denoted u_1''.

2.8.5.4 Other Losses

The components $\zeta_{(i)-(vii)}$ include the main losses in a stage operating with wet steam. Apart from the special losses associated with the initial supersaturation and nucleation of fog, which will be considered later, other minor losses can occur. Gyarmathy [2-9] has shown that losses due to turbulence, centrifuging of fog droplets, and increased blade-surface friction are likely to be small compared with the main losses. A probably significant loss, also not included, will be produced by the coarse water centrifuged to the moving blade tip and there re-entrained into the steam flow. The acceleration of this water by high-velocity steam issuing from the "leakage gap" between blade tip and casing will result in a local loss in available energy. However, the resulting reduction in leakage-steam velocity may improve the aerodynamic performance of the following stage, by improving the uniformity of the inlet flow. This loss component must therefore not be overlooked in a detailed analysis, but is beyond the scope of the present exercise.

2.8.6 Losses Due to Supercooling and Nucleation

Particular losses arise from the initial spontaneous condensation in a turbine. Considering first the nucleation process, the available output is limited by the finite rate at which the latent heat released by condensation is transferred to the steam flow. Initially the heat released is further reduced by the high surface energy (capillary energy) of the extremely small droplets in the flow.

In addition to these liquid-phase effects, there may also be viscous and heat transfer losses in the vapor phase, as a result of the latent heat addition and corresponding change in steam velocity. This latter loss is similar to the reduction in total pressure through a shock wave. In fact, in [2-9], the spontaneous condensation is considered to occur instantaneously—the so-called condensation shock—and losses are calculated accordingly. However, from the calculated examples of subsonic flow with nucleation through a turbine, Fig. 2.5.1, or supersonic nucleation in a nozzle, Fig. 2.4.3, it can be seen that the main heat addition following nucleation is by no means a discontinuous process. Hence the loss component due to "condensation shock" will generally be overestimated by the method of [2-9]. Loss is best calculated by using a detailed one-dimensional wet-steam flow method (e.g., Sec. 2.3) in conjunction with the work-loss formulas presented in the previous section.

A significant reduction in available energy occurs if steam expands to a supercooled condition before nucleation. As no water is present, the flow inefficiency is produced by aerodynamic losses only. However, supersaturated steam has a lower potential work capability than wet steam, as latent heat of vaporization is not released. Compared with the isentropic wet-steam stage therefore, a stage operating with supersaturated steam has an apparently lower efficiency. The magnitude of the loss depends upon the degree of supercooling achieved and can be calculated as follows.

For a stage with equal inlet and exit velocities $(C_1 = C_3)$ the maximum output

$$H_1 - H_3 = h_1 - h_3 = \frac{k}{k-1}\left(\frac{p_1}{\rho_1} - \frac{p_2}{\rho_2}\right)$$

where from Eq. (2.1.3), the isentropic index $k = 1.30$ for supersaturated steam. The isentropic output for wet steam can be calculated similarly, but using an effective index from Eq. (2.1.11) of $n = 1.135$. Inserting the usual polytropic relationship

$$\frac{p}{\rho^k} = \text{constant}$$

the approximate ratio of stage outputs for a common inlet condition is given by

$$\eta = \frac{H_1 - H_3}{H_1 - H_{3ss}} = \frac{k}{k-1} \frac{n-1}{n} \left(\frac{1 - r^{(k-1)/k}}{1 - r^{(n-1)/n}}\right)$$

where $r = p_3/p_1$.

The magnitude of the apparent loss has been calculated for the particular case of an expansion from dry saturated conditions at 60 bar static pressure and is tabulated in Table 2.8.2.

TABLE 2.8.2 Apparent inefficiency of supersaturated expansion (from 60 bar sat.)

Outlet pressure p_3 (bar)	48	36	24	12
Outlet temperature T_3 (°C)	248	215	171	105
Saturation temperature T_{3sat} (°C)	261	244	222	188
Subcooling ΔT_{sub} (°C)	13	29	51	83
Efficiency η	0.970	0.954	0.932	0.895
Work-loss coefficient ζ	0.030	0.046	0.068	0.105

In the example of a multistage wet-steam turbine described in Sec. 2.5, a subcooling of approximately 12°C was obtained before spontaneous nucleation (Fig. 2.5.1). The theoretical equilibrium wetness at the nucleation position was approximately 2.6% and the apparent loss in efficiency from Table 2.8.2 would be ∼3%. The loss is therefore at twice the Baumann rate, being 2.3% loss per 1% mean wetness. This corresponds well with the results

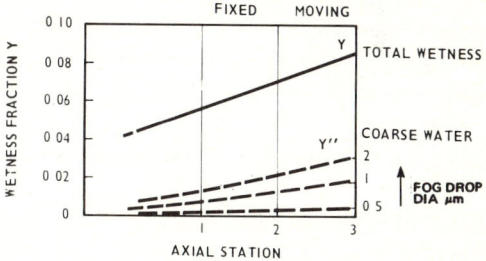

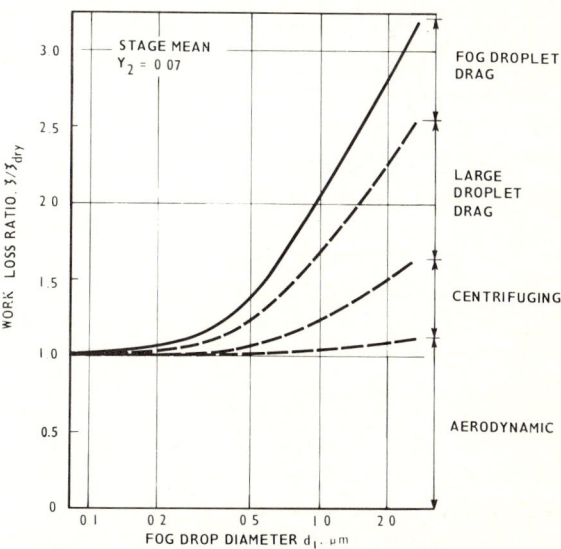

FIG. 2.8.7 Calculated LP-stage loss components (stage wetness assumed for loss calculation).

of [2-34], shown in Fig. 2.8.1, and may explain the proportionately higher losses at low mean wetness.

2.8.7 Example Calculation of Losses

Until adequate data are available to provide the relevant input quantities, detailed wetness-loss calculations are somewhat academic. However, it is of interest to examine the effect of the main variables on the calculated loss. A sample calculation has therefore been made for an LP turbine final stage, and results are summarized in Fig. 2.8.7.

Losses are expressed as a multiple of the "dry" stage work-loss coefficient, where

$$\zeta_{dry} \cong 0.5(\xi_n + \xi_m)$$

and calculations are shown for a stream tube at the mean radius of the stage. Briefly, the results show the considerable influence of the size of fog droplets on the wetness losses. Each of the main loss components increases with increase in fog-drop diameter, the total loss being approximately twice the dry-stage loss at a fog-drop diameter of 1 μm.

Using such methods, the effect of detailed turbine design on wetness loss can ultimately be assessed and the inefficiency minimized. Gyarmathy [2-9], for example, compares turbines with small and large fog droplets and with varying degrees of water extraction from the flow. He concludes that the minimum loss requires both small fog-drop size and effective coarse-water extraction. The feasibility of controlling the nucleation process to produce a fine fog will follow from the studies of spontaneous condensation to be described in Chap. 3.

NOMENCLATURE

A	cross-sectional area of stream tube
a	speed of sound
a_e	equilibrium speed of sound
a_f	frozen speed of sound
$a-j$	mass fractions of water flow (Sec. 2.9)
C	absolute velocity of gaseous phase
C_D	drag coefficient
c_p	specific heat at constant pressure
c_v	specific heat at constant volume
D	drag force on a droplet
d	water drop diameter
d_{32}	Sauter mean diameter
H	specific enthalpy, total

h	specific enthalpy, static
J	nucleation rate per unit mass of wet steam
Kn	Knudsen number (Sec. 2.3)
k	isentropic expansion index
L	reference length
l	molecular mean free path length
M	Mach number (Sec. 2.4)
m	mass of droplet
N	number of droplets per unit length of stream tube
$N^*(d)$	cumulative number fraction of drops below diameter d in a droplet spectrum
n	index of polytropic expansion or number of drops per unit volume
n_m	number of drops per unit mass of wet-steam mixture
P_I	inertia parameter (Sec. 2.2)
P_T	thermal parameter (Sec. 2.2)
P_Y	phase ratio (Sec. 2.2)
p	pressure
$\dot{Q}$	rate of mass increase of a droplet by condensation
$\dot{Q}_Y$	rate of increase of wetness fraction
R	gas constant
Re	Reynolds number (Sec. 2.7)
S	specific entropy
s	streamwise distance
T	absolute temperature
t	time
U	moving blade circumferential velocity
u	droplet velocity
V	volume
v	specific volume
W	velocity of gaseous phase relative to moving blades
We	Weber number (Sec. 2.4)
X	dryness fraction
x	special variable (Sec. 2.4)
x, y	Cartesian coordinates
Y	wetness fraction (by mass flow rate unless indicated)
Z	compressibility (Sec. 2.1)
z	length in direction parallel to turbine axis
α	heat transfer coefficient; flow absolute yaw angle (Sec. 2.8)
$\alpha, \beta, \theta, \lambda, \xi$	flow angles (Sec. 2.7)
β	flow yaw angle relative to moving blades (Sec. 2.8)
δ	thickness of shock wave
ζ	work-loss coefficient
η	isentropic efficiency
λ	thermal conductivity

μ	dynamic viscosity
ξ	enthalpy loss coefficient
ρ	density
σ	surface tension
τ	relaxation time
χ	ratio of specific heats
ω	angular velocity; angular frequency

Subscripts

A	inlet to rotor control volume
B	exit to rotor control volume
e	value for equivalent monodispersion
f	liquid phase—"frozen" value of sound speed
f, d	value for water drops of diameter d
fg	phase transition
g	gaseous phase
m	wet-steam mixture; moving blade
n	nozzle or fixed blade
r, θ, z	radial, circumferential, axial components in a turbine
rel	relative value
s	isentropic
ss	stage isentropic exit condition (Fig. 2.8.3)
*	critical size (Sec. 2.4)

Superscripts

$^-$	mean value
$'$	mass fraction as fine fog
$''$	mass fraction as coarse water

Units

Basic SI units are used unless otherwise stated.

REFERENCES

2-1 "National Engineering Laboratory Tables," H. M. Stationery Office, Edinburgh, 1964.
2-2 Callendar, H. L.: "Properties of Steam," Edward Arnold, London, 1920.
2-3 Horlock, J. H.: Approximate Equations for the Properties of Superheated Steam, *Proc. Inst. Mech. Engrs.*, vol. 173, no. 33, 1959.
2-4 Dzung, L. S.: "Thermostatische Zustandsanderungen des trockenen und des nassen Dampfes," *Z. Angew*, vol. 6, 1955.
2-5 Vukalovich, M. P., Aleksandrov, A. A., and Trakjtengerts, M. S.: A Corrected Equation of State for Superheated Steam, *Teplonenergetika*, vol. 14, no. 10, pp. 65-68, 1967.

2-6 Nicola, M. C.: Der Exponent der isentropischen Expansion für Nassdampf und Sattdampf, *Wärme*, Band 71, Heft 2, 1971.
2-7 Rankine, W. J. M.: "A Manual for the Steam Engine," 1859.
2-8 Zeuner, G.: "Technical Thermodynamics," vol. 2, 83.
2-9 Gyarmathy, G.: "Grundlagen einer Theorie der Nassdampfturbinen," Juris-Verlag, Zurich, 1962.
2-10 Kirillov, I. I., and Yablonik, R. M.: Fundamentals of the Theory of Turbines Operating on Wet Steam, *NASA TT F-611*; translation of "Osnovy teorii viazhnoparovykh turbin," Mashinostroyeniye Press, Leningrad, 1968.
2-11 Cole, J. E., and Dobbins, R. A.: Propagation of Sound Through Atmospheric Fog, *J. Atmos. Sci.*, vol. 27, May 1970.
2-12 Shapiro, A. H., and Hawthorne, W. R.: The Mechanics and Thermodynamics of Steady One-Dimensional Gas Flow, *ASME J. Appl. Mech.*, Dec. 1947.
2-13 Rogers, G. F. C., and Mayhew, Y. R.: "Engineering Thermodynamics," Longmans, London, 1960.
2-14 Modern Computing Methods, *Notes on Applied Science*, no. 16, 2nd ed., National Physical Laboratory, H. M. Stationery Office, London, 1961.
2-15 Mugele, R. A., and Evans, H. D.: Droplet Size Distribution in Sprays, *Ind. Eng. Chem.*, vol. 43, no. 6, pp. 1317–1324, 1951.
2-16 Crane, R. I., and Moore, M. J., Interpretation of Pitot Pressure in Compressible Two-Phase Flow, *J. Mech. Eng. Sci.*, vol. 14, no. 2, 1972.
2-17 Davidson, B. J.: To be published.
2-18 Eschenroeder, A. Q., Boyer, D. W., and Hall, G. J., Non-equilibrium Expansions of Air with Coupled Chemical Reactions, *Phys. Fluids*, vol. 5, no. 5, p. 615, 1962.
2-19 Glauz, R. D.: Combined Subsonic–Supersonic Gas–Particle Flow, *A.R.S. J.*, vol. 32, no. 5, p. 773, 1962.
2-20 Pouring, A. A.: Thermal Choking and Condensation in Nozzles, *Phys. Fluids*, vol. 8, no. 10, 1965.
2-21 Barschdorff, D.: Droplet Formation, Influence of Shock Waves and Instationary Flow Patterns by Condensation Phenomena at Supersonic Speeds, *Proc. Third Int. Conf. Rain Erosion*, Aug. 1970.
2-22 Liepmann, H. W., and Roshko, A.: "Elements of Gasdynamics," Wiley, New York, 1957.
2-23 Moore, M. J., Walters, P. T., Crane, R. I., and Davidson, B. J.: Predicting the Fog-Drop Size in Wet-Steam Turbines, *Inst. Mech. Eng. Conf. Publ.* 3, 1973.
2-24 Jacobs, D.: Private Communication, 1973.
2-25 Deich, M. E.: Investigation of the Velocity of Sound and the Decrement of Attenuation of Sound-Waves in Two-Phase Media, *Prace Instytutu Maszyn Przeplywowych* (*Proc. Inst. Fluid Flow Mach.*), nos. 29–31, pp. 141–159, 1966 (Polish publication).
2-26 Kalinin, A. V.: Progagation and Damping of Small Disturbances in a Two Phase Relaxing Medium, *Heat Transfer–Soviet Research*, vol. 2, no. 5, Sept. 1970.
2-27 Rudinger, G.: Relaxation in Gas-Particle Flow, in P. P. Wegener (ed.), *Non-Equilibrium Flows*, pt. 1, Dekker, New York, 1969.
2-28 Dejong, V. J., and Firey, J. C.: Effect of Slip and Phase Change on Sound Velocity in Steam-Water Mixtures and the Relation to Critical Flow, *I. & E.C. Process Des. Dev.*, vol. 7, no. 3, 1968.
2-29 Petr, V.: The Propagation of Small Disturbances in Wet-Steam, *Fifth Nat. Conf. Czech. Sci–Eng. Soc., Mech. Eng. Grp., Pilsen*, Oct. 1972.
2-30 Shapiro, A. H.: "The Dynamics and Thermodynamics of Compressible Fluid Flow," vol. 1, Ronald, New York, 1953.
2-31 Churchill, S. W., and Usagi, R.: A General Expression for the Correlation of Rates of Transfer and Other Phenomena, *AIChE Jour.*, vol. 18, no. 6, 1972.

2-32 Baumann, K.: Some Recent Developments in Large Steam Turbine Practice, *Eng.*, vol. 111, p. 435, 1921.
2-33 Craig, H. R. M., and Cox, H. J. A.: Performance Estimation of Axial Flow Turbines, *Proc. Inst. Mech. Eng.*, vol. 185, pp. 407–424, 1970–1971.
2-34 Miller, E. H., and Schofield, P.: The Performance of Large Steam Turbine Generators with Water Reactors, *ASME Wint. Ann. Mtng., New York*, Nov. 26–30, 1972.
2-35 Stodola, A., and Loewenstein, L. C.: "Steam and Gas Turbines," Peter Smith, New York, 1945.
2-36 Gardner, G. C.: Events Leading to Turbine Blade Erosion, *Proc. Inst. Mech. Eng.*, vol. 178, pt. 1, no. 23, pp. 593–624, 1964.
2-37 Horlock, J. H.: "Axial Flow Turbines," Butterworths, London, 1966.
2-38 Crane, R. I.: Deposition of Fog Drops on Low Pressure Steam Turbine Blades, *Int. J. Mech. Sci.*, vol. 15, pp. 613–631, 1973.
2-39 Moore, M. J., Langford, R. W., and Tipping, J. C.: Research at CERL on Turbine Blade Erosion, *Proc. Inst. Mech. Eng.*, vol. 182, pt. 3H, p. 61, 1968.
2-40 Puzyrewski, R., and Krzeczkowski, S.: Some Results of Investigations on the Break-up of a Film of Water and the Motion of Drops of Water in the Aerodynamic Wake, *Prace Instytutu Maszyn Przeplywowych* (*Proc. Inst. Fluid Flow Mach.*) nos. 29–31, pp. 21–43, 1966 (Polish publication).
2-41 Laws, J. O.: Measurements of the Fall Velocity of Waterdrops and Raindrops, *Trans. Amer. Geophys. Union*, vol. 22, pt. 3, p. 709, 1940.

CHAPTER 3

Condensation in Flowing Steam

G. Gyarmathy

3.1 INTRODUCTION

3.1.1 Historical Review

3.1.1.1 Experimental Work

The first known reference to supersaturation in steam dates back to 1887, when H. von Helmholtz noted that steam jets issuing into the atmosphere showed delayed fogging. In the late 1890s, C. T. R. Wilson made direct use of supersaturation in his famous cloud chamber experiments, where ionized molecules were made "visible" by water droplets forming along their paths. Pioneering experiments with steam flow in Laval nozzles were made by Stodola in 1913. After about 1930, systematic studies of condensation in high-speed flow were started due to two main reasons. The growing interest in erosion problems of steam turbines inspired Yellot and his co-workers (1934, 1937) to locate the limits of supersaturation (which are called the "Wilson Line," cf. Sec. 3.1.2); Binnie and his co-workers (1938, 1943) further refined these measurements. Since about 1960, the influence of parameters such as pressure, Mach number, and rate of expansion has also been studied in detail.

The other line of experimentation concerned atmospheric moisture condensing in wind-tunnel flows. Here condensation was a nuisance that affected the accuracy of aerodynamic data. After initial observations made by Prandtl, Wieselsberger, and Hermann (1930, 1935, 1942), Oswatitsch presented a thorough theoretical analysis of the condensation process. The practical problem was finally solved by eliminating condensation through drying and preheating the air.

The author thanks his colleague Dr. H. Gallant for discussions concerning Chapters 1 and 3.

The advent of hypersonic wind tunnels in the 1950s soon evoked interest in the condensation of nitrogen and air at very low temperature. Some other mainly space-flight-oriented research touched upon the condensation of metal vapors and organic compounds. However the importance of steam turbines for terrestrial power generation has kept the main interest focused on water, on which a large body of experimental data is now available.

3.1.1.2 Theoretical Work

Most thermodynamic ideas that were of fundamental importance to the explanation of condensation, such as the equilibrium of small droplets with the vapor surrounding them (Thomson 1870, Helmholtz 1886, Gibbs 1878), and the kinetic theory of gases (Boltzmann 1895, Einstein 1910), were formulated almost a hundred years ago. However the first theory describing the formation of droplets in supersaturated vapors, now called the classical nucleation theory, was born much later [Volmer and Weber 1926, Farkas (and Szilárd) 1927, Becker and Döring 1935, and others]. Oswatitsch in 1942 was the first to incorporate nucleation theory into the equations of gas dynamics and thereby achieve a logically complete theory of condensing flows.

Later refinements of various aspects of theory concerned the surface tension of minute droplets (Tolman 1949, Kirkwood and Buff 1949), time lag effects in the nucleation process (Kantrowitz 1951, Probstein 1951, Wakeshima 1954), and the growth rate of droplets (Oswatitsch 1942, Gyarmathy 1963, Hill 1965, Kang 1967).

Since about 1950, some basic premises of classical nucleation theory have been called into question, and a revised nucleation theory has been formulated (Kuhrt 1952, Dunning 1955, Lothe and Pound 1962, and others). The controversy between classical and revised theory is still unsettled [3-1]; the revised theory predicts condensation at much lower values of supersaturation than the classical one. In the case of water vapor, experimental data are definitively in favor of the classical theory.

3.1.1.3 Current Research

More recent investigations into the condensation of flowing steam include the effects of the expansion rate on fog droplet size (Gyarmathy and Meyer 1964, Petr 1969), the appearance of shock waves and pulsating phenomena during condensation at low supersonic Mach numbers (Schmidt 1962, Pouring 1965, Barschdorff 1967), and the influence of foreign nuclei such as dust or entrained water drops (Deich and co-workers, since about 1965). Condensation at high pressure is of interest for nuclear power plants (Gyarmathy et al. 1973) and is receiving increased interest at present. Theoretical research, especially regarding nucleation theory, is being continued at numerous places.

3.1.2 Basic Concepts

In the following some of the basic notions used to describe the process of condensation in expanding vapors will be summarized.

Spontaneous condensation Condensation in the absence of foreign nuclei and/or wall surfaces.*
Wilson point Thermodynamic condition pertaining to maximum supersaturation before spontaneous condensation occurs.
Wilson line Locus of Wilson points obtained with expansions starting from different initial conditions, in a thermodynamic chart. In Fig. 3.1.1 the Wilson line is sketched in the log p vs. T, T vs. s, and h vs. s charts.
Effect of time scale (i.e., of nozzle length) Maximum supersaturation occurring as influenced by rate of expansion.
Wilson zone Region embracing Wilson lines pertaining to expansion rates of practical interest.

---

*Also called homogeneous condensation, in contrast to heterogeneous condensation occurring on foreign nuclei like dust, ions, etc.

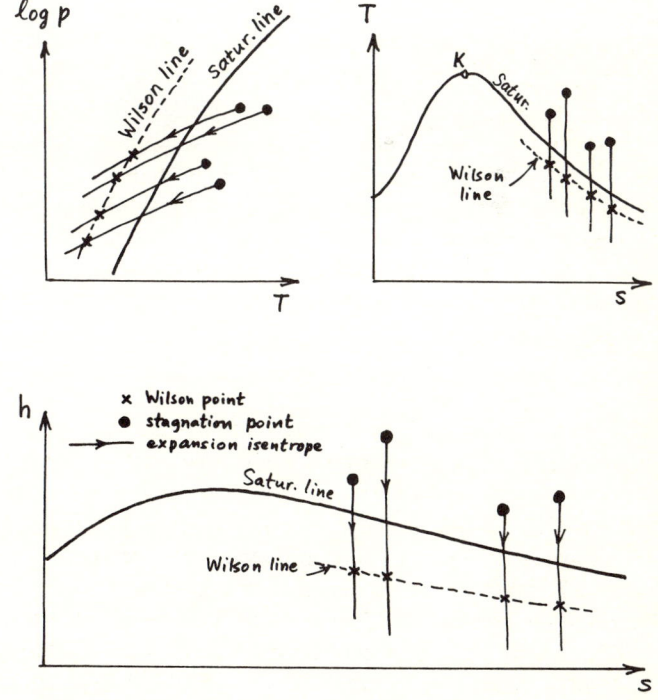

FIG. 3.1.1 Representation of a Wilson line in various thermodynamic charts.

Expansion rate Logarithmic rate of decrease of static pressure in a fluid element (usually referred to noncondensing flow)

$$\dot{P} = -\frac{d \ln p}{dt} = -\frac{1}{p}\frac{dp}{dt} = -\frac{C}{p}\frac{dp}{dx} \tag{3.1.1}$$

It can be shown that $\dot{P}$ in ideal gases is related to the rate of change of other parameters as follows

$$\dot{P} = -\frac{\kappa}{\rho}\frac{d\rho}{dt} = \frac{-\kappa}{(\kappa-1)T}\frac{dT}{dt} = \frac{\kappa M}{1+(\kappa-1)M^2/2}\frac{dM}{dt} = \frac{M^2 C\kappa}{(M^2-1)A}\frac{dA}{dx} \tag{3.1.2}$$

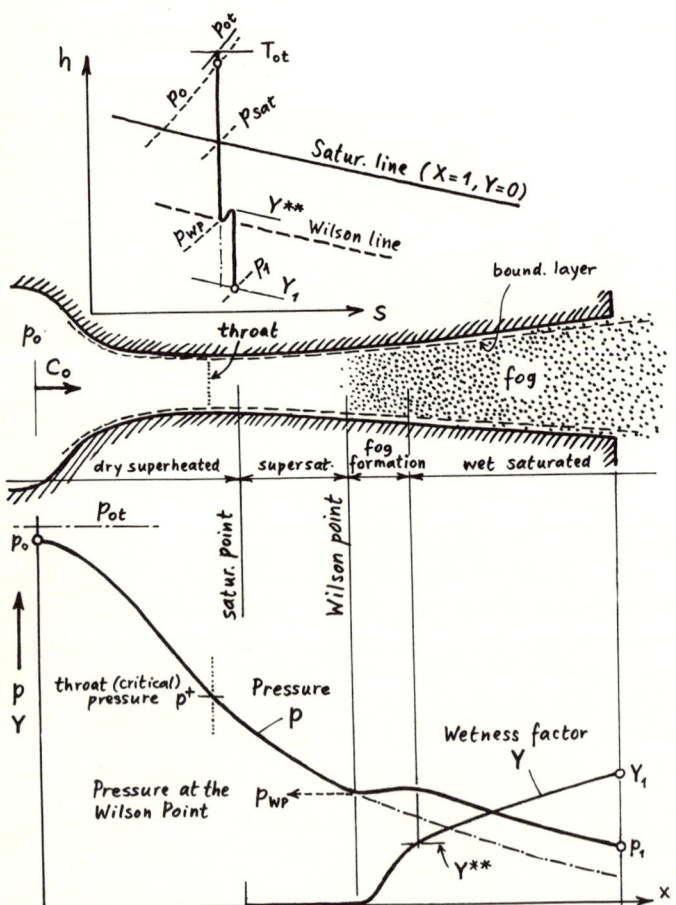

FIG. 3.1.2 Characteristic events in condensing Laval-nozzle flow, with definitions used.

Cloud-chamber expansion Observations indicate the occurrence of supersaturation up to about $S = p/p_s(T) = 4$ in water vapor free of dust particles. Further expansion causes sudden collapse of supersaturation and fogging.

Laval nozzle flow Here observation is easier, because sequences in time are displayed in space. Condensation is manifested by a change of pressure distribution as compared to dry flow. The sequence of events is shown in Fig. 3.1.2.

3.1.3 Steady Gas Flow in Laval Nozzles

3.1.3.1 Main Features

Laval nozzles are convergent-divergent channels, in which a gas or vapor flow is accelerated from subsonic to supersonic speed (Fig. 3.1.3a). Depending on

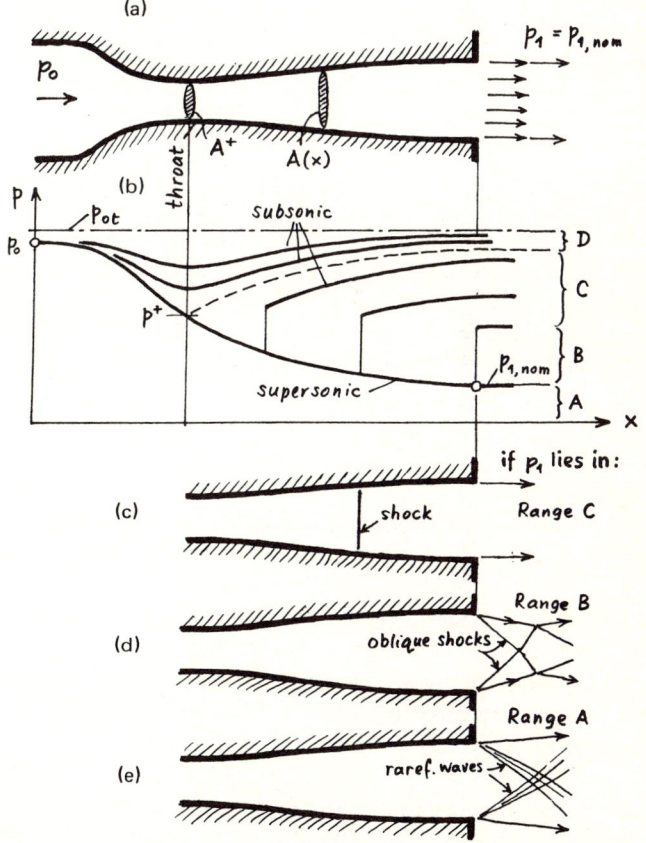

FIG. 3.1.3 Laval-nozzle flow of a dry gas at various back pressure ratios p_1/p_{0t}.

the ratio of back pressure p_1 to inlet stagnation pressure p_{0t}, several operational regimes are known to occur. These lead to characteristically different wall-pressure distributions, as sketched in Fig. 3.1.3b:

Shock-free supersonic flow in divergent part (Fig. 3.1.3a, d, e), if p_1/p_{0t} is lower than nominal (range A) or only slightly higher (range B).
Supersonic flow with shock wave in divergent part (Fig. 3.1.3c), if p_1/p_{0t} lies in range C. Shock causes return to subsonic flow.
All subsonic flow ("Venturi nozzle" mode), if p_1/p_{0t} is in range D.

The exit jet is influenced by oblique shock waves (Fig. 3.1.3d) or by rarefaction waves (Fig. 3.1.3e), if p_1/p_{0t} is in range B or A, respectively. The mass flow through a Laval nozzle is determined by inlet conditions and by throat size, and is independent of back pressure ("choked" flow), except if flow is all-subsonic.

3.1.3.2 Theoretical Description of Idealized Flow

Theoretical description of steady, noncondensing, compressible flow in slender Laval nozzles is usually based on the following assumptions:

One-dimensional
Steady
Adiabatic
Frictionless (boundary layers are taken into account by a correction to passage area only)
Absence of body forces
Perfect-gas fluid
Known variation of effective passage area A with streamwise coordinate x, i.e., exclusion of shock/boundary-layer interaction; $A(x)$ = geometric area minus boundary-layer displacement area

The resulting equations for shock-free flow are summarized in Table 3.1.1; for derivation see, for example, [3-2].

3.1.3.3 Frictional Effects

Friction is usually negligible, except within the wall boundary layers. Figure 3.1.4 shows total pressure traverses made in two rather slender steam nozzles of sizes typical for condensation studies at low pressure [3-3]. The traverses were made at three positions I, II, and III downstream of the throat. Nozzle width b, throat area A^+, and diffuser length-to-width ratio l_0/b are: for nozzle a, $b = 14$ mm, $A^+ = 140$ mm^2, $l_0/b = 8.2$; for nozzle b, $b = 24$ mm, $A^+ = 410$ mm^2, and $l_0/b = 14.4$. Both the total-pressure profiles p_t/p_{0t} and the Mach number profiles M indicate that the flow has a frictionless core.

Such measurements can be used to evaluate the magnitude of deviations from isentropic conditions at various distances y from the nozzle wall, as

TABLE 3.1.1 Gas dynamic equations for shock-free monophase nozzle flow

Conservation of mass

$\rho C A = \dot{m}_0$ (3.1.T1) $\dfrac{d\rho}{\rho} + \dfrac{dC}{C} + \dfrac{dA}{A} = 0$ (3.1.T2)

Conservation of momentum

$\rho C\, dC = -dp$ (3.1.T3)

Conservation of energy

$h + \tfrac{1}{2} C^2 = h_{ot}$ (3.1.T4) $dh + C\, dC = 0$ (3.1.T5)

Equations of state

thermal $\rho = \dfrac{p}{RT}$ (3.1.T6) $\dfrac{d\rho}{\rho} = \dfrac{dp}{p} - \dfrac{dT}{T}$ (3.1.T7)

caloric $h = c_p T$ (3.1.T8) $dh = c_p\, dT$ (3.1.T9)

entropy s $T\, ds = dh - \dfrac{dp}{\rho}$ (3.1.T10)

Definitions

specific heat ratio $\kappa = \dfrac{c_p}{c_p - R}$ (3.1.T11)

acoustic speed $a = \left[\left(\dfrac{dp}{d\rho}\right)_{\text{adiab.}}\right]^{1/2} = (\kappa R T)^{1/2}$ (3.1.T12)

Mach number $M = \dfrac{C}{a}$ (3.1.T13)

Differential equation of motion, from Eqs. (3.1.T2, T3, T5, T7, and T9)

$$\dfrac{dC}{dx} = \dfrac{-C}{1 - (C/a)^2} \dfrac{1}{A(x)} \dfrac{dA}{dx} \quad (3.1.\text{T14})$$

Inlet stagnation state: $p_{ot}, T_{ot}, h_{ot}, \rho_{ot}$

Poisson relations

$$\dfrac{p}{p_{ot}} = \left(\dfrac{\rho}{\rho_{ot}}\right)^{\kappa} = \left(\dfrac{T}{T_{ot}}\right)^{\kappa/(\kappa-1)} \quad (3.1.\text{T15})$$

Expressions in terms of Mach number

$$\dfrac{T}{T_{ot}} = \left(1 + \dfrac{\kappa - 1}{2} M^2\right)^{-1} \quad (3.1.\text{T16})$$

$$\dfrac{p}{p_{ot}} = \left(1 + \dfrac{\kappa - 1}{2} M^2\right)^{-\kappa/(\kappa-1)} \quad (3.1.\text{T17})$$

$$\dfrac{\rho}{\rho_{ot}} = \left(1 + \dfrac{\kappa - 1}{2} M^2\right)^{-1/(\kappa-1)} \quad (3.1.\text{T18})$$

Critical (throat) speed $C^+ = a^+ = \left(\dfrac{2\kappa}{\kappa + 1} R T_{ot}\right)^{1/2}$ (3.1.T19)

Nozzle area relation

$$\dfrac{A}{A^+} = \left[\dfrac{2}{\kappa + 1}\left(1 + \dfrac{\kappa - 1}{2} M^2\right)\right]^{(\kappa+1)/[2(\kappa-1)]} M^{-1} \quad (3.1.\text{T20})$$

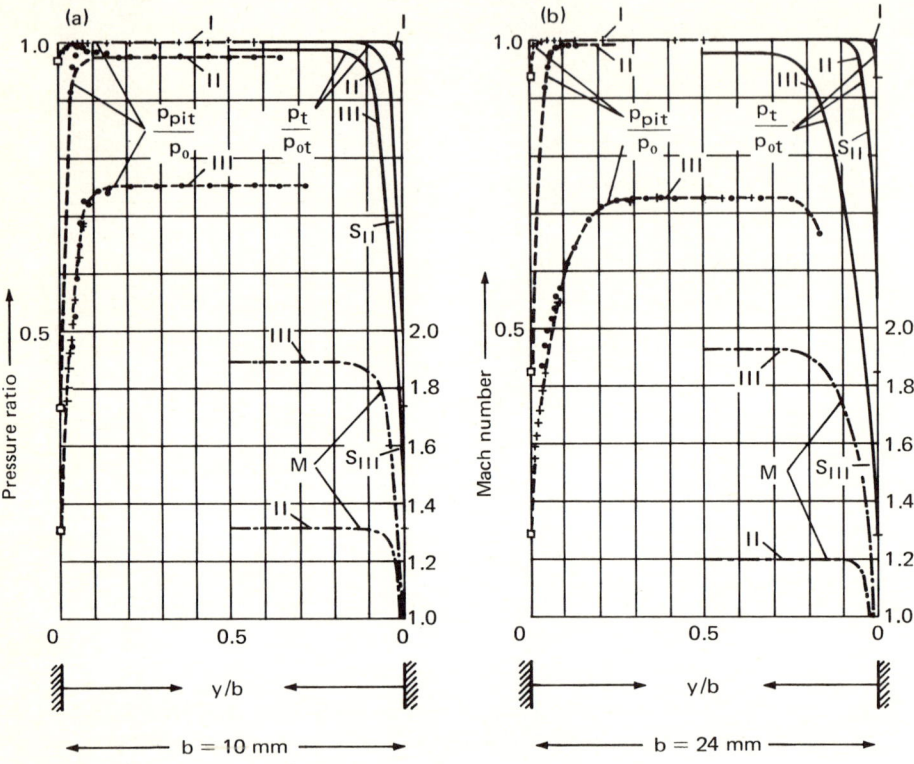

FIG. 3.1.4 Pressure and Mach number profiles measured in two Laval nozzles of different size [3-3].

shown in the Mollier diagram (Fig. 3.1.5). It is seen that, down to pressures of 0.3 p_{ot} and less, isentropic conditions exist in most parts of the flow cross section. In nozzles having smaller relative length, boundary layer effects are of course even less important. In view of this situation, frictional effects are better accounted for by a reduction of the geometric flow area according to the displacement thickness of the boundary layers, than by corrections in which frictional losses are "smeared out" uniformly over the entire flow area.

Shock waves, if present, interact with the boundary layers ("lambda shocks"). Therefore wall-pressure measurements do not indicate clearly the discontinuity associated with the shock; optical methods allow a better identification and localization of shocks.

3.1.3.4 Two-Dimensional Effects

If the nozzle channel is not sufficiently slender, conditions vary across the flow even within the frictionless core. The magnitude of these effects is illustrated in Fig. 3.1.6 (from [3-4]). Here the curves of constant speed (and constant conditions) are shown for a nozzle bounded by two circular profiles and flat side walls.

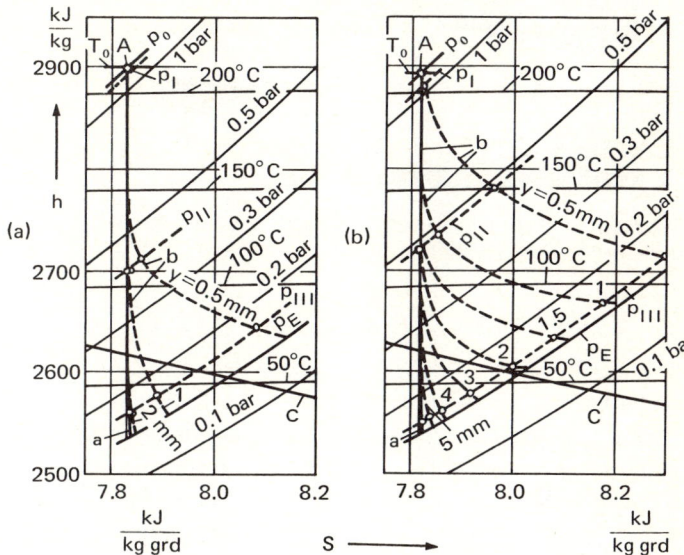

FIG. 3.1.5 Fluid states at various distances from the nozzle wall (cf. Fig. 3.1.4).

In nozzles of strongly two-dimensional shape (an extreme case being turbine cascades), condensation does not occur simultaneously across the entire flow cross section, but rather forms an oblique or V-shaped front, cf. Fig. 3.1.7 (from [1-17]). The condensation front roughly follows the lines of constant conditions.

Disturbances of a two-dimensional nature can be observed in nozzles with profile contours not sufficiently even. Abrupt changes of wall curvature (e.g., at the transition from a highly curved throat portion to a straight diverging portion) may lead to undulations in the pressure distribution measured at the nozzle centerline (see Fig. 3.1.8, after [3-6]). Such parasitic effects can render the interpretation of pressure measurements very difficult.

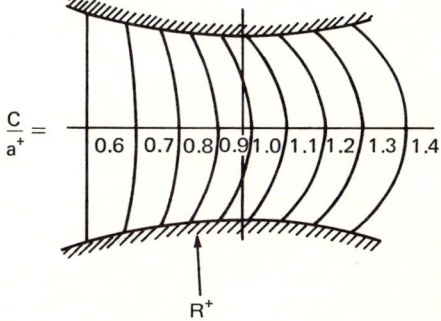

FIG. 3.1.6 Curves of constant flow velocity C near the throat of a "two-dimensional" nozzle ($R^+ = 2.5\ H^+$), from [3-4].

FIG. 3.1.7 V-shaped condensation "shock" in a Laval nozzle having highly curved walls [1-17].

3.1.3.5 Shocks with Heat Release

Chemical reactions (e.g., detonations or burning) or phase transitions (like condensation) involve the addition of a certain amount of latent heat to the flowing gas. (In some cases heat absorption may occur.) The process can, in a simplified way, be looked upon as being instantaneous, in which case the heat release is represented by a discontinuity in the flow conditions. Properties of such discontinuities in one-dimensional subsonic or supersonic flows of ideal gases, when gas mass and number of moles stay constant during the addition of a certain amount of heat (energy Q added per unit mass of gas), are plotted in Fig. 3.1.9 (from [3-5]). The Mach numbers ahead of and behind the discontinuity, M_1 and M_2, are coupled by the curves shown for various values of the parameter $Q' = Q/c_p T_{0t}$. Discontinuities in supersonic flow ($M_1 > 1$) may lead to subsonic

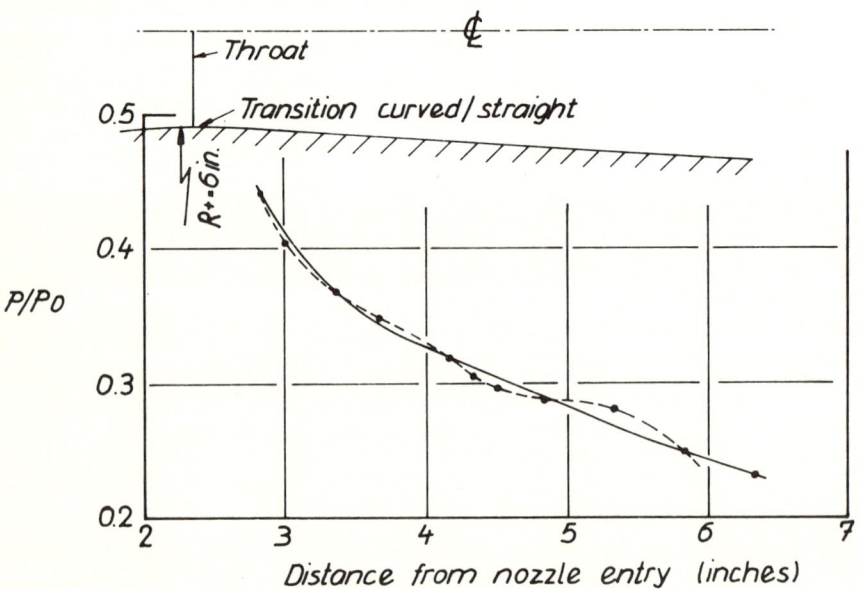

FIG. 3.1.8 Undulating pressure distribution at the nozzle centerline, after [3-6].

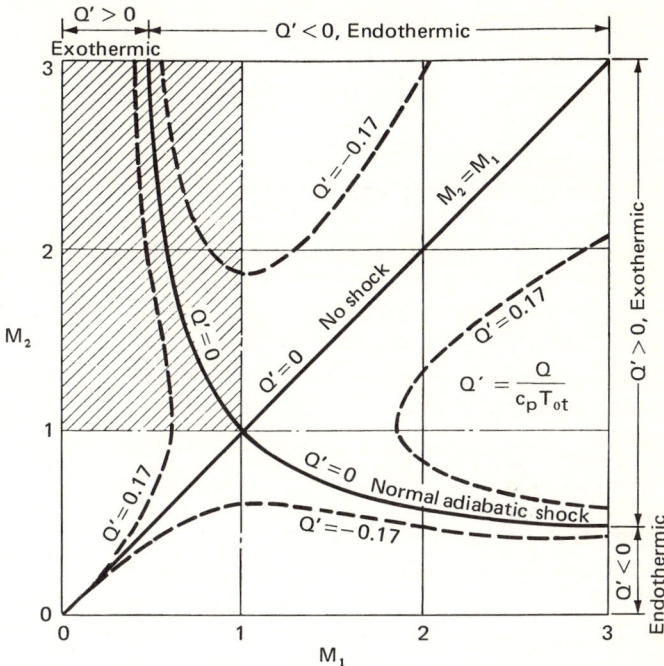

FIG. 3.1.9 Mach number change in shocks with heat addition, from [3-5].

or supersonic downstream flow ("strong detonation": $M_2 < 1$; "weak detonation": $M_2 > 1$).

For a certain positive value of Q' (e.g., for $Q' = 0.17$) there is, around $M_1 = 1$, a region where no solution exists. This means that release of this amount of heat in a steady flow is not possible at such M_1 values. If a chemical reaction or phase transition occurs in this "prohibited" range, the flow has to adjust itself. Subsonic flows will reduce their mass flow rate to produce a sufficiently low M_1 value at the point of heat release ("thermal choking"), while heat addition in supersonic flow will lead to nonsteady (pulsating) phenomena.

3.2 EXPERIMENTAL EVIDENCE IN NOZZLES

3.2.1 Test Equipment Used

3.2.1.1 Nozzle Apparatus Design

In steam work, test rigs capable of continuous operation are commonly used. Superheated steam provided by a boiler is desuperheated either by water injection or by expansion through a turbine stage, and is expanded in a Laval nozzle to a low (e.g., condenser) pressure. In the nozzle, pressure distribution and/or optical and other measurements and observations are made. Inlet stagnation conditions are usually superheated; in the case of steam, this means that condensation begins downstream of the throat. The following aspects merit particular attention and care:

Ensure complete evaporation of injected water ahead of test section. (Provide sufficient steam flow, if required by using by-pass lines.)
Use steam and water of high purity; keep system internally clean.
Keep inlet conditions constant during measurement (precise control of pressure and temperature; sufficient waiting time between tests made at different inlet temperatures).
Avoid nozzle contours with sudden transitions from curved to straight; machine nozzle parts with high precision (smooth walls, minimized leakage gaps).
Ensure that boundary layers are thin with respect to nozzle width and height.

In order to decouple the effects of the inlet conditions from those of the expansion rate, it is advisable to employ nozzle wall contours that lead, in the absence of condensation, to a constant value of the expansion rate $\dot{P}$ in the diverging part of the nozzle. The area distribution of such nozzles can be calculated [3-3] from the equations given in Table 3.2.1. Circular-arc contours [3-7] are cheaper to manufacture and yield almost constant values of $\dot{P}$. Curved-throat contours followed by straight walls can lead to highly variable $\dot{P}$, and cause wavy pressure distributions (Fig. 3.1.8).

3.2.1.2 Instrumentation for Pressure Measurement

In order to measure the subcooling ΔT_{WP} at the Wilson point with sufficient accuracy, great care has to be devoted to the measurement of both the inlet

TABLE 3.2.1 Pressure and nozzle area variation for constant expansion rate

The relation $\dot{P}(x) = -\dfrac{C(x)}{p(x)} \dfrac{dp}{dx} = \text{constant}$ is fulfilled in ideal-gas flows by the following pressure distribution $p(x)$:

$$\ln\left[\frac{1+\sqrt{1-\left(\dfrac{p(x)}{p_{ot}}\right)^{(\kappa-1)/\kappa}}}{1+\sqrt{(\kappa-1)/(\kappa+1)}}\right] - \frac{1}{2}\ln\left[\frac{\left(\dfrac{p(x)}{p_{ot}}\right)^{(\kappa-1)/\kappa}}{2/(\kappa+1)}\right] - \sqrt{1-\left(\dfrac{p(x)}{p_{ot}}\right)^{(\kappa-1)/\kappa}}$$

$$+ \sqrt{\frac{\kappa-1}{\kappa+1}} = \frac{\kappa-1}{2\kappa}\frac{\dot{P}}{\sqrt{2c_p T_{ot}}}(x-x^+) \quad (3.2.\text{T1})$$

from which $A(x)/A^+$ follows as

$$\frac{A(x)}{A^+} = \left(\frac{2}{\kappa+1}\right)^{1/(\kappa-1)} \sqrt{\frac{\kappa-1}{\kappa+1}} \left(\frac{p(x)}{p_{ot}}\right)^{-1/\kappa} \left[1-\left(\frac{p(x)}{p_{ot}}\right)^{(\kappa-1)/\kappa}\right]^{-1/2} \quad (3.2.\text{T2})$$

TABLE 3.2.2 Error propagation factors in nozzle experiments

Assumed original error[a]	Resulting error of ΔT_{WP} (°K)
+1% in p_{0t}	+1.0
+1°K in T_{0t}	−1.0
+1% in $p(x)$	−0.6
+$\Delta x_{0.01}$ in x	+0.6

[a] $p_{0t} = 2$ bar, $T_{0t} = 150°C$, and $\Delta x_{0.01}$ = distance along which static pressure decreases by 1%.

stagnation conditions (pressure p_{0t} and temperature T_{0t}) and the static-pressure distribution $p(x)$. Error propagation factors for typical inlet conditions are summarized in Table 3.2.2. Typically it is desirable to measure ΔT_{WP} with an accuracy better than ±1°K. This requires, e.g., as a combination

Errors of p_{0t} to be $< 0.2\%$
Errors of T_{0t} to be $< 0.5°K$
Errors of $p(x)$ to be $< 0.5\%$
Errors of x to be $\ll \Delta x_{0.01}$

The severity of these limits may partially explain some of the scatter of the experimental data. If the inlet flow is not completely dry, prohibitive errors are likely to be incurred in the measurement of T_{0t}.

For measuring $p(x)$, the following arrangements have been used:

Nozzle wall equipped with a *series of* closely spaced *pressure taps* [3-9, 3-12]
One pressure tap in fixed plane wall, with *movable contoured wall element* on opposite side [3-8, 3-10, 3-11]
One pressure tap in a *movable rail*, which is accommodated in a plane nozzle wall [3-3]
One pressure tap in *movable tube* spanned along nozzle centerline [3-7]

Best accuracy of $p(x)$ seems to be obtained with the last method, provided the flow cross section is large as compared to the tube. A view of this arrangement is shown in Fig. 3.2.1. Scavenging of pressure lines with air is recommended in all cases. Fast-response pressure transducers are required in order to detect flow oscillations [3-7, 3-15].

3.2.1.3 Other Instrumentation

These include:

Interferometers for measuring density distributions [3-7]
Optical arrangements for light scattering or attenuation, in order to measure droplet size [3-10, 3-11, 3-12, 3-13]

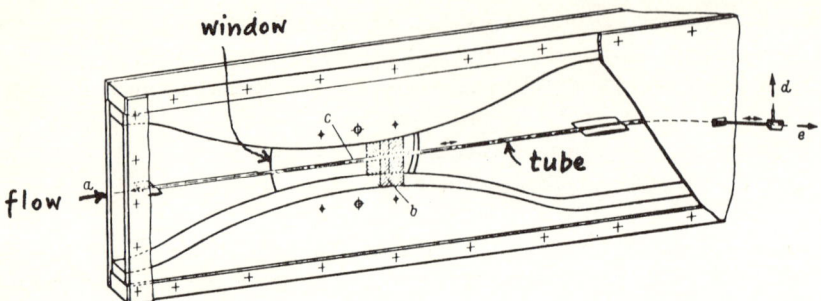

FIG. 3.2.1 Central tube used for measuring the static pressure distribution (front side wall not shown in drawing), from [3-7].

Thin-wire resistance thermometers for detecting the presence of wetness in the flow [3-14, 3-3]

Pitot probes used for boundary-layer traverses [3-3] or for along-the-axis measurements [3-16]

An optical arrangement for simultaneous measurement of light attenuation and scattering is depicted in Fig. 3.2.2 [3-12].

3.2.2 Phenomenology of Condensation in Nozzles

In the following, a brief description is given of the phenomena observed in nozzles in the absence of foreign condensation nuclei. Condensation of humid air and of steam shows qualitatively the same behavior. In nozzles having uneven or highly curved contours, some of the phenomena described can be overshadowed by disturbances or modified by two-dimensional effects, and their observation may become impossible.

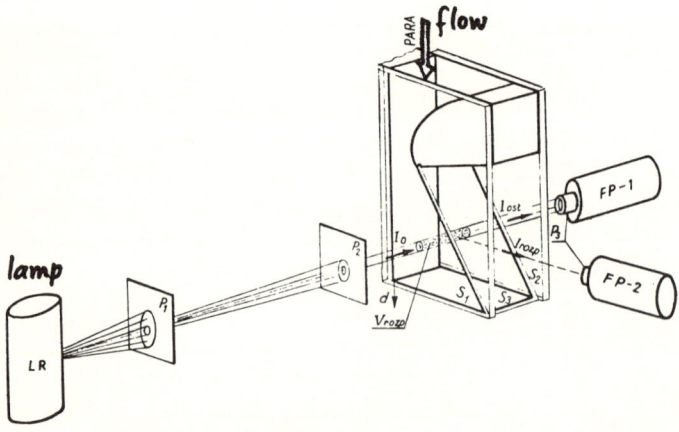

FIG. 3.2.2 Optical arrangement for droplet size measurement based on scattered and attenuated light [3-12].

We consider a Laval nozzle expanding superheated dry steam from a constant inlet stagnation pressure p_{0t} to a back pressure p_1 sufficiently low to ensure shock-free supersonic flow. The inlet stagnation temperature T_{0t} shall be varied, starting from a high level (A in Fig. 3.2.3) for which superheated conditions prevail down to the nozzle exit. Only the flow outside the boundary layers (e.g., along the centerline) will be considered; here the expansion shall be regarded as isentropic.

Figure 3.2.3 shows a series of typical inlet conditions $A, B, \ldots, E$ with pertinent expansion lines (top part) and static pressure distribution (bottom part). The main features for each case are

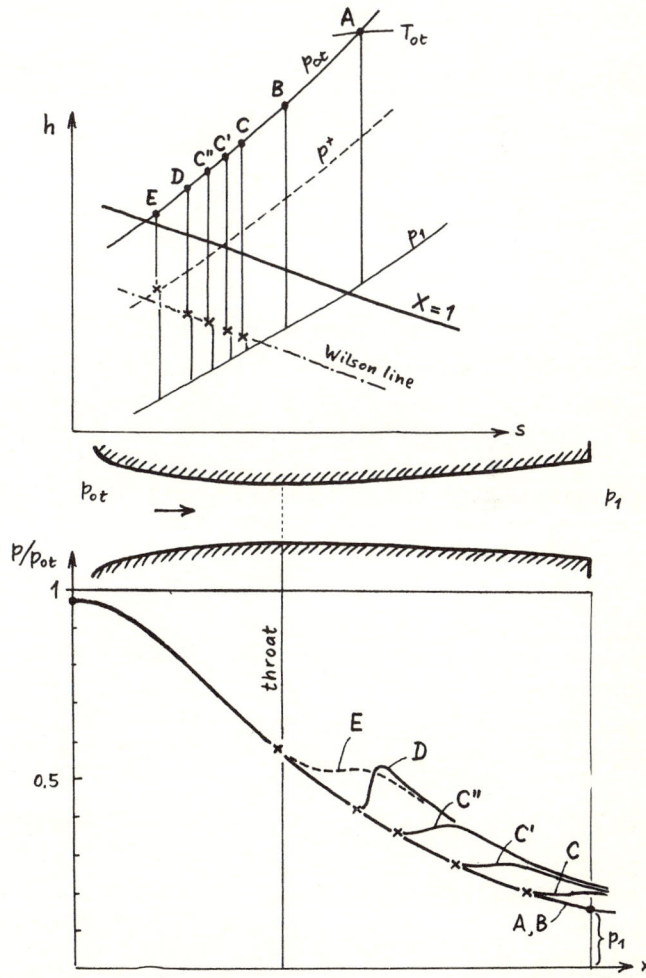

FIG. 3.2.3 Condensing flow with various values of the inlet superheat $T_{0t} - T_s(p_{0t})$.

- A All-superheated expansion (steam behaving as gas)
- B Supersaturated near downstream end; flow as in A
- C Pressure "bump" near nozzle exit; indicates attainment of Wilson line (Wilson point: x)
- C', C'' Wilson point is shifted upstream; pressure bump becomes steeper
- D Pressure bump is very steep (condensation is combined with shock)
- E Pressure bump begins in or near throat; high-speed instrumentation reveals pressure fluctuations, while slow-response pressure measurement yields a time-averaged ("smeared-out") pressure distribution

Optical interferometry, which has a sharper spatial and time resolution than usual pressure measurement, reveals additional details (Fig. 3.2.4). The

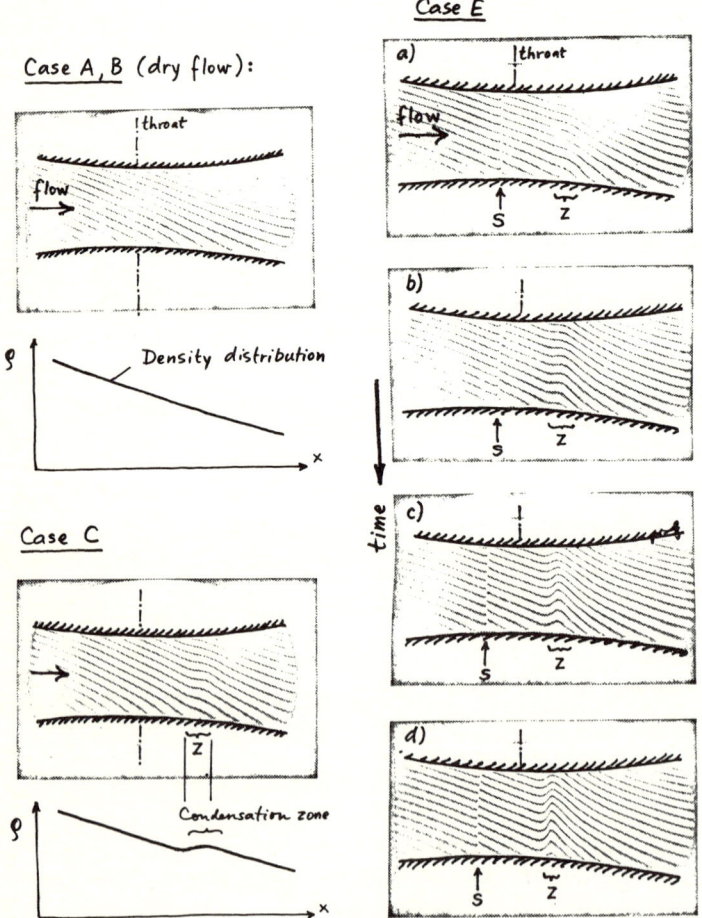

FIG. 3.2.4 Interferograms of condensing flow with different inlet conditions, from [3-17].

pictures shown were made with humid air flow [3-17], but could also be reproduced with steam flow [3-7]. In conditions A and B, flow density varies smoothly. For case C the interferogram reveals a continuous disturbance of the density field downstream of the throat (at the location where pressure measurement shows the "bump"); the disturbance spreads all across the flow. This disturbed region can be shown to be identical with the zone of fog formation. For case D (not depicted), there appears a shock wave (density jump) within this zone. Finally, for case E, interferograms reveal a periodic process of high frequency: the condensation zone appears at Z in frame a, shifts upstream and gets steeper in the process (frames b, c, d), until a shock is formed that travels into and through the throat (S in frames a, b, c, d), and decays in the subsonic flow. As soon as the shock has traveled a slight distance, a new disturbance begins to form behind it, as seen in frame a. Frequencies of shock formation were found to be of the order of 500 to 1000 Hz. The frequency depends on the inlet conditions [3-7] and on nozzle geometry [3-18]. These periodic phenomena seem not to appear in nozzles having highly curved throat contours.

Condensation in the subsonic part of the nozzle in low-pressure steam nozzles can only be realized with subcooled inlet stagnation conditions. No such experiments have been reported, however. This might be due to the difficulty of generating, in flow systems, subcooled vapor free of condensation nuclei.

If a light beam is transmitted across the nozzle (a typical arrangement is shown on top of Fig. 3.2.5 [3-11]) and its distance from the throat varied, the transmitted intensity I exhibits a sudden decrease in the region where the pressure bump is observed. As made visible by its scattered light, a fine fog consisting of tiny droplets is being formed in this region across the entire flow cross section. The appearance of wetness in this region can also be demonstrated by the use of thin-wire thermometer probes [3-14, 3-3] (Fig. 3.2.6). By suitable illumination of the nozzle channel, the fog formed in the flow can be made clearly apparent. Figure 3.2.7, taken from [3-3], shows a nozzle illuminated from the upstream direction.

To round off the phenomenological picture of spontaneous condensation in nozzles, the influence of a shock wave imposed by high back pressure (range C in Fig. 3.1.3) should be mentioned. Since pressure and temperature are made to increase by the shock, the steam becomes drier. Under certain conditions (high initial superheat, strong shock), complete evaporation of the fog can be caused. Curves c, b, and e in Fig. 3.2.6 pertain to such a situation.

At inlet conditions of low superheat, when the condensation occurs close to the throat, the location of the condensation zone is very sensitive to variations of inlet conditions. In such situations, pressure, etc., in the region of the "bump" can oscillate, and the position of a shock, if present, will also fluctuate.

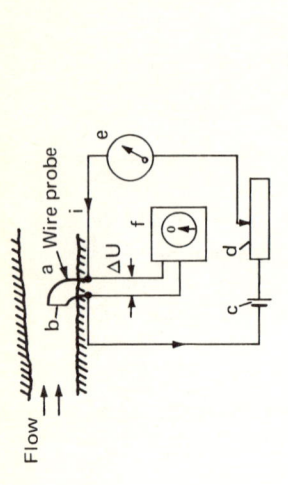

a Schematic illustration of the photometric device.
b A sectional view of the nozzle.

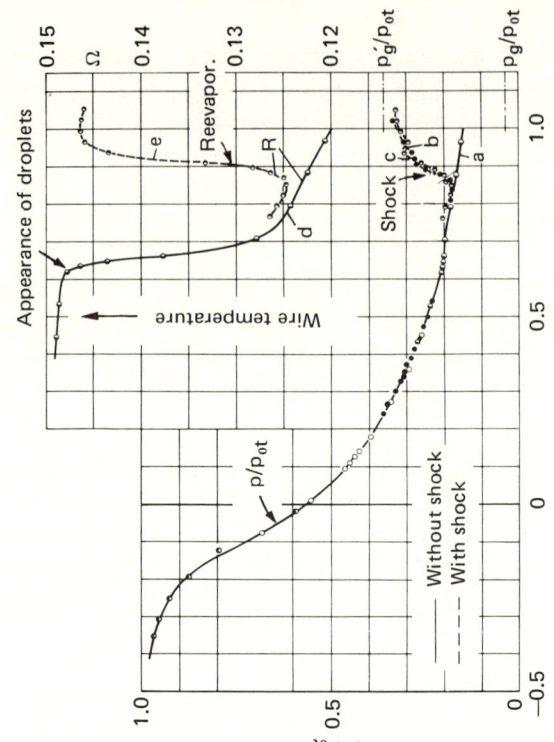

FIG. 3.2.6 Variation of pressure ratio and wire-probe temperature along a nozzle [3-3].

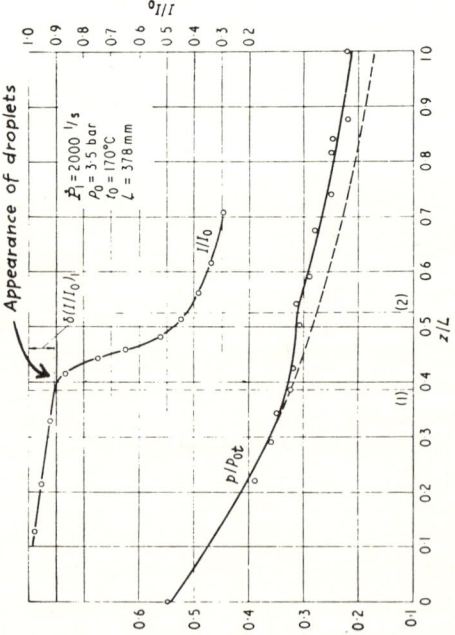

FIG. 3.2.5 Variation of the pressure ratio and light intensity attenuation along the nozzle axis for $\dot{P} = 2000$ l/s and $p_{ot} = 1.15$ bar. Reproduced by courtesy of the Council of the Institution of Mechanical Engineers, from [3-11].

↑ throat   ↑ fog formation

FIG. 3.2.7 Fog formation made visible by illumination [3-3].

3.2.3 Synopsis of Experimental Data

The aim of nozzle experiments mostly consists in obtaining data on one or more of the following:

Position of Wilson line (i.e., supercooling at the Wilson point for various inlet conditions)
Size and number of fog droplets formed
Effects of heat addition on the flow (appearance of shocks, periodic phenomena)
Detailed sequence of events, for sake of comparison with theory

Many investigators have alternatively used several nozzles of different length in order to study the effects of the expansion rate. Some experiments were carried out in configurations different from usual Laval nozzles. Such cases will be treated in conjunction with turbine cascades.

Expansion rate values $\dot{P}$ cited in the following are based on noncondensing flow and refer to the region of the nozzle where condensation occurred in the experiment.

3.2.3.1 Pressure Distributions in Various Nozzles

Figures 3.2.8 and 3.2.9 show a variety of static pressure distributions obtained in various nozzles and with various inlet conditions [3-3, 3-7, 3-10, 3-19].

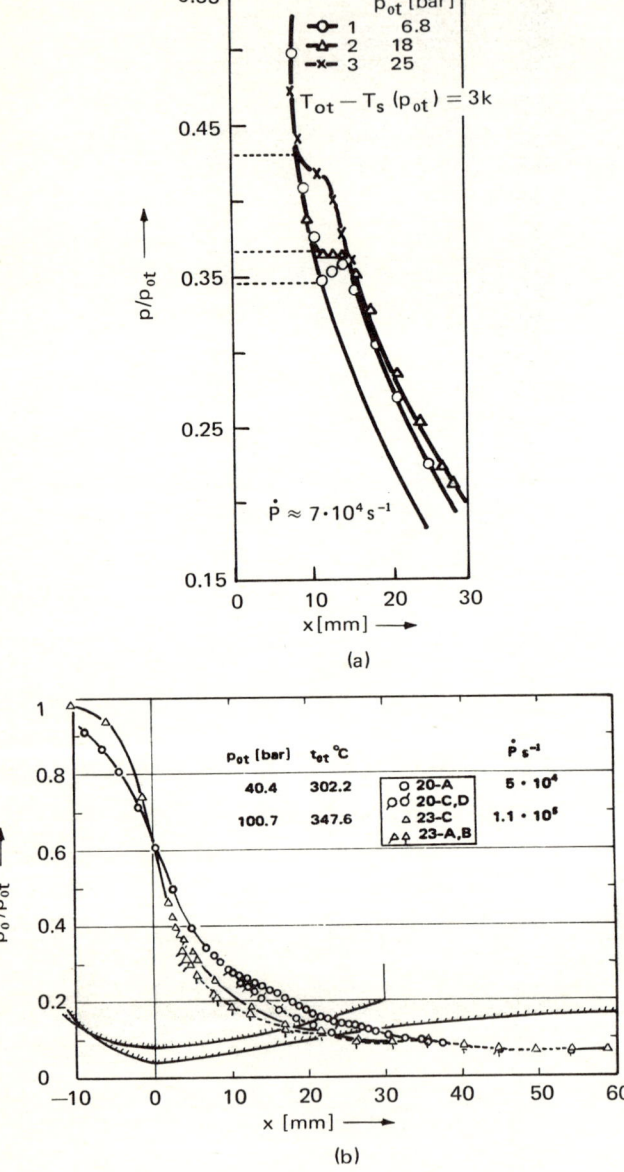

FIG. 3.2.8 Pressure distributions measured at high inlet pressure, (a) from [3-19], (b) from [3-10].

Figure 3.2.8 presents high-pressure experiments, of which few are available in the literature. Figure 3.2.9 gives a selection of low-pressure results. The quality of the nozzle contours can be judged on basis of the shape of the dry-superheated pressure curves. The curves of Fig. 3.2.9b are of excellent quality, while the nozzles of Fig. 3.2.9a were obviously very poorly shaped.

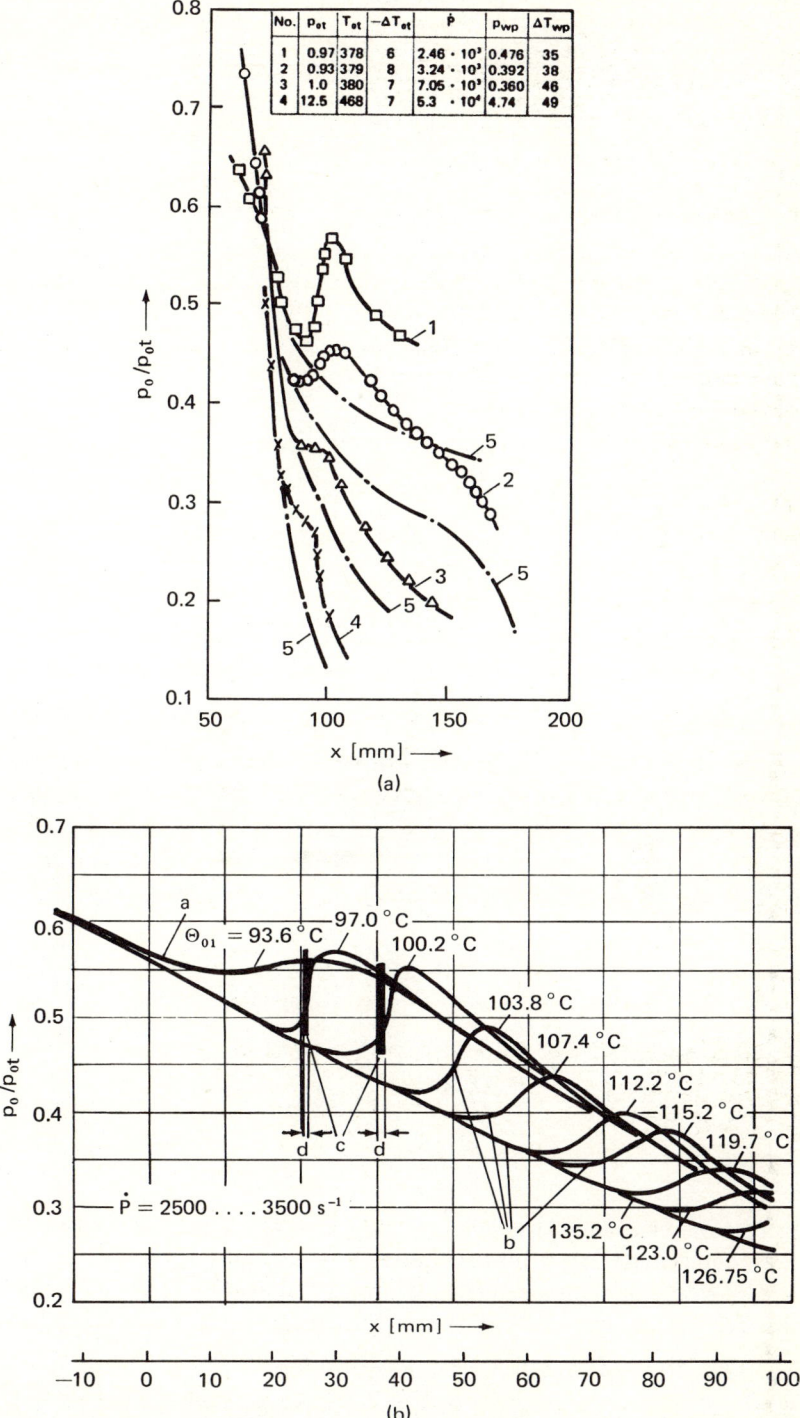

FIG. 3.2.9 Pressure distributions measured in low-pressure nozzles: (*a*) from [3-19] for four different nozzles; (*b*) from [3-7].

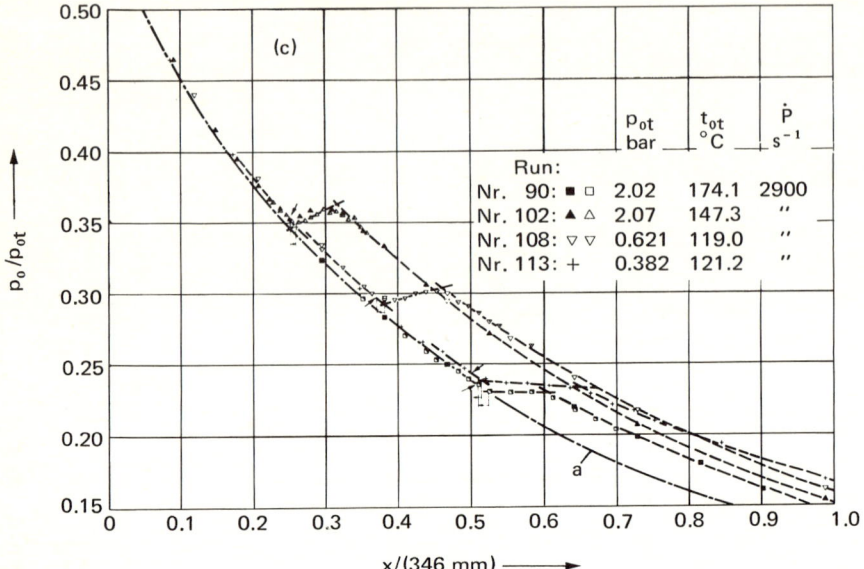

FIG. 3.2.9 (*continued*) Pressure distributions measured in low-pressure nozzles: (c) from [3-3].

The Wilson point is usually identified by the pressure p_{WP} at the point of departure of the S-shaped "bump" curve from the dry-superheated pressure curve (Fig. 3.1.2). Other parameters of interest in the Wilson point can be inferred from p_{WP}, p_{0t}, and T_{0t} using the equations of state of dry steam as follows (isentropic core flow is assumed)

$$s_{WP} = s_g(p_{0t}, T_{0t}) \tag{3.2.1}$$

$$h_{WP} = h_g(p_{WP}, s_{WP}) \tag{3.2.2}$$

$$T_{WP} = T_g(p_{WP}, s_{WP}) \tag{3.2.3}$$

$$\Delta T_{WP} = T_s(p_{WP}) - T_{WP} \tag{3.2.4}$$

$$S_{WP} = \frac{p_{WP}}{p_s(T_{WP})} \tag{3.2.5}$$

In case of thick boundary layers, two-dimensional effects, or nonisentropic core flow, appropriate corrections are necessary [3-3].

The Wilson lines obtained from the measured distributions will be discussed in Sec. 3.2.3.3. Comparison with theory will be dealt with in Sec. 3.3.

3.2.3.2 Size and Growth of Fog Droplets

Mean droplet radii $\bar{r}$ obtained by various authors [3-10, 3-13, 3-19, 3-20] using optical methods are assembled in Fig. 3.2.10. Figures 3.2.10*a*, *b*, and *d* refer to

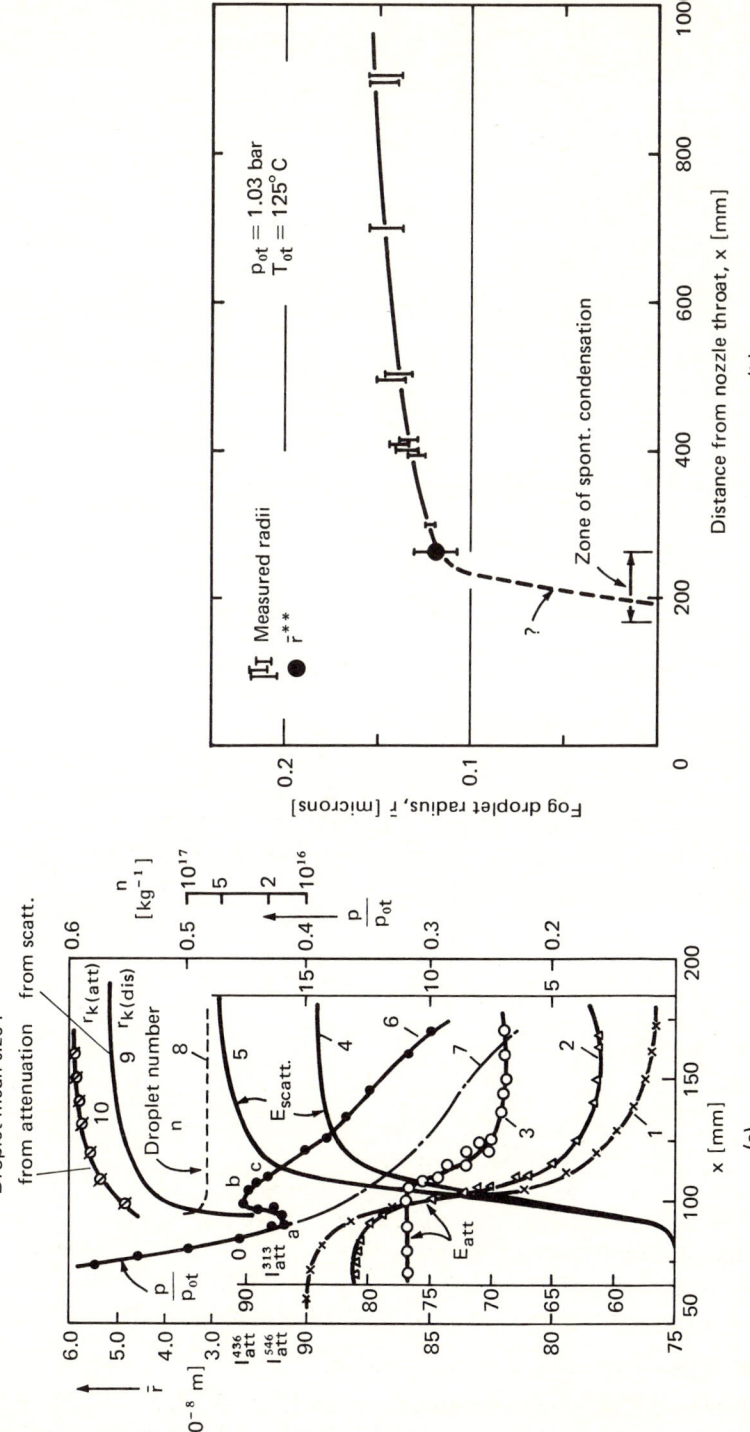

FIG. 3.2.10 Variation of average fog-droplet size along diverse nozzles (light attenuation and light scattering measurements): (a) $\dot{P} \approx 12{,}000 \text{ s}^{-1}$, from [3-19]; (b) $\dot{P} = 1000 \text{ s}^{-1}$, reproduced by courtesy of the Council of the Institution of Mechanical Engineers, from [3-13].

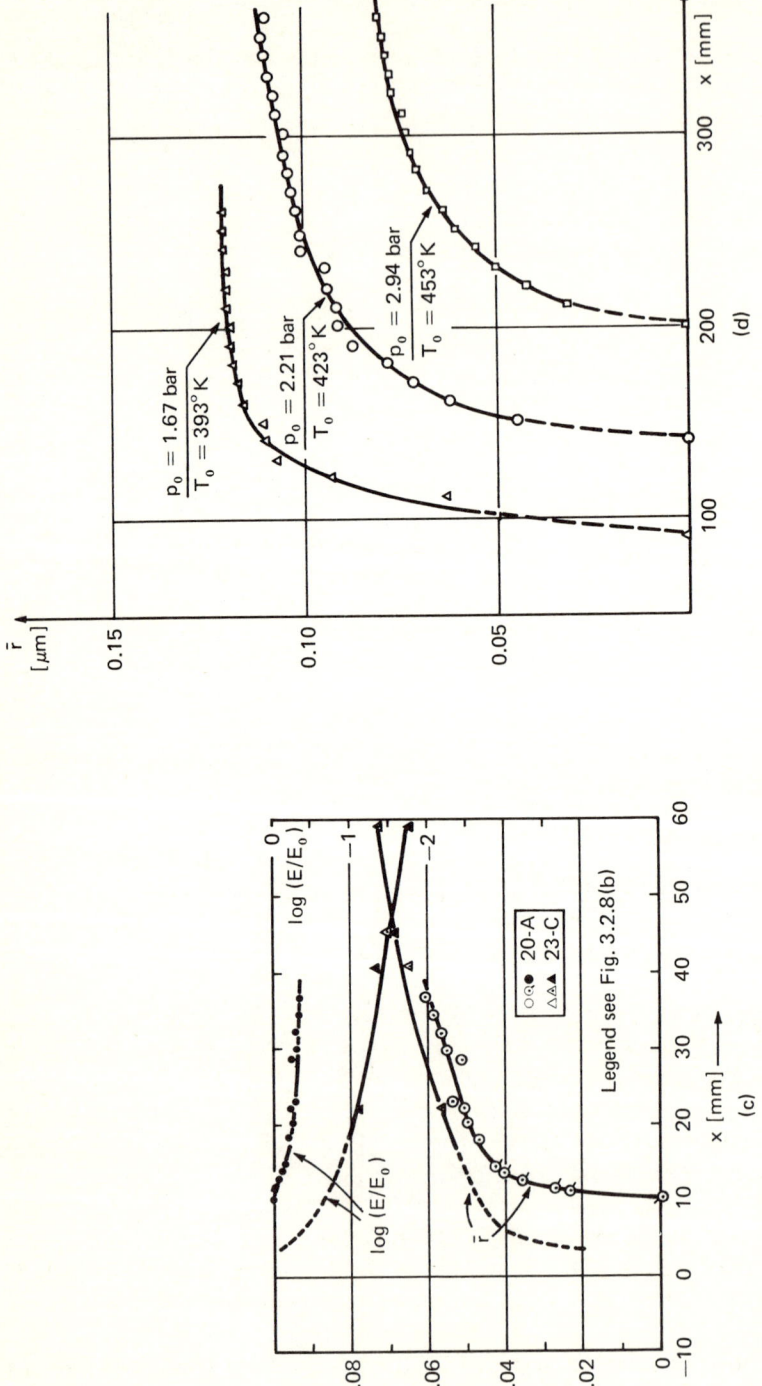

FIG. 3.2.10 (*continued*) Variation of average fog-droplet size along diverse nozzles (light attenuation and light scattering measurements): (*c*) $\dot{P} = 5 \times 10^4$, 1.1×10^5 s^{-1}, from [3-10]; (*d*) $\dot{P} \approx 2000$–4000 s^{-1}, from [3-20].

low pressure, c to high pressure. Expansion rates differ considerably from case to case. However the evolutions of mean droplet size along the nozzle axis reveal some common features. These are

Rapid growth of droplets in the initial stage of condensation
Slower growth further downstream
Final sizes of droplets in the range 0.05–0.2 μm in the nozzles used

The slowing down of droplet growth indicates that supersaturation has nearly disappeared in the latter part of the nozzle. The growth of droplets here reflects the increase of wetness due to continued expansion.

Both direct measurement of specific droplet number [3-19, 3-20] and theoretical estimates regarding the coagulation of droplets indicate that the number of droplets, N per unit mass of steam stays essentially constant along the flow (Fig. 3.2.10a).

In order to express the quality of the fog (i.e., the specific number of droplets) by a representative droplet size, it is convenient to use the size that corresponds to the amount of moisture theoretically present at the Wilson line, i.e., about $Y = 0.03$ wetness. This representative mean droplet size

$$\bar{r}^{**} \equiv \bar{r}_{Y=0.03} \tag{3.2.6}$$

is approximately equal to the size at which the $\bar{r}$ curves reach their plateau (Fig. 3.2.10b). Experimental data on $\bar{r}^{**}$ will be discussed in Sec. 3.2.3.4.

3.2.3.3 Wilson Lines

Wilson points obtained from a number of experiments by various authors are plotted in Figs. 3.2.11 to 3.2.14. The graphical symbols indicate the magnitude of the expansion rate. Tails attached to the symbols identify references.

The low-pressure data (Fig. 3.2.11) indicate excellent agreement between the results of Barschdorff [3-7] and Gyarmathy and Meyer [3-3]. For expansion rates typical for steam turbine blading ($\dot{P} = 1000$–5000 s^{-1}), the Wilson lines at low pressure are found to lie in a range corresponding to 2.9–3.3% equilibrium wetness. Points lying at high wetness ($> 4\%$) are probably in error due to wet inlet flow conditions.

Data obtained in nozzles having profile contours that lead to strong variations of $\dot{P}$ along the nozzle axis often show considerable scatter, the magnitude of which is illustrated in Fig. 3.2.12, taken from [3-21]. It is unlikely however that all of this scatter can be explained by nozzle shape alone.

At very low pressure ($p_{WP} = 0.015$–0.07 bar ≈ 10–50 Torr) a continuous extension of low-pressure Wilson lines is found to exist, as shown by Fig. 3.2.13, which is a p vs. T diagram, taken from [3-22]. It is believed that

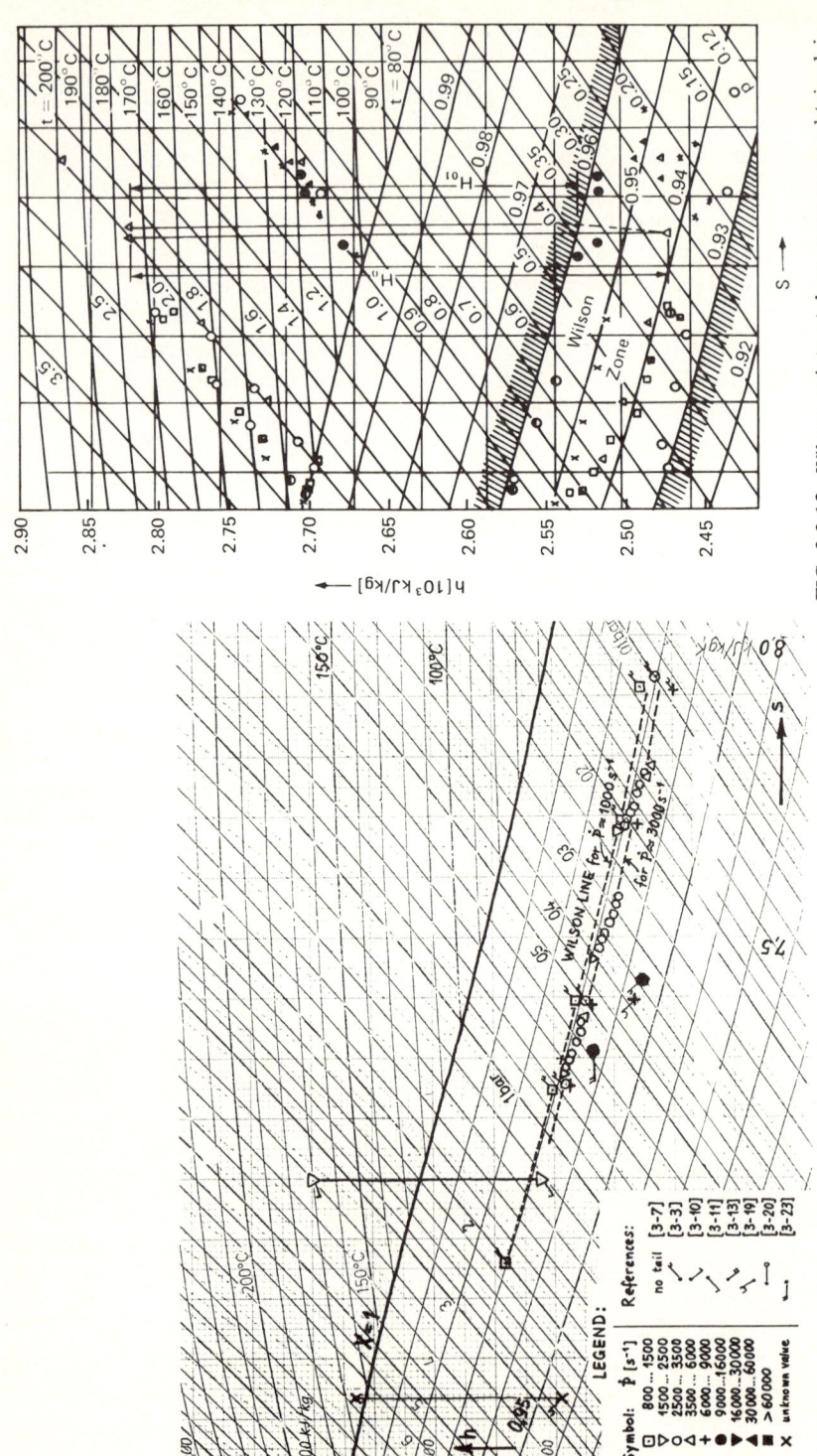

FIG. 3.2.12 Wilson points at low pressure as obtained in [3-21].

FIG. 3.2.11 Wilson points measured at low pressure.

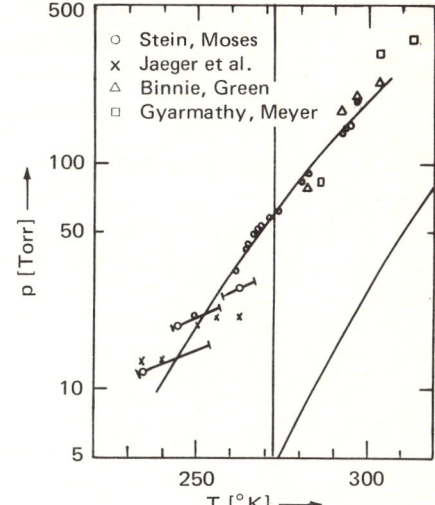

FIG. 3.2.13 Wilson points of steam at very low pressure [3-22].

liquid fog droplets are formed during spontaneous condensation of steam even if local flow temperatures lie as far as 30°K below the freezing point [3-22].

At high pressure, few published data exist [3-10, 3-19] (Fig. 3.2.14). The disagreement is large, and further data are necessary before Wilson lines can be localized with any degree of reliability. Some of the measurements were taken

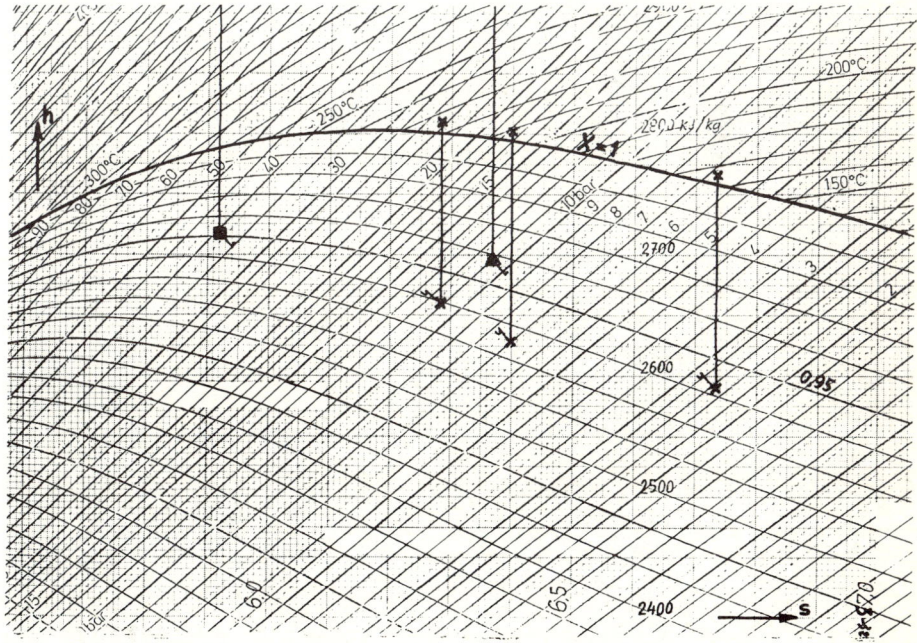

FIG. 3.2.14 Wilson points measured at high pressure. For explanation of symbols, see Fig. 3.2.11.

with low-superheat inlet conditions, as indicated in the figure. It is likely that these results were falsified by wet inlet flow.

3.2.3.4 Fog Droplet Sizes

Representative mean-fog droplet sizes, as defined by Eq. (3.2.6) and evaluated from measurements made by various authors, are plotted in Fig. 3.2.15. The pressure at the Wilson point p_{WP} is used as abscissa. The expansion rate values are indicated by the same symbol shapes as in Figs. 3.2.11 and 3.2.14.

As seen from Fig. 3.2.15, there is agreement that droplets formed in Laval nozzles at expansion rates of 1000 s^{-1} and above have radii of the order of 10^{-7} m (i.e., 0.1 μm) and smaller. There seems to be a definite effect of expansion rate on fog droplet size, as first predicted in [1-18] and verified in [3-3], although results obtained in some nozzles [3-19, 3-20, 3-23] do not fit into the general trend. Optical measurements (point symbols) and radii values inferred from the time duration of rapid droplet growth (continuous lines) show agreement at least in the trends. The latter values were calculated [3-3] from the time elapsing between the formation of nuclei at the Wilson point and the attainment of equilibrium conditions further downstream by using the growth law of droplets (Sec. 3.3). The duration of growth can be inferred from the width of the pressure bump. At high pressure, only few data are available [3-10, 3-19]. The results of Saltanov et al. [3-19], obtained in nozzles with unidentified $\dot{P}$ values, seem to suggest that low-pressure trends are maintained.

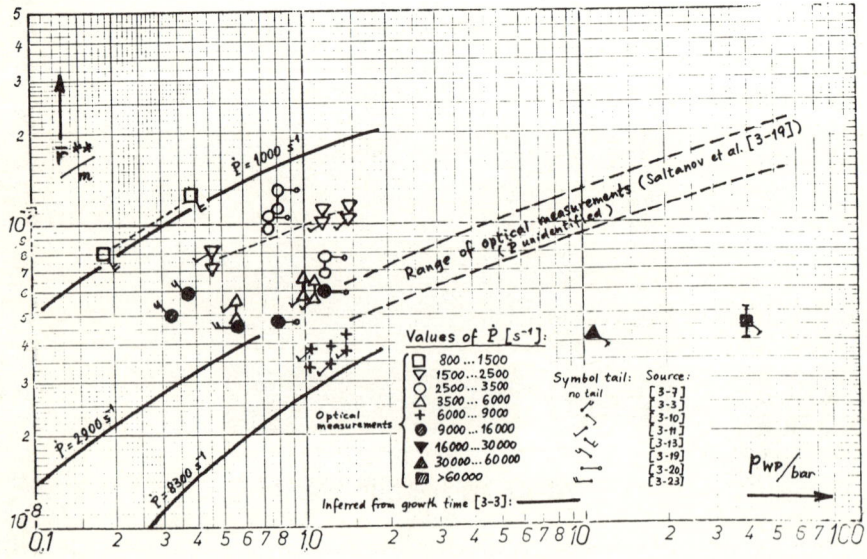

FIG. 3.2.15 Fog droplet size in Laval nozzles.

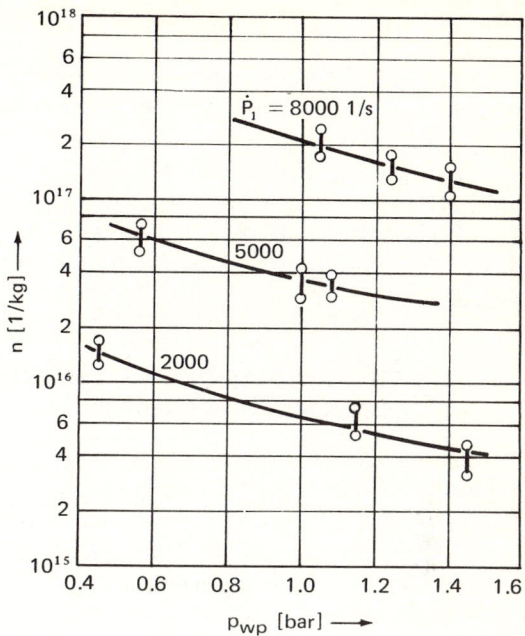

FIG. 3.2.16 Variation of the number of droplets formed at the Wilson line with pressure and expansion rate. Reproduced by courtesy of the Council of the Institution of Mechanical Engineers, from [3-11].

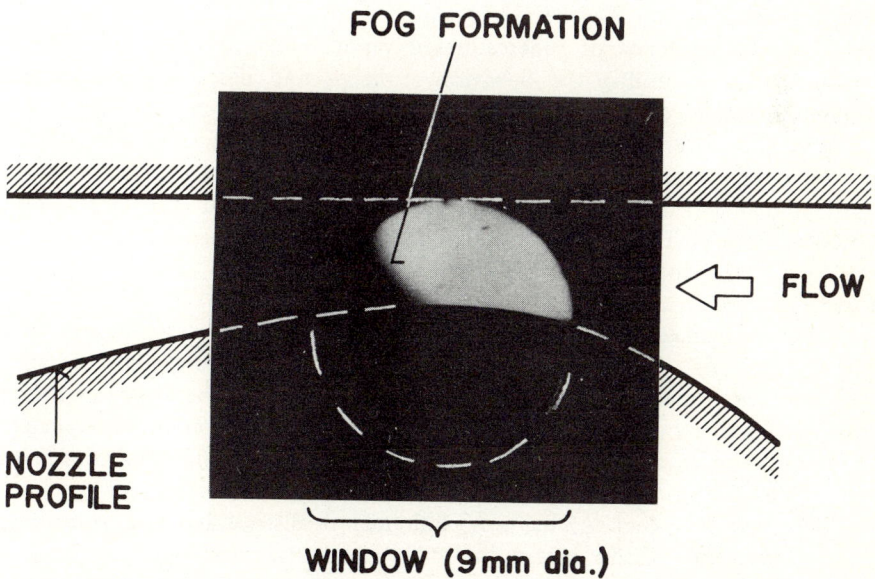

FIG. 3.2.17 Visual appearance of high-pressure steam fog (lower half of window field is obstructed by nozzle profile; in upper half fog appears as dark brown region covering the white illuminated background), from [3-10].

Data on the specific number of droplets, determined by purely optical methods [3-11] are shown in Fig. 3.2.16. Per kilogram steam there are 10^{15} to 10^{18} fog droplets being formed in typical low-pressure nozzles.

The visual appearance of nozzle fog depends considerably on droplet size and concentration. Low-pressure fogs formed at high $\dot{P}$ values are only visible if well illuminated; coarser low-pressure fogs resemble milky water [3-3, 3-12] (Fig. 3.2.7). At very high pressure (high steam density), fog in nozzles is virtually opaque, cf. Fig. 3.2.17, from [3-10].

3.3 THEORETICAL DESCRIPTION OF THE CONDENSATION PROCESS

In this section the theoretical equations describing the spontaneous condensation of vapors will be developed and discussed. First, nucleation and droplet growth are treated. The equations governing these processes will then be used, in conjunction with the basic gas dynamic equations, to describe one-dimensional condensing steam flow.

3.3.1 Nucleation Theory

3.3.1.1 Basic Ideas

Discussions in this section rely mainly on the excellent reviews of theory recently published by Wegener [3-1] and Feder et al [1-19]. As mentioned in Sec. 3.1, the existence of supersaturated vapor states, and also the thermodynamic laws governing the behavior of drops and describing the critical (minimum stable) droplet size, were discovered in the last decades of the last century. Soon afterwards, the collapse of supersaturation and formation of fog in clean vapors were observed experimentally. This was contradictory to thermodynamic theory, which suggested that small droplets were unstable in saturated steam. The mechanisms by which the spontaneous formation of small droplets could eventually be explained remained hidden until the mid-1920s.

It was recognized very early that the kinetic theory of gases could possibly offer a clue; first attempts, in which triple and multiple collisions of vapor molecules were made responsible for the process, failed. The correct solution was visualized by Volmer and Weber in 1926: relying on thermodynamic statistics, they postulated that in vapors the probable number of molecular clusters (consisting of two or more vapor molecules which coincidentally stuck together) can be calculated by using Boltzmann's distribution law. According to this law, the relative probability of a certain thermodynamic state in a constant-volume constant-energy system, e.g., a state in which a certain number of molecules are forming a cluster instead of individually flying around, is proportional to the factor $e^{-\Delta S/k}$, where $-\Delta S$ is the entropy decrease associated with the formation of such a cluster from vapor. Boltzmann's constant has the value $k = 1.38 \cdot 10^{-23}$ J/K. Using the cluster distribution thus obtained, Volmer and Weber were able to calculate the equilibrium (i.e., most probable) concentration of critical-sized droplets pertaining to a given supersaturation. From this, later workers (Farkas in 1927 and others) arrived at expressions giving the production rate of supercritical-sized droplets, i.e., the so-called nucleation rate.

The nucleation rate I (number of supercritical-sized droplets produced per unit mass* of vapor per unit time) is given in the form

$$I = Ke^{-\Delta G_{\text{crit}}/kT_g} \qquad (3.3.1)$$

where ΔG_{crit} is the change of Gibbs free enthalpy† upon formation of a critical-size cluster. K is a proportionality factor to be determined from theoretical considerations. Further below an expression for K will be derived on the basis of the theory developed by Volmer-Weber, Farkas, Becker-Döring, Zeldovich, Frenkel, and others [3-1]. Later, recent amendments to this classical theory will be briefly mentioned.

All considerations presented here are based on the model that clusters down to the critical size and even below (containing, e.g., only 50 molecules) can be treated as liquid drops. The surface tension value in the thermodynamic equations need not be set equal to that of a plane surface. However little is known about the proper variation of σ with r. Derivation of the nucleation rate on the basis of purely molecular concepts has not yet been successful.

3.3.1.2 Critical Droplet Size

The expression for r_{crit}, cf. Eq. (1.4.2), can be derived from a mental experiment devised by Thomson (Lord Kelvin). The critical radius is given for ideal-gas vapors as

$$r_{\text{crit}} = \frac{2\sigma/\rho_f RT_g}{\ln S} = \frac{2\sigma \rho_g/\rho_f p}{\ln S} \qquad (3.3.2)$$

The number of molecules in a critical droplet is given, through Eq. (1.3.21), as

---

*Most textbooks define a nucleation rate J per unit *volume*; in flows, however, the use of I is more convenient.

†In Eq. (3.3.1) $-\Delta G/T_g$ has taken the place of the entropy change $-\Delta S$. This modification is not arbitrary. It results from the assumption of boundary conditions more appropriate for the treatment of nucleation in flowing vapors. For given pressure and temperature $\Delta G/T_g$ takes over the role played by ΔS in a constant-volume constant-energy system (see, e.g., [3-24]).

$$i_{\text{crit}} = \frac{4\pi\rho_f}{3m_m} r_{\text{crit}}^3 \tag{3.3.3}$$

3.3.1.3 Equilibrium Population of Critical Droplets

The equilibrium number $N_{i,\text{eq}}$ of clusters of a given size i per unit mass of vapor is determined, according to Sec. 3.3.1.1, by the probability $\exp(-\Delta G_i/kT_g)$. This means that

$$N_{i,\text{eq}} = N_m \exp\left(-\frac{\Delta G_i}{kT_g}\right) \tag{3.3.4}$$

where N_m is the total number of vapor molecules per unit mass. Since the Gibbs free enthalpy change during reversible formation of a droplet of radius r containing a number i of molecules is found to be

$$\Delta G_i = -im_m RT_g \ln S + 4\pi\sigma r^2 \tag{3.3.5}$$

and from this ΔG_{crit} follows by introduction of Eqs. (3.3.3) and (3.3.2) as

$$\frac{\Delta G_{\text{crit}}}{kT_g} = \frac{i_{\text{crit}}}{2} \ln S = \frac{4\pi\sigma}{3kT_g} r_{\text{crit}}^2 \tag{3.3.6}$$

we have

$$N_{\text{crit,eq}} = N_m \exp\left(-\frac{i_{\text{crit}}}{2}\ln S\right) = N_m \exp\left(-\frac{4\pi\sigma}{3kT_g} r_{\text{crit}}^2\right) \tag{3.3.7}$$

It is seen from Fig. 1.4.2, where the relation (3.3.4) is plotted for a typical saturated and supersaturated vapor state, that droplets of critical size are by many orders of magnitude more rare than monomer molecules. A first attempt to derive a nucleation rate from the above expression for $N_{\text{crit,eq}}$ will be discussed in the next section.

3.3.1.4 A Rough Expression for the Nucleation Rate

It is known from kinetic theory that the number β of gas molecules impinging on unit boundary surface per unit time can be calculated from pressure, temperature, and molecular mass of the gas as

$$\beta = \frac{p}{\sqrt{2\pi m_m kT_g}} = \frac{p}{m_m\sqrt{2\pi RT_g}} \tag{3.3.8}$$

In order to obtain a first estimate for the order of magnitude of the factor K in the nucleation rate expression (3.3.1) we assume that

The number of critical-size droplets that are present in unit mass of vapor always equals the equilibrium value $N_{\text{crit,eq}}$

A molecule impinging on the surface of a critical droplet adheres to it, i.e., transforms it into a supercritical stable droplet

The fact that supercritical droplets may become subcritical due to loss (evaporation) of molecules can be disregarded

None of these assumptions is strictly correct, as will be shown. Even so, this simple concept gives a useful rough estimate of I.

$$I_{\text{estim}} \approx 4\pi r_{\text{crit}}^2 \frac{p}{m_m \sqrt{2\pi RT_g}} N_{\text{crit,eq}}$$

$$= \frac{4\pi r_{\text{crit}}^2}{\sqrt{2\pi} \, m_m^2} \rho_g \sqrt{RT_g} \exp\left(-\frac{4\pi\sigma}{3kT_g} r_{\text{crit}}^2\right) \quad (3.3.9)$$

where Eq. (3.3.7), the identity $N_m = 1/m_m$ (Sec. 1.3.2.1) and $p = \rho_g RT_g$ have been used.

Let us estimate the order of magnitude of the pre-exponential factor that corresponds to K in Eq. (3.3.1). With $r_{\text{crit}} = 10^{-9}$ m, $m_m = 3 \cdot 10^{-26}$ kg, $\rho_g = 1$ kg/m³, $\sqrt{RT_g} \approx 500$ m/s, we have

$$K \approx 5 \cdot 10^{-18} \cdot 1/9 \cdot 10^{52} \cdot 1 \cdot 5 \cdot 10^2 \approx 10^{36} \text{ kg}^{-1}\text{s}^{-1}$$

Obviously in order to make I assume physically significant values (e.g., at least 1 nucleus per g/μs, or 10^9 kg⁻¹s⁻¹) the negative exponent may not be too large. If $r_{\text{crit}} = 10^{-9}$ m, the exponent has the following value (assuming $T_g = 300°$K, $\sigma = 0.06$ N/m):

$$\left(-\frac{4\pi\sigma}{3kT_g} r_{\text{crit}}^2\right) \approx -\frac{4\pi \cdot 0.06 \cdot 10^{-18}}{3 \cdot 1.38 \cdot 10^{23} \cdot 300} \approx -60$$

which gives

$$I_{\text{estim}} \approx (10^{36} \text{ kg}^{-1}\text{s}^{-1})e^{-60} \approx 10^{36-26} \approx 10^{10} \text{ kg}^{-1}\text{s}^{-1}$$

For smaller values of r_{crit} (higher supersaturation), much higher values of I_{estim} would result. However let us first derive a more accurate expression for the pre-exponential factor.

3.3.1.5 Classical Nucleation Theory

With increasing cluster size i, the equilibrium number of clusters in supersaturated states, as given by Eqs. (3.3.4) and (3.3.5), decreases fairly rapidly from the high number of monomeric molecules toward its minimum at the critical droplet size (Fig. 1.4.2). Then however it increases again to go to infinity for $i \to \infty$. Since it is physically absurd to form infinitely many infinitely large drops from a finite number of molecules, such equilibrium distributions can never be reached; as soon as they are approached, the supersaturation will disappear. This means that the equilibrium distribution $N_{i,\text{eq}}$ pertaining to a supersaturated state may in reality exist only for very small cluster sizes ($i \ll i_{\text{crit}}$). What is then the *actual* number N_{crit} of critical nuclei, from which a nucleation rate can be calculated?

Farkas (1927) introduced a theoretical artifice by which a nucleation rate close to reality can be formulated. This artifice consists of the assumption of a steady-state hypothetical model system, in which a particular *non*equilibrium distribution of clusters is assumed to exist, such that there is a net flux of clusters growing to and beyond critical size. At some size, clusters are thought to be continually removed from the system and replaced by an equal amount of single molecules. Steady state requires the net flux to be independent of cluster size, and from this condition the distribution function can be elegantly derived by considering the simultaneous decay and build-up of clusters of various sizes by absorption and emission of single molecules. The corresponding net flux of growth yields the nucleation rate.

The result of this calculation is

$$I = \beta 4\pi r_{\text{crit}}^2 N_{\text{crit,eq}} \sqrt{\frac{1}{2\pi k T_g} \left(\frac{\partial^2 \Delta G_i}{\partial i^2}\right)_{\text{crit}}} \qquad (3.3.10)$$

Using Eq. (3.3.8) and differentiating ΔG_i from Eq. (3.3.5), one can transform this to

$$I = \sqrt{\frac{2\sigma}{\pi m_m^3}} \frac{\rho_g}{\rho_f} \exp\left(-\frac{4\pi r_{\text{crit}}^2 \sigma}{3kT_g}\right) \qquad (3.3.11)$$

which is the final expression for the classical nucleation rate.

A numerical estimate, using the same conditions as in Sec. 3.3.1.4 and $\rho_f = 10^3$ kg/m^3, gives the following value

$$I = \sqrt{\frac{2 \cdot 0.06}{\pi \cdot 27}} 10^{-78} \frac{1}{10^3} e^{-60} \approx 0.04 \cdot 10^{36} e^{-60} \approx 0.04 \cdot 10^{10} \text{ kg}^{-1}\text{s}^{-1}$$

Comparison with the value of I_{estim} found in Sec. 3.3.1.4 reveals that the first rough expression differs by only about one order of magnitude from the final one. The behavior of the function

$$I = I(\ln S) \quad \text{or} \quad I = I(\Delta T)$$

is crucial for the process of spontaneous condensation. The curves of I shown in Fig. 3.3.1 for various pressure levels reveal that I rises from negligibly small values at low and moderate subcooling to significant values ($\approx 10^{10}$–10^{20} kg^{-1}s^{-1}) within a relatively slight change of ΔT. Indeed this fast rise of I with ΔT is the explanation for the observed sudden occurrence of spontaneous nucleation. The ΔT values at which a certain value of I is reached depend on pressure. This is due to the variation of crucial material properties (in particular of σ) with temperature.

It is to be noted that the expression for I is very sensitive to the value of surface tension used. (Since $r_{crit} \sim \sigma$, surface tension enters the exponent to the third power!) Deficiencies of the nucleation theory can therefore be "easily" compensated by slightly different assumptions for σ. The curves are based on bulk-liquid surface tension data taken from [1-12].

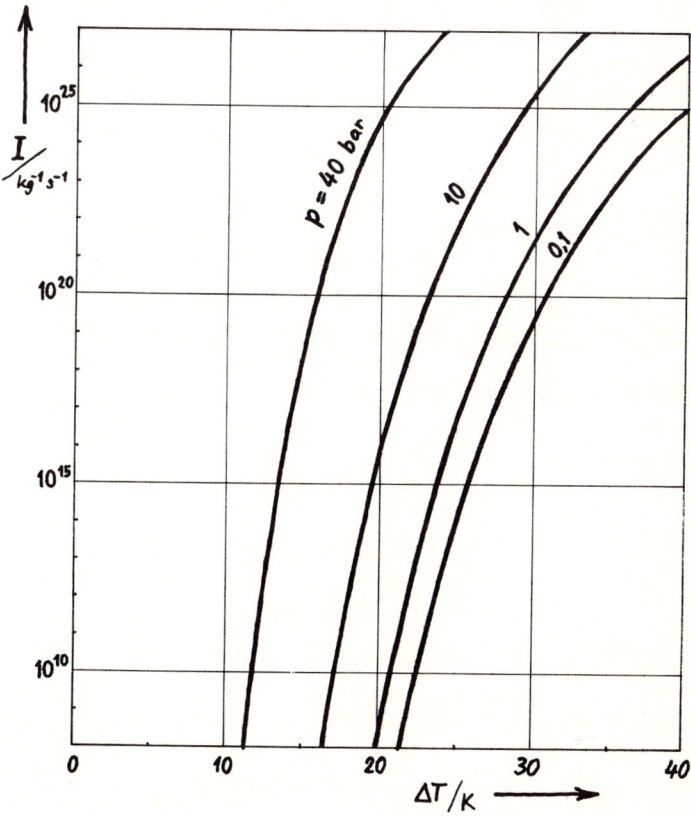

FIG. 3.3.1 Nucleation rate I as a function of subcooling and pressure, calculated from classical theory, Eq. (3.3.11); surface tension as for bulk liquid (see Fig. 1.3.4), density ρ_g as for real-gas steam.

3.3.1.6 Time Delay of Nucleation

In a rapid expansion, supersaturation is being increased at a very fast rate, and time may be insufficient for steady-state nucleation to become established. Several authors [1-19] have estimated the time delay of this build-up process. Wakeshima (1954) finds for the step-response of nucleation rate

$$I(t) = I\left[1 - \exp\frac{-t}{\Delta t_{\text{nucl}}}\right]$$

in which I is given by Eq. (3.3.11), and the delay time of nucleation Δt_{nucl} can be expressed as

$$\Delta t_{\text{nucl}} = \frac{kT_g}{2\beta 4\pi r_{\text{crit}}^2}\left(\frac{\partial^2 \Delta G_i}{\partial i^2}\right)_{\text{crit}}^{-1} \quad (3.3.12)$$

or with (3.3.2), (3.3.5), and (3.3.8)

$$\Delta t_{\text{nucl}} = \sqrt{\frac{\pi}{2}}\frac{\rho_f}{\rho_g}\frac{r_{\text{crit}}}{\ln S \sqrt{RT_g}} \quad (3.3.13)$$

For typical conditions ($\rho_f/\rho_g \approx 10^3$, $r_{\text{crit}} \approx 10^{-9}$ m, $\ln S \approx 2$, $\sqrt{RT_g} \approx 500$ m/s) the nucleation delay time has the value

$$\Delta t_{\text{nucl}} \approx 10^3 \cdot \frac{10^{-9}}{2 \cdot 500} = 10^{-9}\,\text{s}$$

which is negligibly small as compared to the usual duration of the expansion process. Therefore the nucleation rate given by Eq. (3.3.11) may be used in nozzle-flow calculations.

3.3.1.7 Surface Tension of Droplets

As noted in Sec. 1.4.1.1, the value of σ for highly curved surfaces may be considerably different from the σ value for a plane liquid surface. No conclusive theories exist however, so we will omit corrections to nucleation theory arising from the $\sigma(r)$ dependence. The same applies for other property data, like ρ_f and Δh_{fg}, where the effects of a dependence are less significant anyway.

3.3.1.8 Modified Nucleation Theories

Feder et al. [1-19], reviewing classical theory, introduced a correction due to differences between the temperature of clusters and of vapor. Their correction factor

$$\frac{I_{\text{Feder}}}{I} = \left[1 + \frac{2(\kappa - 1)}{\kappa + 1}\left(\frac{L}{RT_g} - \frac{1}{2}\right)^2\right]^{-1} \tag{3.3.14}$$

is on the order of 1/50, i.e., it *reduces* the nucleation rate as compared to Eq. (3.3.11).

Reexamination of classical theory in the last ten years [1-19, 3-1] has led, as mentioned in Sec. 3.1.1.2, to an ardent discussion about the correct formulation of cluster thermodynamics. The revised theory results in nucleation rates that are about 10^{17} times higher than the classical rate! Such theories predict condensation to take place at supersaturations that are in the case of steam much smaller than observed. Therefore it makes sense to stay with classical theory in the calculation of steam nozzle flow.

A different argument than the one outlined in Sec. 3.3.1.5, based on a nonsteady physical model, was used by Deich et al. [3-25], to derive a nucleation rate expression. This expression is cited by Moore et al. [3-26], who find that the nucleation rate values given by this expression fit steam experiments quite satisfactorily. For the analytic expression the reader is referred to [3-26].

3.3.2 Droplet Growth Theory

3.3.2.1 Growth Rate Equation

Nuclei that emerge from the nucleation process as supercritical, stable droplets will further grow by condensation of vapor on their surface and by coagulation with other droplets.

Coagulation (Sec. 1.4.3.2) is generally thought to play an unimportant role in nozzle and low-pressure turbine flow [3-27, 1-30]. In the high-pressure blading of nuclear turbines coagulation between droplet streams moving at different speeds may possibly be important, but this problem seems not to have been analyzed in detail. Coagulation will therefore not be further considered here.

The rate of condensation on a drop is governed by the rate at which the latent heat can be carried away from the surface into the cooler vapor. (If the vapor is hotter than the drop, evaporation occurs.) In treating heat transfer, relative velocity between droplet and vapor will be neglected (Sec. 1.4.3.1), and the droplet will be assumed to be spherical and to be surrounded by an infinite vapor space. In the heat balance of small ($r \leqslant 1$ μm) droplets the heat capacity $c_f \rho_f (4\pi/3) r^3$ can be shown to be negligible compared with the amounts of latent heat released by condensation. The heat transfer is driven by the temperature difference $T_r - T_g$. According to Eq. (1.4.3), the validity of which will be scrutinized in Sec. 3.3.2.2, the surface temperature T_r depends also on the radius of the droplet and approaches $T_s(p)$ for large sizes ($r \gg r_{\text{crit}}$). The driving temperature difference for droplet growth equals, according to Eq. (1.4.3),

$$T_r - T_g = \Delta T \left(1 - \frac{r_{\text{crit}}}{r}\right) \qquad (3.3.15)$$

According to the concepts of heat conduction, around the drop there is a vapor region in which temperature T varies continuously from the value T_g far from the drop to the value T_r at the drop surface (Fig. 3.3.2). This continuum concept becomes invalid however if the droplet size is comparable to or smaller than the mean free path of vapor molecules [Knudsen number $\text{Kn} = \bar{l}/2r > 1$, Eq. (1.4.7)]. Then heat transfer can be described only by molecular laws.

Stodola [1-1] was the first to investigate these questions in depth, and his analysis of droplet behavior in steam has remained one of the most instructive treatises written on the subject. Oswatitsch [3-8], pioneer of condensing-flow theory, has utilized molecular equations for small droplets and an expression based on heat conduction for large ones. Molecular heat transfer from spheres is extensively treated in [3-27a]. An equation for the heat transfer coefficient α_r embracing continuum and molecular regimes [1-18, 1-20, 3-28] is given by Eq. (1.4.10).

The rate of latent heat liberation $\dot{Q}_l$ is connected to droplet growth rate $\dot{r} = dr/dt$ by

$$\dot{Q}_l = \Delta h_{fg} \frac{dm}{dt} = \Delta h_{fg} 4\pi r^2 \rho_f \frac{dr}{dt} \qquad (3.3.16)$$

Since the heat capacity of the droplet is negligible, $\dot{Q}_l$ equals at any time the heat $\dot{Q}$ carried away by heat transfer into the (colder) vapor. Equating $\dot{Q}_l$ to the heat transfer rate $\dot{Q}$ given by Eq. (1.4.11) and considering Eq. (1.4.10), we get

$$\frac{dr}{dt} = \frac{\lambda_g}{\rho_f \Delta h_{fg}} \frac{1}{1 + (2\sqrt{8\pi}/1.5 a_{\text{th}} \text{Pr}_g)[\kappa/(\kappa+1)] \text{Kn}} \frac{T_r - T_g}{r} \qquad (3.3.17)$$

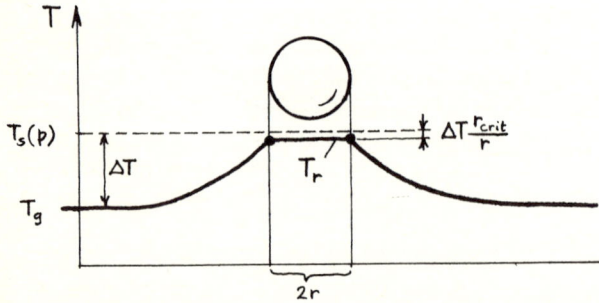

FIG. 3.3.2 Temperature field in subcooled vapor in the vicinity of a droplet.

or using (3.3.15)

$$\dot{r} = \frac{\lambda_g}{\rho_f \Delta h_{fg}} \frac{1 - r_{crit}/r}{1 + (2\sqrt{8\pi}/1.5 a_{th} Pr_g)[\kappa/(\kappa + 1)] \text{Kn}} \frac{\Delta T}{r} \quad (3.3.18)$$

For steam, if the thermal accommodation of molecules hitting the drop is assumed to be perfect ($a_{th} = 1$), we can write, using (1.4.10a)

$$\dot{r} = \frac{\lambda_g}{\rho_f \Delta h_{fg}} \frac{1 - r_{crit}/r}{r + 1.59 \bar{l}} \Delta T \quad (3.3.19)$$

In Fig. 3.3.3 a few growth curves $r(t)$ computed from this equation are shown. For sake of simplicity, steam conditions (pressure and subcooling) were assumed to remain constant. At $t = 0$, $r = r_0 = 1.1 r_{crit}$ was arbitrarily assumed.

It is seen that

Growth is slower at low pressure than at high
Growth is slow while drop size is still near critical
A tenfold increase of diameter (1000-fold increase of mass!) occurs, even at low pressure, in a matter of microseconds.

At lower values of subcooling, growth is slower, as evident from Eq. (3.3.19). For large drop sizes ($r > 1$ μm) relative velocity between drop and vapor may be significant and require corrections to Eq. (3.3.19), e.g., [1-18].

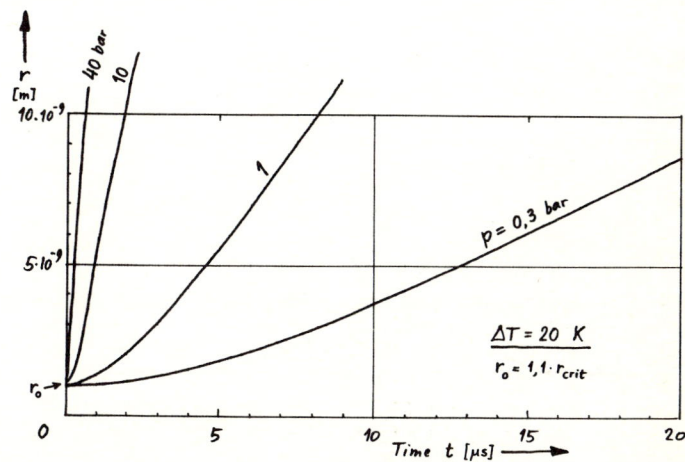

FIG. 3.3.3 Growth of a near-critical droplet of various pressure levels, as calculated from Eq. (3.3.19).

3.3.2.2 Kinetic Expression for Droplet Growth Rate

In numerous publications on condensing flow, low-pressure situations were of primary interest and therefore purely molecular growth laws have been utilized, e.g., [3-12, 3-29]. In this approach both molecular heat transfer and mass transfer are simultaneously considered, Eq. (1.4.3) is discarded, and droplet temperature T_r is regarded as unknown. In the following discussion this molecular approach, which is relevant for cases with $\bar{l} \gg r$, i.e., $\mathrm{Kn} \gg 1$, will be outlined and its results compared to the more general growth law expressed by Eq. (3.3.18).

The molecular heat transfer coefficient of a sphere is the limit case of Eq. (1.4.10) for $\mathrm{Kn} \to \infty$, giving

$$\alpha_r = \frac{\lambda_g}{r} \frac{1.5 \, \mathrm{Pr}_g}{2\sqrt{8\pi}} \frac{\kappa + 1}{\kappa} \frac{a_{\mathrm{th}}}{\mathrm{Kn}}$$

Taking material property values at saturation conditions pertaining to vapor pressure p, we may write $\mathrm{Pr}_g = \mathrm{Pr}'' = \mu'' c_p / \lambda'' = (\mu'' R / \lambda'') \kappa / (\kappa - 1)$ and set, according to Eqs. (1.4.7) and (1.3.14)

$$\mathrm{Kn} = \frac{1.5 \mu'' \sqrt{RT_g}}{2rp}$$

The heat transfer coefficient thus obtained

$$\alpha_r = a_{\mathrm{th}} \frac{\kappa + 1}{\kappa - 1} \frac{Rp}{\sqrt{8\pi RT_g}}$$

can be used in Eq. (1.4.11) to express the heat transfer rate from the droplet under molecular conditions as

$$\dot{Q} = 4\pi r^2 a_{\mathrm{th}} \frac{\kappa + 1}{\kappa - 1} \frac{Rp}{\sqrt{8\pi RT_g}} (T_r - T_g) \tag{3.3.20}$$

The heat capacity of the droplet being neglected, the heat balance requires $\dot{Q} = \dot{Q}_l$, where the latter is given by Eq. (3.3.16). From this we can express droplet growth rate as

$$\frac{dr}{dt} = \frac{a_{\mathrm{th}}}{\Delta h_{fg} \rho_f} \frac{\kappa + 1}{\kappa - 1} \frac{Rp}{\sqrt{8\pi RT_g}} (T_r - T_g) \tag{3.3.21}$$

where T_r is unknown. For the thermal accommodation coefficient the empirical value $a_{\mathrm{th}} \approx 1$ may be used.

The mass balance of the droplet can be formulated by starting from the number of vapor molecules impinging upon the drop. The total mass of the molecules impinging in unit time, according to Eq. (3.3.8), is

$$\dot{m}_{\text{imp}} = m_m \beta 4\pi r^2 = \frac{4\pi r^2 p}{\sqrt{2\pi R T_g}}$$

Of these impinging molecules a certain fraction ξ is assumed to adhere to the drop, at least for a short time, while the fraction $1 - \xi$ immediately rebounds. The factor ξ is called the condensation coefficient. Its value must be deduced from measurements. These indicate that for water vapor molecules at water surfaces, ξ is close to unity [3-29, 3-30]. The mass of vapor molecules adhering to the drop in unit time is thus written as

$$\dot{m}_{\text{adh}} = \xi \frac{4\pi r^2 p}{\sqrt{2\pi R T_g}} \qquad (3.3.22)$$

How many molecules are evaporating from the drop in unit time? In order to answer this question the hypothesis must be made that the number of molecules leaving the drop depends only on the size and temperature of the drop itself and not on the conditions in the vapor. Then the number of molecules leaving (evaporating) must equal the number of vapor molecules impinging and adhering under equilibrium conditions, i.e., when $T_g = T_r$ and $p = p_s(T_r)$ exp $(2\sigma/r\rho_f R T_r)$ hold; cf. Eq. (1.4.1). Denoting the value of ξ pertaining to equilibrium vapor conditions* by ξ_{eq} we have, by analogy with Eq. (3.3.22)

$$\dot{m}_{\text{evap}} = \xi_{\text{eq}} \frac{4\pi r^2 p_s(T_r)}{\sqrt{2\pi R T_r}} \exp\left(\frac{2\sigma}{r\rho_f R T_r}\right) \qquad (3.3.23)$$

The mass balance of the drop is then expressed as

$$\dot{m}_{\text{adh}} - \dot{m}_{\text{evap}} = \frac{dm}{dt} = 4\pi r^2 \rho_f \frac{dr}{dt}$$

from which, by inserting Eqs. (3.3.22) and (3.3.23) and by setting $\xi = \xi_{\text{eq}}$, we have

$$\frac{dr}{dt} = \xi \frac{p}{\rho_f \sqrt{2\pi R T_g}} \left[1 - \frac{p_s(T_r)}{p} \sqrt{\frac{T_g}{T_r}} \exp\left(\frac{2\sigma}{r\rho_f R T_r}\right)\right] \qquad (3.3.24)$$

*It is likely that ξ is fairly independent of temperature, i.e., that $\xi \approx \xi_{\text{eq}}$.

Thus a second expression for the droplet growth rate given by Eq. (3.3.21) is obtained. Equating the two expressions, one gets an equation from which the unknown temperature T_r of the drop can be determined

$$\frac{a_{th}R}{2\Delta h_{fg}} \frac{\kappa + 1}{\kappa - 1} (T_r - T_g) = \xi \left[1 - \frac{p_s(T_r)}{p} \sqrt{\frac{T_g}{T_r}} \exp\left(\frac{2\sigma}{r\rho_f RT_r}\right) \right] \quad (3.3.25)$$

Droplet growth can now be calculated by assigning numerical values to the accommodation and condensation coefficients (e.g., $a_{th} = 1$ and $\xi = 1$), and by first using Eq. (3.3.25) to determine T_r and then calculating dr/dt either from Eq. (3.3.21) or from Eq. (3.3.24), both procedures being exactly equivalent. By integration of dr/dt the growth history $r(t)$ can be obtained.

Such a calculation of $r(t)$ has been made for the constant vapor conditions of $p = 1$ bar, $\Delta T = 20°$K, assuming $a_{th} = 1$ and trying the values $\xi = 0.3$ and 1. The results are plotted in Fig. 3.3.4, together with the solution obtained from Eq. (3.3.19). It is seen that the two methods yield very nearly identical curves $r(t)$ and $T_r(t)$, especially if $\xi = 1$ is used.*

The growth law expressed by Eq. (3.3.18 or 3.3.19) offers the advantage that the (iterative) calculation of T_r is made unnecessary. Therefore preference will be given to the method outlined in Sec. 3.3.2.1.

---

*It can be shown that Eqs. (1.4.3) and (3.3.25) in fact lead to nearly identical values of T_r if the left side of Eq. (3.3.25) is negligibly small as compared to the value of ξ, a condition that is fulfilled by steam in all cases of practical interest.

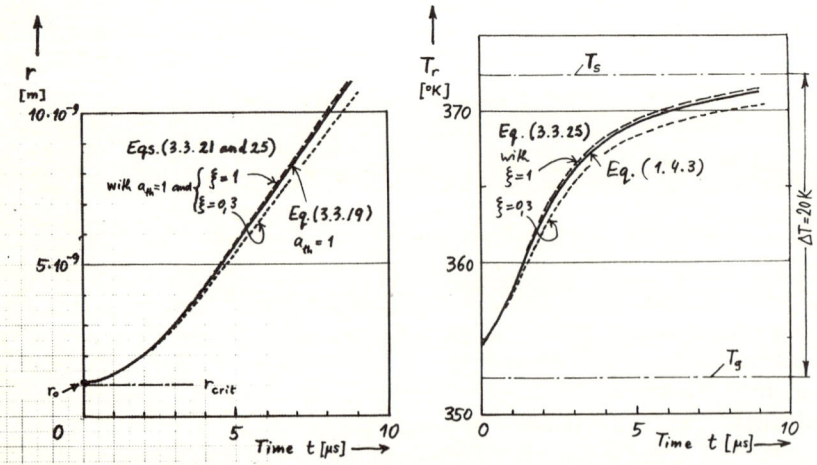

FIG. 3.3.4 Droplet radius and droplet temperature during growth in a constant-subcooling steam environment ($p = 1$ bar, $\Delta T = 20°$K, $r_0 = 1.1\, r_{crit}$).

3.3.3 One-Dimensional Flow with Condensation

The aim of this theory is to describe the flow of condensing steam through a slender Laval nozzle. Inlet stagnation conditions and nozzle area distribution are supposed to be known.

Oswatitsch [3-8] was the first to describe condensing flows by a logical and complete theory, using the basic equations of compressible flow and taking condensation into account with the help of nucleation theory and of growth laws. All subsequent calculations were in essence based on his approach.

3.3.3.1 System of Equations

If condensation occurs, the equations describing the flow of a noncondensing gas, as presented in Table 3.1.1, must be modified in order to account for the fact that a varying part of the fluid (mass fraction Y) may be in liquid form. The equations for two-phase flow are listed in Table 3.3.1 and further below. These equations are based on the following assumptions:

One-dimensional
Steady
Adiabatic
Frictionless
Free of body force fields
Vapor phase: thermally and calorically perfect
Liquid phase: consists of droplets, is incompressible, occupies negligible volume, has same speed as surrounding vapor ("zero slip"), and its enthalpy corresponds to saturated liquid ($h_f = h''$)
No droplets are present other than the ones formed by spontaneous nucleation
Subcritical droplets are not to be attributed to the liquid phase

It is seen from Table 3.3.1 that the conservation equations are unchanged as compared to monophase flow (cf. Table 3.1.1). In the equations of state, vapor and liquid phases are considered side by side (subscripts g and f respectively). It is seen that in Eq. (3.3.T8) the liquid phase as a whole is assumed to have the temperature T_s, which corresponds to setting $h_f = h''$ (see 7th assumption above). The change of T_s is rigidly coupled to the change of p by Eq. (3.3.T10).

From these basic equations a differential equation (3.3.T13) can be deduced that gives flow velocity gradient dC/dx in terms of nozzle area distribution $A(x)$. This equation differs from the corresponding equation for monophase flow, Eq. (3.1.T14) in Table 3.1.1, in two respects. First, on the right side there appears a new term that contains dY/dx, the wetness gradient

TABLE 3.3.1 Gas dynamic equations for two-phase flow

Conservation of mass:

$$\rho C A = \dot{m}_0 \quad (3.3.\text{T1}) \qquad \frac{d\rho}{\rho} + \frac{dC}{C} + \frac{dA}{A} = 0 \quad (3.3.\text{T2})$$

Conservation of momentum:

$$\rho C\, dC = -dp \quad (3.3.\text{T3})$$

Conservation of energy:

$$h + \tfrac{1}{2} C^2 = h_{ot} \quad (3.3.\text{T4}) \qquad dh + C\, dC = 0 \quad (3.3.\text{T5})$$

Equations of state:

thermal:
$$\rho = \frac{\rho_g}{1-Y} = \frac{p}{(1-Y)RT_g} \quad (3.3.\text{T6})$$

$$\frac{d\rho}{\rho} = \frac{dp}{p} + \frac{dY}{1-Y} - \frac{dT_g}{T_g} \quad (3.3.\text{T7})$$

caloric:
$$h = (1-Y)[h_{go} + c_p(T_g - T_0)] + Y[h_{fo} + c_f(T_s - T_0)] \quad (3.3.\text{T8})$$

$$dh = -dY[\Delta h_{fg} - c_p(T_s - T_g)] + (1-Y)c_p\, dT_g + Yc_f\, dT_s \quad (3.3.\text{T9})$$

where $\Delta h_{fg} = h_{go} - h_{fo} + (c_f - c_p)(T_0 - T_s)$

Clausius-Clapeyron (1.3.9):
$$dT_s = \frac{RT_s^2}{\Delta h_{fg}} \frac{dp}{p} \quad (3.3.\text{T10})$$

Rearrangement to obtain dC/dx in terms of dA/dx:

Eliminate dp from (T3) using (T6, T7)
Express therefrom dT_g/T_g as $f_1(dC, dY, dp, \ldots)$ (3.3.T11)
Eliminate dp/p from (T10) using (T7)
Eliminate dh from (T5) using (T9, T10)
Express therefrom dT_g/T_g as $f_2(dC, dY, dp, \ldots)$ (3.3.T12)
Equate (T11) and (T12)
Eliminate $d\rho/\rho$ using (T2)
Rearrange to obtain the following expression:

$$\frac{dC}{dx} = \frac{C}{1-(C/C^*)^2}\left[-\frac{1}{A(x)}\frac{dA}{dx} + \frac{\Delta h_{fg} - c_p T_s}{c_p T_g}\frac{1}{1-Y}\frac{dY}{dx}\right] \quad (3.3.\text{T13})$$

where

$$C^{*2} = \kappa R T_g (1-Y)\left[1 + \frac{Y}{1-Y}\frac{\kappa-1}{\kappa}\frac{c_f T_s^2}{\Delta h_{fg} T_g}\right]^{-1} \quad (3.3.\text{T14})$$

$$\kappa = c_p/(c_p - R) \quad (3.3.\text{T15})$$

---

in the flow. The magnitude of this term depends on the detailed structure of the fluid and will be treated below. Second, the so-called critical speed C^* given by Eq. (3.3.T14), in which the denominator of the first factor becomes zero, is no longer identical with the acoustic speed a. This latter can be shown to vary in steam/droplet mixtures with the frequency of the sound waves; its highest value is $a_{\max} = \sqrt{(1-Y)\kappa R T_g}$ and pertains to high frequencies (cf. Chap. 2). At the location where $C = C^*$ is reached, Eq. (3.3.T13) has a singularity, and the bracketed expression must obviously assume the value zero. In condensing flow ($dY/dx > 0$), critical conditions are therefore attained not at the throat but further downstream (at positive dA/dx).

As a general comment it may be added that the form of the expression defining C^*, Eq. (3.3.14), is dependent on the particular assumptions made in deducing the differential equation, (3.3.T13). In Chap. 2, an equivalent differential equation is established by treating T_f as a free parameter (like Y), the change of which is described by a separate differential equation, instead of being coupled to the pressure by the assumption $T_f = T_s(p)$. As a result, a different expression is obtained for C^*. Simultaneously there appears a third term, containing dT_f/dx on the right side of the equation. Despite this formal difference, both yield nearly identical solutions $C(x)$, provided that the assumption $T_f = T_s$ is sufficiently well justified. We may therefore conclude that the critical speed C^* is, in contrast to the acoustic speed a, not a physical property of the fluid but merely a matter of mathematical formulation.

The computation of condensing flow requires an additional equation, which expresses the wetness gradient dY/dx in terms of known parameters. Since all liquid is present in the form of droplets, and the droplets in general have various sizes depending on their "age" (i.e., on the location $x = \xi$ at which they were "born" as stable nuclei), one can write for any location x in the flow

$$Y = Y(x) = \frac{4\pi}{3} \rho_f \int_{x_s}^{x} [r(x, \xi)]^3 \nu(\xi)\, d\xi \qquad (3.3.26)$$

where $r(x, \xi)$ = radius of a droplet born at ξ, at the location x in the nozzle ($x \geq \xi$)

$\nu(\xi)d\xi$ = number of droplets per unit mass of medium born between $x = \xi$ and $x = \xi + d\xi$

x_s = location where droplet formation begins; theoretically this is the point where saturation conditions are crossed

Differentiation yields for the wetness gradient dY/dx

$$\frac{dY}{dx} = \frac{4\pi}{3} \rho_f \left[r(x,x)^3 \nu(x) + \int_{x_s}^{x} 3[r(x,\xi)]^2 \frac{\partial r(x,\xi)}{\partial x} \nu(\xi)\, d\xi \right] \qquad (3.3.27)$$

where $\nu(x)$ and $\nu(\xi)$ can be expressed on the basis of the nucleation rate I as $\nu\, dx = I\, dt = I\, dx/C$ so that

$$\nu(x) = \frac{I(x)}{C(x)} \qquad (3.3.28a)$$

$$\nu(\xi) = \frac{I(\xi)}{C(\xi)} \qquad (3.3.28b)$$

Since the droplets are considered to be born with a radius equaling the local critical size, $r(x, x) = r_{\text{crit}}(x)$. The radius gradient $\partial r/\partial x$ can be expressed by growth rate $dr/dt = \dot r$ as

$$\frac{\partial r(x, \xi)}{\partial x} = \frac{\dot r(x, \xi)}{C(x)} \tag{3.3.29}$$

Here $\dot r$ is known from Eq. (3.3.19), in which $r = r(x, \xi)$ is to be set. By integration along x, one obtains $r(x, \xi)$ as

$$r(x, \xi) = r_{\text{crit}}(\xi) + \int_{\xi}^{x} \frac{\partial r(x', \xi)}{\partial x'} \, dx' \tag{3.3.30}$$

where x' is the variable of integration running from $x' = \xi$ to $x' = x$. Thus Eq. (3.3.27) obtains the final form

$$\frac{dY}{dx} = \frac{4\pi}{3} \rho_f r_{\text{crit}}^3(x) \, \nu(x)$$

$$+ \frac{4\pi \rho_f}{C(x)} \int_{x_s}^{x} \left[r_{\text{crit}}(\xi) + \int_{\xi}^{x} \frac{\dot r(x', \xi)}{C(x')} \, dx' \right]^2 \dot r(x, \xi) \nu(\xi) \, d\xi \tag{3.3.31}$$

The first term represents condensation due to the appearance of new droplets of critical size, the second the condensation on all droplets formed at previous locations.

Equation (3.3.31) and Eqs. (1.4.2), (3.3.11), (3.3.19), and (3.3.28) specifying r_{crit}, I, $\dot r$, and ν, express dY/dx as a known function of local conditions at x, so that the system of equations is complete. These equations are summarized in Table 3.3.2.

3.3.3.2 Numerical Solution

The system of equations given in Table 3.3.2 cannot be solved in closed form. Numerical procedures employing the Runga-Kutta or other similar methods for integration along x are required [3-7, 3-12, 3-26, 3-30].

There are two possibilities in making the expression of dY/dx, Eq. (3.3.31), numerically tractable. In the first method the integral over $d\xi$ is approximated by a sum comprising discrete groups of droplets, each group being constituted of equal-sized droplets formed in a given step of the integration process. (Depending on the step width used in the nucleation zone, a number of 20 to 100 droplet categories are usually taken into account.) Droplet numbers in each category are calculated as $N_k = \nu(\xi_k)\Delta\xi_k$, where $\Delta\xi_k = \xi_k - \xi_{k-1}$ is the step in which the kth droplet group is formed. Parallel to the integration of Eq. (3.3.T13) for obtaining $C(x)$ at the next x step, Eq. (3.3.30) has to be

TABLE 3.3.2 System of equations describing condensing flow

Two differential equations for $C(x)$ and $Y(x)$:

$$\frac{dC}{dx} = \frac{C}{1-(C/C^*)^2}\left[-\frac{1}{A(x)}\frac{dA}{dx} + \frac{\Delta h_{fg} - c_p T_s}{c_p T_g}\frac{1}{1-Y}\frac{dY}{dx}\right] \quad (3.3.\text{T}13)$$

$$\frac{dY}{dx} = \frac{4\pi}{3}\rho_f r_{\text{crit}}^3 \, \dot{\nu} + \frac{4\pi\rho_f}{C}\int_{x_S}^{x} r(x,\xi)^2 \dot{r}(x,\xi)\nu(\xi)\,d\xi \quad (3.3.31)$$

One integral equation for $r(x,\xi)$:

$$r(x,\xi) = r_{\text{crit}}(\xi) + \int_{\xi}^{x} \frac{\dot{r}(x',\xi)}{C(x')}\,dx' \quad (3.3.30)$$

Fourteen algebraic equations for the unknown functions $\dot{r}(x,\xi)$, $C^*(x)$, $\Delta T(x)$, $S(x)$, $r_{\text{crit}}(x)$, $I(x)$, $\nu(x)$, $\bar{l}(x)$, $\rho(x)$, $h(x)$, $p(x)$ $\rho_g(x)$, $T_g(x)$ and $T_s(x)$:

$$\dot{r} = \frac{\lambda_g}{\rho_f \Delta h_{fg}} \frac{1 - r_{\text{crit}}/r}{r + 1.59\bar{l}} \Delta T \quad (3.3.19)$$

$$C^* = \sqrt{\kappa R T_g (1-Y)\left[1 + \frac{Y}{1-Y}\frac{\kappa-1}{\kappa}\frac{c_f T_s^2}{\Delta h_{fg} T_g}\right]^{-1}} \quad (3.3.\text{T}14)$$

$$\Delta T = T_s(p) - T_g \quad (1.3.22)$$

$$\ln S = \frac{4740°\text{K}}{T_s} \frac{\Delta T}{T_g} \quad (1.3.24)$$

$$r_{\text{crit}} = 2\sigma/\rho_f R T_g \ln S \quad (3.3.2)$$

$$I = \sqrt{\frac{2\sigma}{\pi m_m^3}}\frac{\rho_g}{\rho_f}\exp\left(-\frac{4\pi r_{\text{crit}}^2 \sigma}{3kT_g}\right) \quad (3.3.11)$$

$$\nu = I/C \quad (3.3.28)$$

$$\bar{l} = \frac{1.5\mu'\sqrt{RT_s}}{p} \quad (1.3.14)$$

$$\rho = \frac{\dot{m}_0}{CA(x)} \quad (3.3.\text{T}1)$$

$$h = h_{ot} - \frac{C^2}{2} \quad (3.3.\text{T}4)$$

$$p = (1-Y)\rho R T_g \quad (3.3.\text{T}6)$$

$$\rho_g = (1-Y)\rho \quad (3.3.\text{T}6)$$

$$h = (1-Y)[h_{go} + c_p(T_g - T_0)] + Y[h_{fo} + c_f(T_s - T_0)] \quad (3.3.\text{T}8)$$

$$T_s = \frac{2061°\text{K}}{5.52 - \log(p/1\,\text{bar})} \quad \text{from } (1.3.11)$$

Initial conditions at nozzle inlet ($x = x_0$):

$$Y_0 = 0 \qquad C_0 = \frac{\dot{m}_0}{A(x_0)\rho_0}$$

where ρ_0 is determined by C_0 and by inlet stagnation conditions evaluated for each droplet category. Thus a complete size distribution of droplets at each location x is being obtained, from which mean droplet radii are calculated by averaging.

A second method, which is more economical in terms of computer time and storage requirements but does not yield droplet size distributions, consists of describing the droplet population by total number, total radius sum, and total surface area sum [3-7], from which the droplet radius can be calculated directly.

A difficulty in the integration process arises due to the singularity of Eq. (3.3.T13) at the location where

$$C = C^* = \sqrt{\frac{(1-Y)c_p R T_g}{c_p - R + \frac{Y}{1-Y}\frac{c_f T_s}{\Delta h_{fg}}\frac{RT_s}{T_g}}} \qquad (3.3.32)$$

is reached. If the mass flow rate $\dot{m}_0$ used in the equation of continuity does not have the proper numerical value, dC/dx will either become infinite (which means that $\dot{m}_0$ is physically impossible, i.e., too large) or $C = C^*$ will not be reached at all and the solution obtained will remain subsonic ("Venturi" solution, with $\dot{m}_0$ smaller than required for supersonic flow). Iteration procedures may be necessary to find the proper value of $\dot{m}_0$ at which the bracketed term and the denominator in Eq. (3.3.T13) become zero simultaneously.

3.3.3.3 Typical Results

The results of calculations referring to one of the nozzles described in [3-3] and to initial stagnation conditions of $p_{ot} = 1$ bar, $t_{ot} = 140°C$, are shown in detail in Fig. 3.3.5. Results are plotted only for the section of the nozzle where fog formation occurs (50 to 150 mm downstream of the throat in this example). The plotted curves comprise

Nondimensional static pressure	p/p_{ot}
Flow velocity	C
Subcooling	ΔT
Critical droplet radius	r_{crit}
Nucleation rate	I
Droplet radii for various categories k	r_k
Wetness fraction	Y

It is seen that the rise of subcooling and the concomitant decrease of r_{crit} cause a fast rise of nucleation rate to high values on the order of 10^{20} nuclei/kg·s and more. The droplets begin to grow, and wetness begins to appear (Y rises from zero). Simultaneously, as a result of liberation of latent heat, ΔT reverses its trend (collapse of supersaturation). Heat liberation and

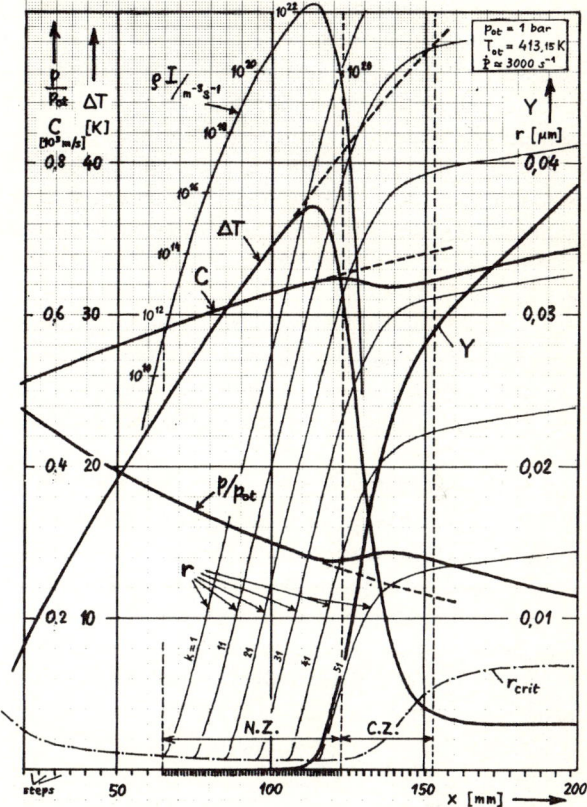

FIG. 3.3.5 Typical numerical solution for condensing nozzle flow.

condensation affect the pressure and velocity distribution and lead to a bump in the p/p_{ot} curve and to a dip in the C curve. The pressure bump is closely similar to those observed in experiments (Sec. 3.2). As soon as ΔT begins to decrease, nucleation is rapidly stopped. The growth of droplets however continues until subcooling disappears. At this point ($x \approx 150$ mm), phase equilibrium is practically reached ($Y \approx 0.03$), and droplet growth is slowed down to the rate required to form additional wetness as the expansion further penetrates the wet-steam region.

The Wilson point, defined in Sec. 3.1.2 as the maximum of ΔT, fairly exactly coincides with the first noticeable departure of pressure from its "dry" trend. Nucleation is seen to be essentially completed before sizeable amounts of moisture (say, $Y \geqslant 0.005$) appear. Therefore the fog formation zone is often subdivided into a nucleation zone (N.Z.) and a condensation zone (C.Z.). Most of the liberation of latent heat takes place in the latter, and the pressure bump is situated here.

On the bottom scale, the step widths used in solving the equations are shown. The width is varied in order to reduce computation time wherever possible without undue loss of numerical accuracy. In the nucleation zone and in the condensation zone, fine steps are used because of the rapid change of the I and r_k values.

Figure 3.3.6 shows mass distribution of droplets (constant $\cdot v \cdot r^3$) at the beginning and the end of the condensation zone. It is seen that the fog produced consists of fairly uniform-sized droplets. The character of the droplet growth curves r_k resembles the measured curves of mean droplet size, e.g., Fig. 3.2.10c and d.

As seen from Fig. 3.3.5, such theoretical calculations shed light on the sequence of mechanisms underlying the observed behavior of condensation and may therefore help the interpretation of experimental data. Quantitative comparisons between theory and experiment will be discussed in the next section.

3.3.3.4 Comparison of Theory and Experiment

Comparison between calculated and measured pressure distributions and droplet sizes have been carried out by numerous authors [1-18, 3-3, 3-7, 3-12, 3-16, 3-26, 3-30]; see Fig. 3.3.7 as an example. Generally it is found that the classical nucleation rate equation predicts condensation at a somewhat early stage if the bulk-liquid value of surface tension is being used. (In terms of ΔT, it predicts a peak subcooling which is 2-4°K too low.) Therefore some authors introduce artificial corrections in order to induce agreement, e.g., by reducing I by a constant factor (about 1000), or by using a higher value (about 10%) for σ than valid for the bulk liquid [3-7]. Other expressions for the nucleation rate [3-26] and various alternatives to the droplet growth laws have also been frequently tried [3-12, 3-30].

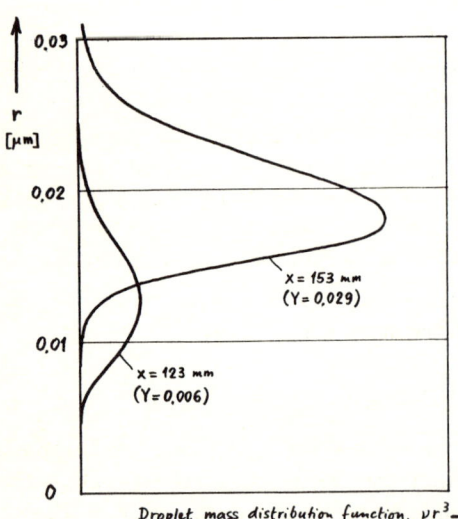

FIG. 3.3.6 Calculated distribution of droplet mass over droplet radius (data as in Fig. 3.3.5).

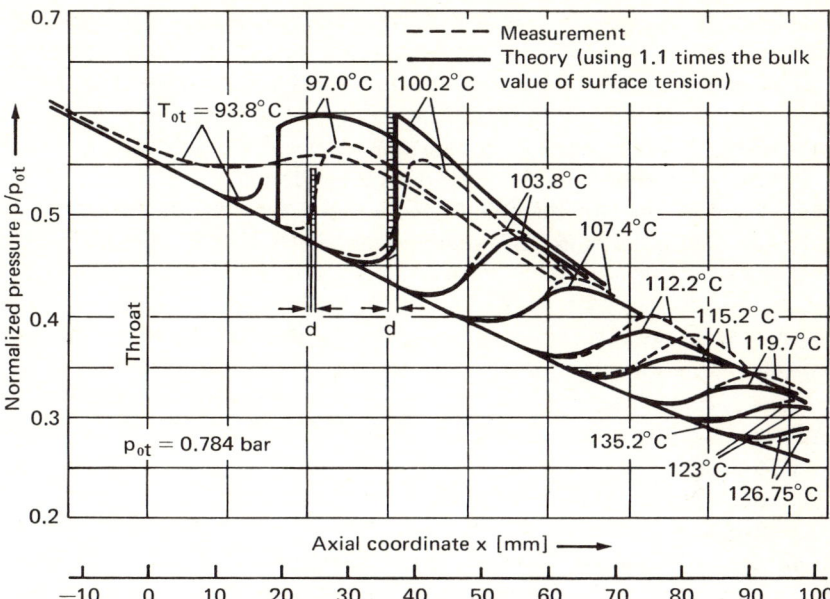

FIG. 3.3.7 Pressure distributions in a nozzle with and without a shock in the condensation zone, with inlet temperature varied [3-7].

It seems to be difficult to decide by such comparisons which correction comes closest to reality. Simultaneous comparison of pressure curves and droplet sizes are more meaningful than comparisons of pressure curve only. In any event, testing of various theories requires measurements of very high quality.

3.3.3.5 Condensation in the Presence of Shock Waves

It has been shown by Pouring [3-16] and later by Barschdorff [3-7] that incorporation of the shock equations into the calculation procedure (which has not been discussed here) can quite accurately predict the presence and location of a shock wave in the condensation zone in cases of low initial superheat. Pressure distributions involving a shock, as measured and calculated by Barschdorff, can be seen in Fig. 3.3.7 for inlet temperatures of 97°C and 100.2°C. At higher initial temperatures shock-free flow was observed and calculated. Quite recently the theory has been extended to unsteady shock configurations observed at very low initial superheat [3-31a].

3.4 INFLUENCE OF FOREIGN NUCLEI

On foreign nuclei (dust particles, ions, and the like), and on droplets present in the inlet flow, condensation can begin at or closely after the crossing of the saturation line. The intensity of this condensation of course depends on various factors (such as number, size, slip speed, material properties, and shape) of these foreign nuclei. Once a foreign particle is covered with condensed water, it behaves very much like a water droplet of equal size, and therefore the influence of foreign nuclei can be theoretically explained by postulating the presence of a certain number of water droplets in the inlet flow.

Such calculations may be performed by including in Eq. (3.3.31) a third term that accounts for the growth of these droplets, but will not be discussed here in detail. Results of such calculations based on three different assumptions with regard to initial foreign droplet size are compared with experiment in Fig. 3.4.1 [3-26]. In the case considered, condensation on these droplets was sufficient to keep subcooling beneath the values required for spontaneous nucleation; therefore no pressure bump appeared, but there was a deviation from the dry isentropic pressure distribution beginning near the saturation point.

If the moisture entrained at the inlet consists of relatively large droplets, the condensation on these droplets is slow and manifests itself only in retarding the increase of subcooling, i.e., in shifting the spontaneous condensation toward lower pressures. Examples of such flows with inlet moisture generated by expansion through a turbine stage positioned ahead of the nozzle are shown in Fig. 3.4.2 [3-21].

Whether salt molecules dissolved in the vapor can also influence the process of nucleation and condensation has not been investigated up to the present. Generally it is assumed that the vapor is chemically clean, although very probably even in high-quality vapor the specific number of salt molecules is comparable to, or even higher than, the number of drops formed (Sec.

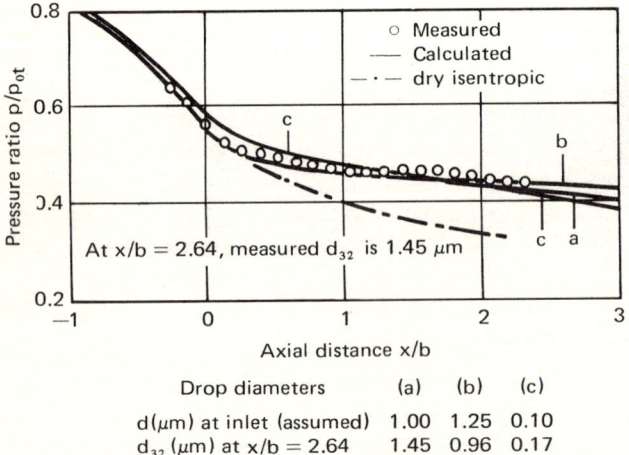

Drop diameters	(a)	(b)	(c)
d (μm) at inlet (assumed)	1.00	1.25	0.10
d_{32} (μm) at x/b = 2.64	1.45	0.96	0.17

FIG. 3.4.1 Complete elimination of spontaneous condensation by finely dispersed inlet wetness; theoretical curves a, b, c based on various assumptions for initial size of droplets [3-26].

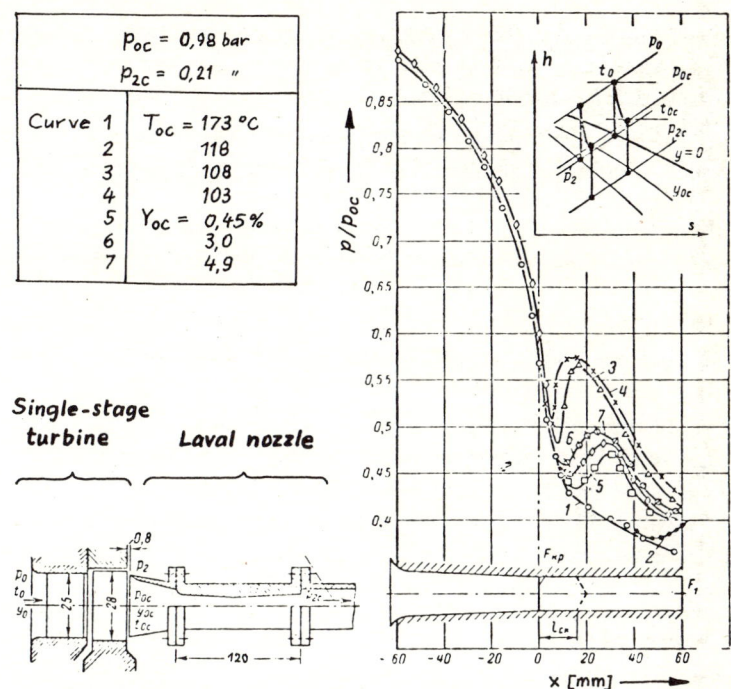

FIG. 3.4.2 Delaying of spontaneous condensation by wetness generated in a turbine stage (cases 5, 6, 7) [3-21].

1.3.2.3). A similar effect on condensation of humid air, that of "smoke" particles produced by salts, has been investigated in [3-31].

3.5 CONDENSATION IN CASCADES AND TURBINES

Condensation in cascades and turbine stages is much more complicated than in straight nozzles because two- and three-dimensional nonuniformities, turbulent fluctuations, and nonsteady flow patterns may have simultaneous influence. It is difficult to understand and predict some of the effects, and a satisfactory and accepted theory comprising all these phenomena has not yet been developed. It is certain however that a simple analogy to nozzle condensation, as hypothesized earlier, e.g., in [1-18], does not exist in general.

Experimental findings have kept accumulating in recent years, and some tentative explanations have been forwarded. Some of the highlights of these developments will be sketched in the following sections.

3.5.1 Experimental Evidence

3.5.1.1 Cascade Flow

In narrow, Laval-nozzle-type blade channels leading to supersonic outlet velocity, pressure bumps can be detected in a fashion similar to straight nozzles [1-17]; see Fig. 3.5.1.

Supersonic expansion fans around corners have been shown to induce oblique condensation shocks; see Fig. 3.5.2, taken from [3-32] and [3-33].

In slightly diverging nozzle channels ending in an oblique expansion fan, unsteady flow effects were observed when inlet superheat was low [3-34]; see Fig. 3.5.3.

Schlieren photographs showing condensation shocks in supersonic cascade flow have been published in [1-17, 3-35]. An influence of the condensation process on the shock pattern was clearly noticeable.

Light-scattering measurement in the outlet flow of a turbine cascade [3-36] has shown that condensation droplets were preferentially being formed in and near the wake region of blades; see Fig. 3.5.4.

The effect of turbulence induced by a grid in the inlet flow on condensation in nozzles was found to be negligible [3-26]; see Fig. 3.5.5.

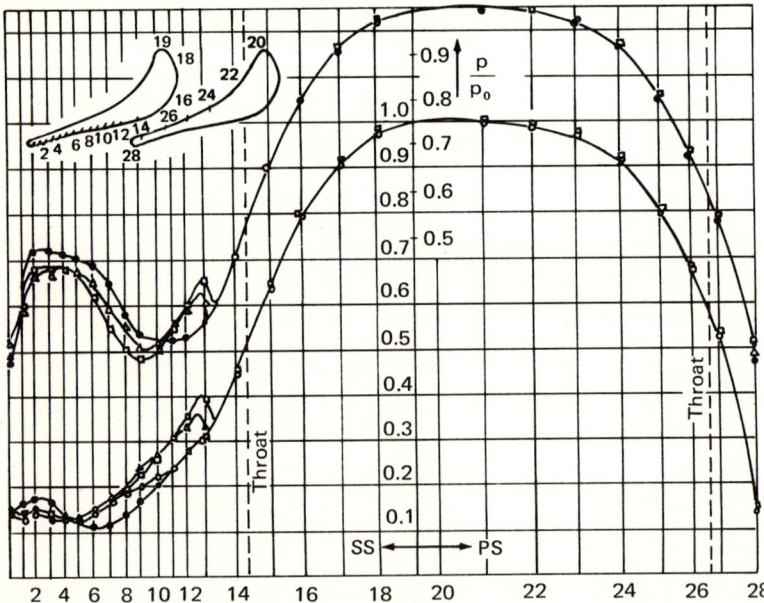

FIG. 3.5.1 Pressure distributions in a turbine cascade, showing condensation. Inlet superheat is varied. Upper curves: $M_1 = 1.4$, lower curves: $M_1 = 1.8$. SS = suction side, PS = pressure side [1-17].

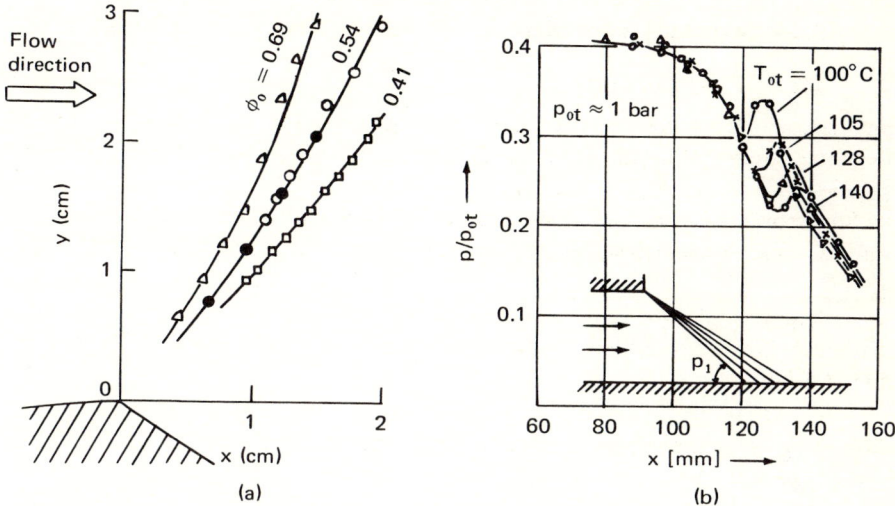

FIG. 3.5.2 Oblique condensation fronts in expansion fans around corners: (*a*) humid air flow at various initial humidities ϕ_0 [3-32]; (*b*) steam flow at various inlet temperatures [3-33].

3.5.1.2 Turbine Flow

Fog droplet-size measurements in the exhaust of a test turbine, by the light attenuation probe depicted in Fig. 3.5.6*a*, have resulted in fog droplet sizes on the order of 0.01 to 0.02 μm; see Fig. 3.5.6*b*. These sizes are about equal to those found in nozzles having comparable mean values of the expansion rate [3-13].

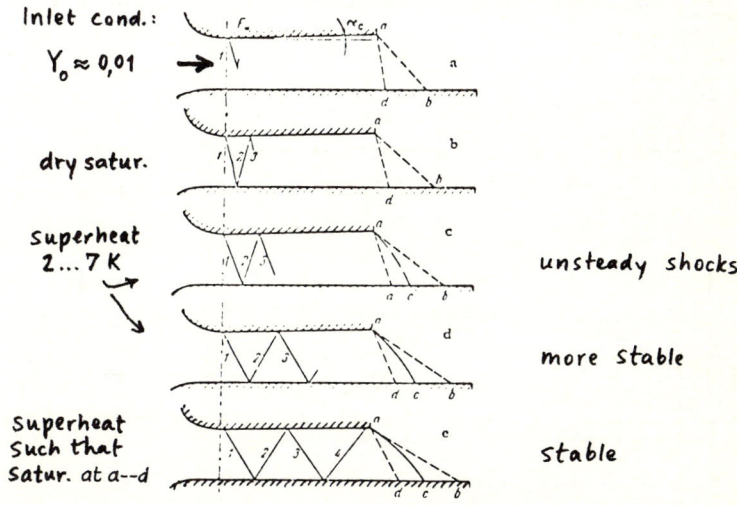

FIG. 3.5.3 Condensation patterns in an overexpanded nozzle [3-34].

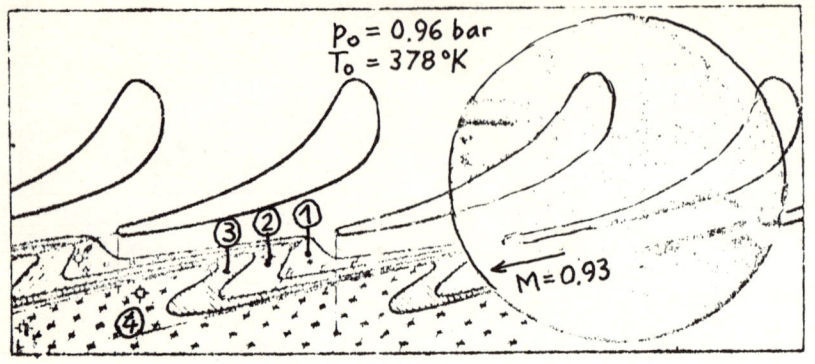

(1) $\bar{r} = 6 \times 10^{-8}$ m; (2) 7×10^{-8} m; (3) 8×10^{-8} m; (4) $(9-10) \times 10^{-8}$ m.

(a)

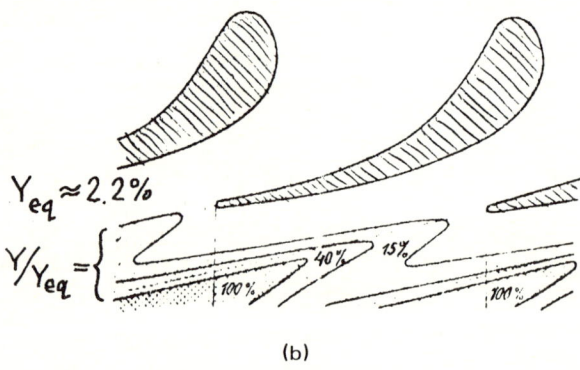

(b)

FIG. 3.5.4 Results of optical measurement behind a cascade: (*a*) droplet sizes; (*b*) wetness amounts (wake regions show more intense condensation than midchannel flow) [3-36].

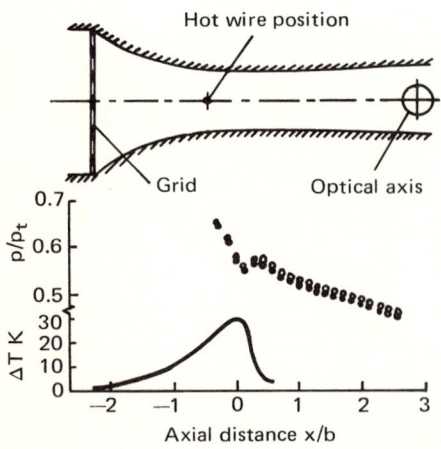

FIG. 3.5.5 Condensation in Laval nozzle with and without grid-generated turbulence; at $x/b = 2.92$, measured droplet dia. is 0.40 μm without grid, 0.37 μm with grid [3-26].

Measured pressure: ○ Without grid
—— Calculated subcooling ● With grid

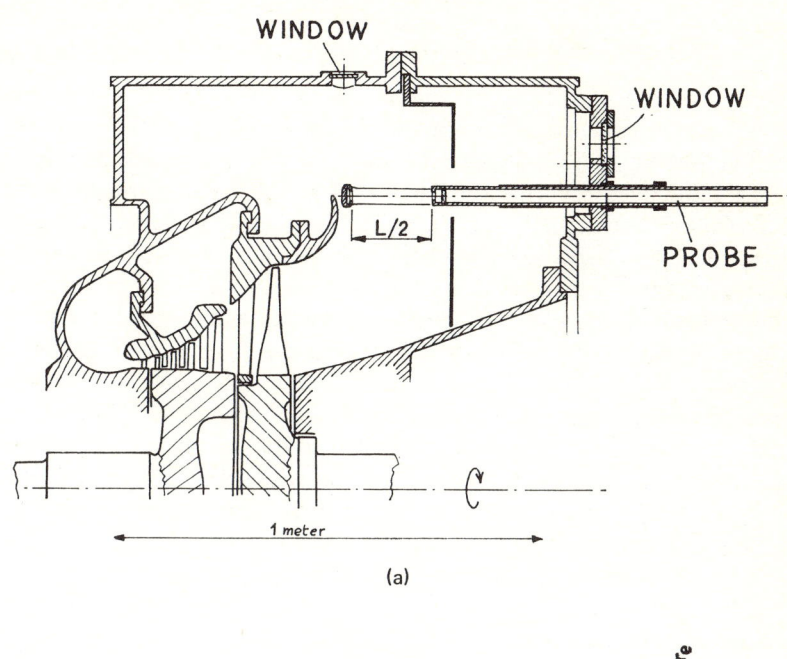

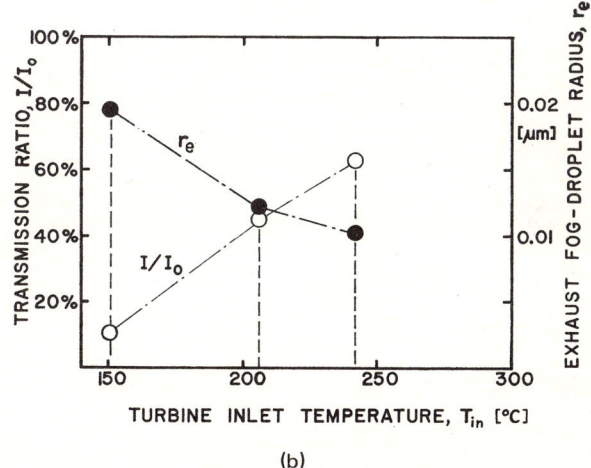

FIG. 3.5.6 Exhaust of a scaled-down model low-pressure turbine: (a) light-attenuation probe arrangement; (b) measured droplet sizes [3-37].

In an experiment made in the flow leaving a turbine stage [3-37], droplets have been detected even at conditions at which the Wilson line was not yet reached; see Fig. 3.4.2. This finding suggests that condensation in turbine stages may begin earlier than in Laval nozzles, provided that the criterion of reaching the Wilson line is based on main-flow average conditions.

3.5.2 Theoretical Explanations

Three tentative explanations have been offered for the difference found between the condensation process in nozzle and turbine flow.

Filippov et al. [3-37] put forward the hypothesis that wake eddies, leaving the trailing edge of blade rows operated at subcooled outlet conditions, were the sites of "premature" nucleation.* This hypothesis, which was based on the experimental observation cited in Fig. 3.5.4, was substantiated by theoretical estimates of the magnitude of the subcooling at the eddy centers. Further downstream the nuclei produced in the wake eddies are supposed to spread out by mixing, and induce condensation in the main flow by seeding.

Moore et al. [3-26] analyzed the condensation in a stream tube running close to the convex side of a turbine bucket; see Fig. 3.5.7. They concluded that condensation may occur in the over-expansion region, even if the conditions at the cascade exit do not yet attain the Wilson line. Such nuclei, once formed, would spread out to seed other regions of the flow.

Gyarmathy, in a discussion contribution to [3-38], pointed to the possible role of static-temperature fluctuations in modifying the nucleation process in turbines. Vehement fluctuations are known to be present in multistage turbines. They were measured by Wood [3-38] and calculated theoretically by Gyarmathy and Spengler [3-39].

---

*The term premature refers here and in the following to predictions based on time-average, main-flow conditions.

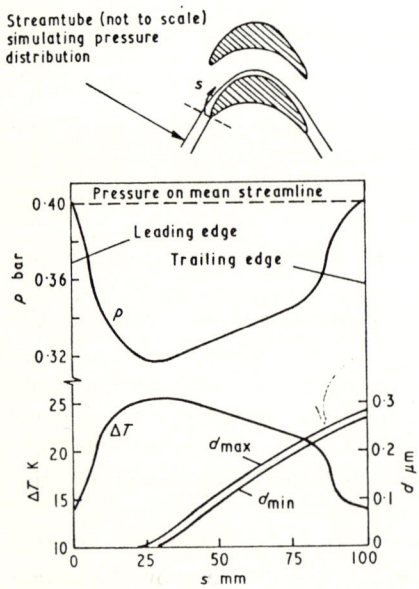

FIG. 3.5.7 Spontaneous condensation near the convex side of a turbine blade [3-26].

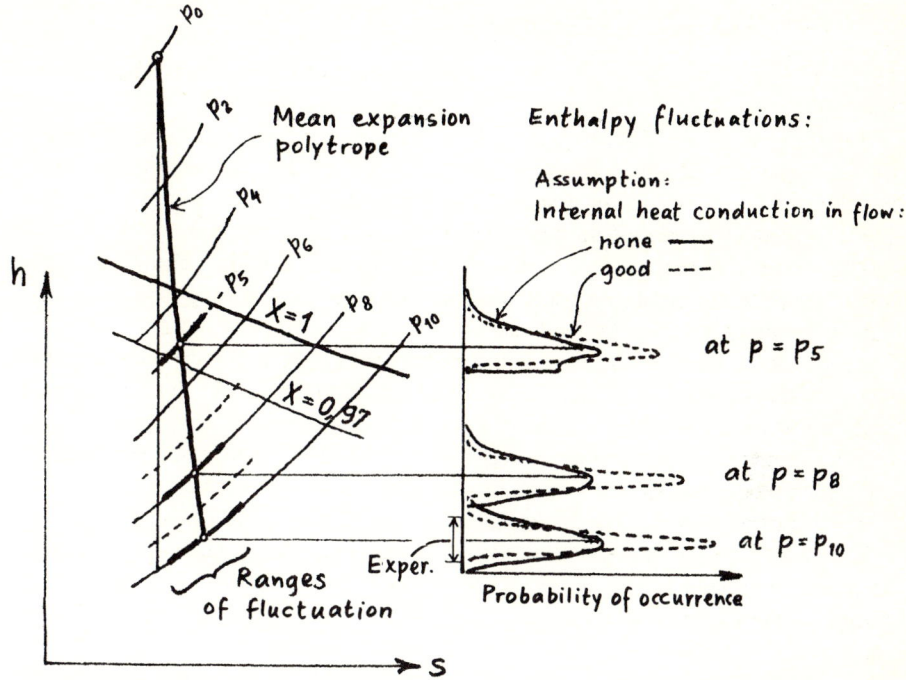

FIG. 3.5.8 Calculated fluctuations of enthalpy at several interstage pressures in a five-stage low-pressure turbine, after [3-39]. Exper.: mean-square fluctuation from hot-wire measurement [3-38].

In turbomachines the static temperature fluctuates more severely than in other flows of high turbulence, because some of the fluid particles may happen to pass through several blade rows without being noticeably affected by friction; i.e., their condition may follow an isentrope, while in other fluid particles more-than-average dissipation may occur. Thus there are colder and hotter flow "packages" intermingled in the main stream. In the cold packages high subcooling values may be attained much before the average flow conditions would reach the Wilson line. Thus these temperature fluctuations are likely to cause premature nucleation. In Fig. 3.5.8 the intensity of these temperature (or static enthalpy) fluctuations, as calculated in [3-39] for the five-stage low-pressure turbine investigated in [3-38], is shown in the Mollier diagram. While static pressure stays fairly constant, fluid conditions are scattered over an enthalpy range almost as large as the heat drop in a stage; see the probability distribution curves shown for three intercascade locations. Comparison with the hot-wire measurement supports the calculated intensity of fluctuations.

In conclusion it is possible, or even likely, that in reality turbine condensation is determined by a combination of the three above-mentioned and perhaps other hitherto undetected phenomena. More detailed observations are required before conclusive statements can be made.

NOMENCLATURE

a	speed of sound
A	cross-sectional area of stream tube or nozzle
a_{th}	coefficient denoting thermal accommodation
b	nozzle width
C	absolute velocity of gaseous flow
C^*	critical velocity (Sec. 3.3.3)
c_p	specific heat of vapor at constant pressure
c_f	specific heat of liquid
d	droplet diameter
ΔG	change of Gibbs free enthalpy
h	specific enthalpy
H	nozzle height
i	number of molecules in a droplet of a given size
I	nucleation rate
k	Boltzmann constant
k	number indicating droplet category (Fig. 3.3.5)
K	proportionality factor
Kn	Knudsen number (Sec. 3.3.2)
l_0	diffusor length
$\bar{l}$	mean free path length
m	mass
M	Mach number
N	number of droplets per unit mass of wet steam
N_m	number of molecules per unit mass of vapor
$N_{i,eq}$	equilibrium number of clusters of a given size i per unit mass of vapor
p	pressure
$\dot{P}$	expansion rate
$\dot{Q}$	heat transfer rate
Q_l	latent heat
r	radius of droplet
R^*	radius of curvature in nozzle throat
R	gas constant
$\bar{r}^{**}$	mean radius of droplet corresponding to wetness on Wilson line ($Y = 0.03$)
$\dot{r}$	droplet growth rate
s	specific entropy
S	supersaturation ratio or entropy
T	temperature
t	time
x	coordinate; distance along nozzle axis

Y	wetness fraction
α	heat transfer coefficient
β	number of droplets impinging on unit boundary surface per unit time
κ	index of isentropic expansion
λ	thermal conductivity
μ	dynamic viscosity
ν	number of droplets per unit mass
ξ	condensation coefficient or value of x where droplet is born
ρ	density
σ	surface tension
ϕ	humidity (Fig. 3.5.2)

Subscripts

f	liquid phase
g	gaseous phase
fg	phase transition
m	molecular
r	referring to droplet
s	saturation
t	stagnation conditions
0	inlet conditions
WP	Wilson point

Superscripts

$^-$	mean value
$'$	saturated liquid
$''$	saturated vapor
$+$	throat
$\cdot$	variation with time

REFERENCES

3-1 Wegener, P. P. (ed.): "Nonequilibrium Flows," pt. I, Dekker, New York, 1969. (Especially chap. 4, by Wegener.)

3-2 Ames Research Staff: Equations, Tables, and Charts for Compressible Flow, *NACA Rept.* 1135, 1953.

3-3 Gyarmathy, G., and Meyer, H.: Spontane Kondensation, *VDI Forschungsheft* 508, VDI-Verlag, Düsseldorf, 1965.

3-4 Oswatitsch, K.: "Gasdynamik," Springer, Wien, 1952.

3-5 Stever, H. G.: Condensation Phenomena in High-Speed Flows, in H. W. Emmons (ed.), "Fundamentals of Gas Dynamics," Princeton University Press, Princeton, N. J., 1958.

3-6 Binnie, A. M.: Notes on Gas Flow through a Nozzle, *Proc. Roy. Soc. (London)*, pp. 492–499, 1950.

3-7 Barschdorff, D.: Verlauf der Zustandsgrössen und gasdynamische Zusammenhänge bei der spontanen Kondensation reinen Wasserdampfes in Lavaldüsen, *Forsch. Ing.-Wes.* vol. 37, no. 5, pp. 146-157, 1971.

3-8 Oswatitsch, K.: Kondensationserscheinungen in Ueberschalldüsen, *Z. Angew. Math. Mech.*, vol. 22, no. 1, pp. 1-14, 1942.

3-9 Binnie, A. M., and Woods, M. W.: The Pressure Distribution in a Convergent-Divergent Steam Nozzle, *Proc. Inst. Mech. Eng. (London)*, vol. 138, pt. I, no. 2, pp. 229-266, 1938.

3-10 Gyarmathy, G., Burkhard, H.-P., Lesch, F., and Siegenthaler, A.: Spontaneous Condensation of Steam at High Pressure: First Experimental Results, *Inst. Mech. Eng. (London), Conf. Publ.* 3, pp. 182-186, 1973.

3-11 Petr, V.: Measurement of an Average Size and Number of Droplets during Spontaneous Condensation of Supersaturated Steam, *Proc. Inst. Mech. Eng., (London)*, vol. 184, pt. 3G(III), pp. 22-28, 1969-1970.

3-12 Puzyrewski, R.: Condensation of Water Vapor in a Laval Nozzle (in Polish), *Inst. Fluid Flow Mach., Gdansk (Poland)*, 1969.

3-13 Gyarmathy, G., and Lesch, F.: Fog Droplet Observations in Laval Nozzles and in an Experimental Turbine, *Proc. Inst. Mech. Eng. (London)*, vol. 184, pt. 3G(III), pp. 29-36, 1969-1970.

3-14 Binnie, A. M., and Green, J. R.: An Electrical Detector of Condensation in High-Velocity Steam, *Proc. Roy. Soc. (London)*, A181, no. 984, pp. 134-154, 1943.

3-15 Yousif, F. H., Campbell, B. A., and Bakhtar, F.: Instability in Condensing Flow of Steam, *Proc. Inst. Mech. Eng. (London)*, vol. 186, no. 37, pp. 439-448, 1972.

3-16 Pouring, A. A.: Thermal Choking and Condensation in Nozzles, *Phys. Fluids*, vol. 8, no. 10, pp. 1802-1810, 1965.

3-17 Barschdorff, D.: "Kurzzeitfeuchtemessung und ihre Anwendung bei Kondensationserscheinungen in Lavaldüsen," Diss. Techn. Hochschule Karlsruhe, 1967.

3-18 Zierep, J., and Lin, S.: Ein Aehnlichkeitsgesetz für instationäre Kondensationsvorgänge in Lavaldüsen, *Forsch. Ing.-Wes.*, vol. 34, no. 4, pp. 97-99, 1968.

3-19 Saltanov, G. A., Seleznev, L. J., and Tsiklauri, G. V.: Generation and Growth of Condensed Phase in High-Velocity Flows, *Int. J. Heat Mass Transfer*, vol. 16, pp. 1577-1587, 1973.

3-20 Król, T.: Results of Optical Measurements of Diameters of Drops Formed Due to Condensation of Steam in a Laval Nozzle (in Polish), *Trans. Inst. Fluid Flow Mech. (Poland)*, no. 57, pp. 19-30, 1971.

3-21 De h, M. E., and Filippov, G. A.: "Gas Dynamics of Two-Phase Media" (in Russ..an), Izd. Energiya, Moscow, 1968.

3-22 Stein, G. D., and Moses, C. A.: Rayleigh Scattering Experiments on the Formation of Water Clusters Nucleated from the Vapor Phase, *J. Colloid. Interfac. Sci.*, vol. 39, no. 3, pp. 504-512, 1972.

3-23 Gardzilewicz, A., and Król, T.: An Attempt to Determine Liquid Phase Dispersion in Condensing Steam from the Measurement of Light Attenuation (in Polish), *Trans. Inst. Fluid Flow Mach. (Poland)*, no. 40, pp. 23-36, 1968.

3-24 Joos, G.: "Lehrbuch der theoretischen Physik," 7. Aufl., Akad. Verl. Geest u. Portig, Leipzig, 1950.

3-25 Deich, M. E., Stepanchuk, V. F., and Saltanov, G. A.: Calculation of the Rate of Formation of Condensation Nuclei in Subcooled Steam (in Russian), *Izv. Akad. Nauk Uzbek SSR, Energetika i Transport*, no. 2, p. 34, 1968.

3-26 Moore, M. J., Walters, P. T., Crane, R. I., and Davidson, B. J.: Predicting the Fog-Drop Size in Wet-Steam Turbines, *Inst. Mech. Eng. (London), Conf. Publ.* 3, pp. 101-109, 1973.

3-27 Wu, B. J.-CH.: "A Study on Vapor Condensation by Homogeneous Nucleation in Nozzles," doctoral dissertation, Yale University, New Haven, Conn., 1972.
3-27a Schaaf, S. A., and Chambré, P. L.: Flow of Rarified Gases, in H. W. Emmons (ed.), "Fundamentals of Gas Dynamics," Princeton University Press, Princeton, 1958.
3-28 Gyarmathy, G.: Zur Wachstumsgeschwindigkeit kleiner Flüssigkeitstropfen in einer übersättigten Atmosphäre, *ZAMP 14*, vol. 3, pp. 280–293, 1963.
3-29 Hill, P. G., Witting, H., and Demetri, E. P.: Condensation of Metal Vapors During Rapid Expansion, *Trans. ASME, J. Heat Transfer*, vol. 85, p. 303, 1963.
3-30 Mills, A. F., and Seban, R. A.: The Condensation Coefficient of Water, *Int. J. Heat Mass Transfer*, vol. 10, no. 12, pp. 1815–1827, 1967.
3-30a Hill, P. G.: Condensation of Water Vapour During Supersonic Expansion in Nozzles, *J. Fluid Mech.*, vol. 25, no. 3, pp. 593–620, 1966.
3-31 Buckle, E. R.: Effects of Seeding on the Condensation of Atmospheric Moisture in Nozzles, *Nature*, vol. 208, p. 367, 1965.
3-31a Filippov, G. A., Povarov, O. A., and Priahin, V. V.: "The Investigation and Analysis of Wet Steam Turbines" (in Russian), Energiya, Moscow, 1973.
3-32 Deich, M. E., Stepanchuk, V. F., Saltanov, G. A., and Tsiklauri, G. V.: Experimental Study of Condensation Jumps, *High Temperature*, vol. 2, no. 5, pp. 710–716, 1964.
3-33 Smith, L. T.: Experimental Investigation of the Expansion of Moist Air around a Sharp Corner, *AIAA J.*, vol. 9, no. 10, pp. 2035–2037, 1971.
3-34 Deich, M. E., Kurshakov, A. V., Saltanov, G. A., and Yatcheni, I. A.: A Study of the Structure of Two-Phase Flow behind a Condensation Shock in Supersonic Nozzles, *Heat Transfer–Soviet Res.*, vol. 1, no. 5, pp. 95–105, 1969.
3-35 Ikeda, T., and Suzuki, A.: Some Findings on the Flow Behaviour of Last-Stage Turbine Buckets by Linear Cascade Tests in Steam, *Inst. Mech. Eng. (London), Conf. Publ.* 3, paper C 26/73, pp. 46–55, 1973.
3-36 Deich, M. E., Filippov, G. A., Saltanov, G. A., and Yatcheni, I. A.: Study of Steam Condensation in Swirled-Flow Regions of Turbine Cascades, *Heat Transfer–Soviet Res.*, vol. 5, no. 1, pp. 62–71, 1973.
3-37 Filippov, G. A., Saltanov, G. A., and Ignatevskii, E. A.: Analysing the Condensation of Supersaturated Steam in Turbine Stages, *Thermal Eng.*, vol. 17, no. 12, p. 26, 1970.
3-38 Wood, N. B.: Flow Unsteadiness and Turbulence Measurements in the Low Pressure Cylinder of a 500 MW Steam Turbine, *Inst. Mech. Eng. (London), Conf. Publ.* 3, paper 54/73, pp. 115–121, 1973.
3-39 Gyarmathy, G., and Spengler, P.: "Ueber die Strömungsfluktuationen in mehrstufigen thermischen Turbomaschinen, Traupel–Festschrift, Juris-Verlag, Zürich, pp. 95–141, 1974.

CHAPTER 4

Instrumentation for Wet Steam

4.1 A REVIEW OF INSTRUMENTATION FOR WET STEAM

M. J. Moore

The development of suitable instrumentation is an important part of the study of wet steam, determining to a large extent the rate at which new knowledge is obtained. The development problems lie in two main categories: the familiar problem of making a measurement without distorting the two-phase flow being measured, and the particular problem of restricted access to the steam space in a turbine. Both aspects are discussed in the following pages, and a range of instrumentation is described for research applications or for performance measurements in operating plants.

4.1.1 Pressure Measurement

For pressure measurement in wet-steam flows, we require a miniature pressure transducer giving a stable signal proportional to absolute pressure and capable of continuous operation in a hot wet environment. As such a transducer is not yet available, the pressure-recording device is generally remote from the measuring point, the pressure being transmitted via small-diameter tubing. As will be shown the arrangement of the sensing point in the wet-steam flow may be similar to the commonly used geometries of probes for aerodynamic measurements in single-phase flows (e.g., Prandtl tubes and Kiel probes). The main problems of pressure measurement in wet steam concern the transmission of pressure through tubing to the measuring instrument.

4.1.1.1 Water-Filled Lines

An arrangement used for many years is shown in Fig. 4.1.1, where the lines connecting sensing point and manometer are filled with water. Condensing

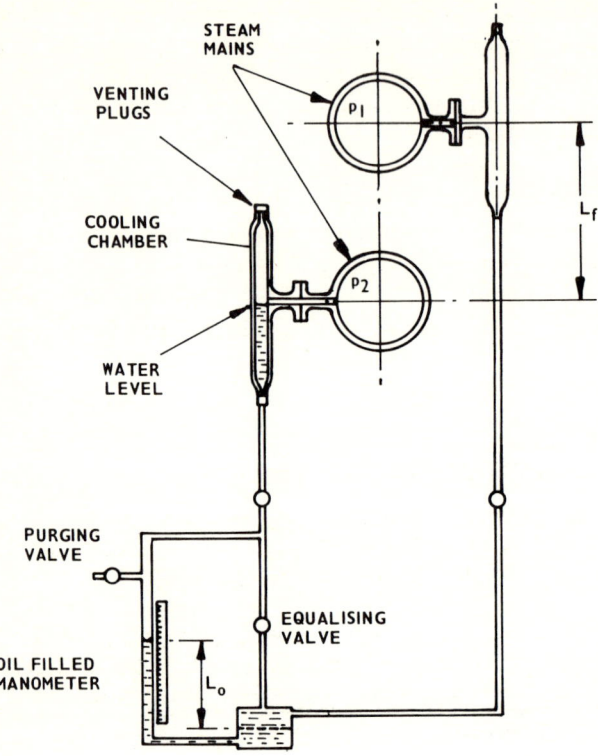

$(p_1 - p_2) = L_o p_o - L_f p_f$

FIG. 4.1.1 Pressure measurement using water-filled lines.

pots at the measuring point ensure that the steam/water surface is maintained at a known position, and a correction to the observed manometer reading is made to compensate for the different heights of the water columns. Manometer fluids must therefore be immiscible with water (e.g., mercury, various heavy organic liquids such as dibromoethyl benzene, "Alkazene" spec.gr. 1.74). In practice the restrictive geometry of the arrangement limits its application, and of course the system is difficult to operate at low steam pressures. For accurate measurements, the lines need occasional purging with water to remove accumulations of gases that may leak into or be released from the water.

4.1.1.2 Continuously Purged Air Lines

Elaborate arrangements would be necessary to ensure a continuous steam path from sensing point to recording device. Ingested water or condensation in the pressure lines tends to accumulate in slug form, the surface tension effects distorting the pressure transmitted. To keep lines free from water, a small flow of dry gas can be admitted near the transducer to pass down the line and flow into the steam at the sensing point. The purge flow must be small to (a) avoid significant distortion of the steam flow field, and (b) minimize the frictional pressure drop from the injection point to the sensing point in the steam flow.

An arrangement for continuous air purging of a system for measuring subatmospheric pressure in wet steam is shown in Fig. 4.1.2a. Atmospheric air is induced via a reservoir that is maintained at a constant pressure of, say, 0.9 bar. The purge flow rate is then adjusted by a fine-control needle valve. The system is calibrated to determine the frictional loss in pressure through the system, giving typical characteristics shown in Fig. 4.1.2b. Purge flow rate can

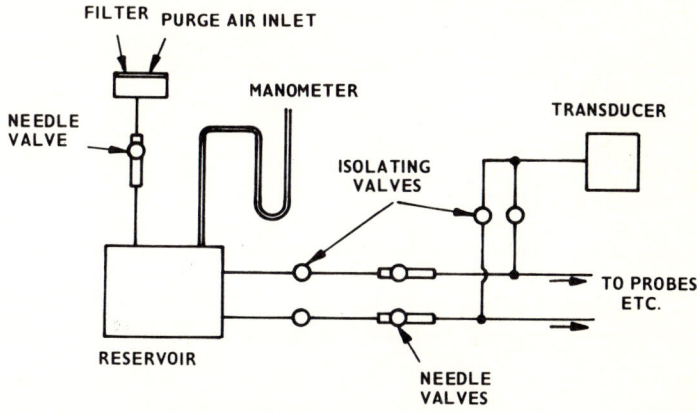

(a) ARRANGEMENT OF SYSTEM

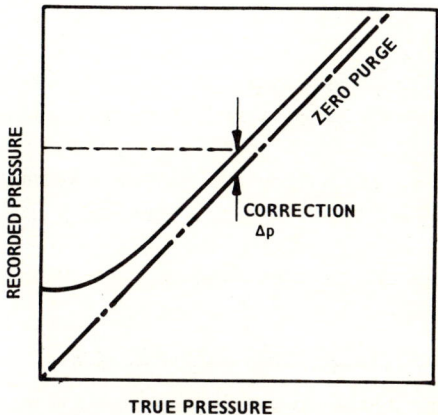

(b) TYPICAL CALIBRATION CURVE

FIG. 4.1.2 System using continuous purging.

be seen to be excessive when the correction is a major proportion of the recorded pressure or where choking conditions occur, usually at the sensing point, making recorded pressure independent of steam pressure.

With care the continuously purged system can provide accurate pressure measurement [4-1] and is particularly useful for motorized instrument traversing. An example is shown in Fig. 4.1.3 of a Pitot-tube traverse across a blade wake. In the region of falling total pressure where steam flows out of the pressure line, the unpurged system adequately records the pressure variation. For increasing pressure, condensation of the inflowing steam in the line results in an unacceptable lag in recorded pressure.

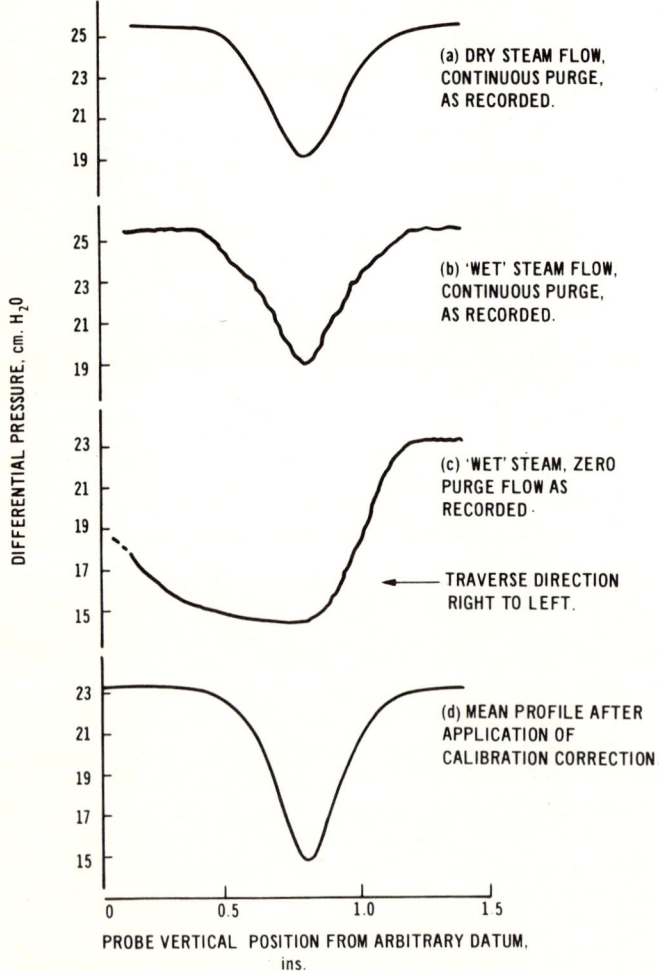

FIG. 4.1.3 Measurements of a particular blade wake profile showing the effect of various factors on result obtained.

An alternative method of control/calibration has been developed by Smith [4-2], where a range of orifices of decreasing diameter is used in sequence to control purge flow, instead of the needle valve in Fig. 4.1.2. The recorded pressures are then extrapolated to give the true value at zero orifice diameter (zero purge flow).

4.1.1.3 Intermittent Purging

As an alternative to the relatively complicated arrangements for continuous purging, the intermittent purge arrangements offer a simple reliable technique particularly suited to measurements in difficult environments, such as in the hot, noisy, often dusty atmosphere of a turbine hall. As before, an accurate pressure recording is ensured by filling the lines to the sensing point with air immediately prior to making the measurement. A sufficient quantity of air is admitted to blow through any steam and/or water in the system, and care is taken in designing the system to ensure that no pockets of water can remain. Purging raises the pressure in the line above the value at the sensing point, and when purge air is shut off the pressure reduces to the appropriate value.

A suitable system is shown in Fig. 4.1.4, together with a typical pressure/time variation recorded at the measuring instrument. For accuracy and reliability, the response time t_* must be short—preferably less than 10 s—and in practice this requirement places definite restrictions on probe and pressure line geometry. In particular, for a given length of line L and volume V of the pressure transducer, the response time t_* is determined mainly by the internal diameter D of the tube. An approximate expression for t_* can be obtained from [4-3] as follows

$$t_* = \frac{2L}{\sqrt{3}a} - \ln\left(\frac{p_* - p}{p_i - p}\right) \frac{32\mu}{\chi p} \left(\frac{L}{D}\right)^2 \left[\frac{V}{\frac{\pi}{4} D^2 L} + \frac{1}{2}\right] \quad (4.1.1)$$

where p is the pressure at the sensing point and p_i the initial value of pressure at the transducer at the time of shutting off the purge flow ($t = 0$). Typical values of t_* are plotted in Fig. 4.1.5.

4.1.1.4 System Response to Fluctuating Pressures

The response of the pressure lines and transducer to a fluctuating pressure at the sensing point is particularly relevant to the measurement of conditions in the vicinity of steam-turbine moving blades. This subject has been studied extensively for small sinusoidal pressure fluctuations [4-4 to 4-7], the theory of [4-7] giving the following expression for amplitude ratio for a uniform tube

$$\frac{\Delta p_*}{\Delta p} = R(\cosh \phi L)^{-1} \quad (4.1.2)$$

where

$$\phi = \frac{\omega}{a_0} \sqrt{\frac{\chi}{n}} \sqrt{\frac{J_0(\alpha)}{J_2(\alpha)}}$$

$$\alpha = i^{3/2} \frac{D}{2} \sqrt{\frac{\rho\omega}{\mu}}$$

$$n = \left[1 + \frac{\chi - 1}{\chi} \frac{J_2(\alpha\sqrt{\mathrm{Pr}})}{J_0(\alpha\sqrt{\mathrm{Pr}})}\right]^{-1}$$

and J_0, J_2 are Bessel functions of order zero and two, respectively.

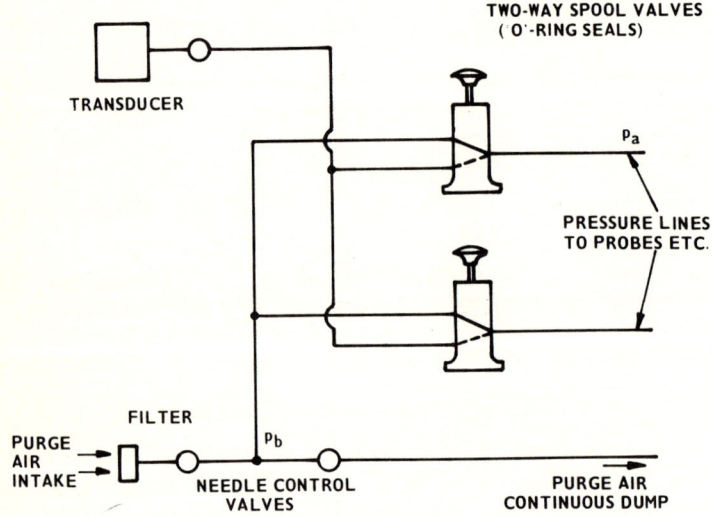

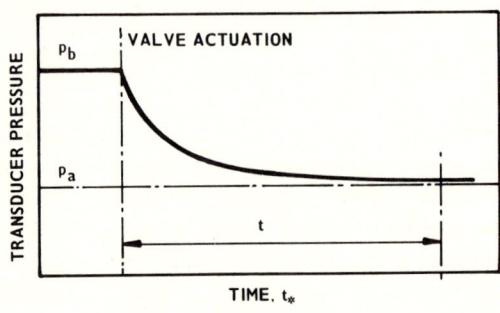

FIG. 4.1.4 System for intermittent purging.

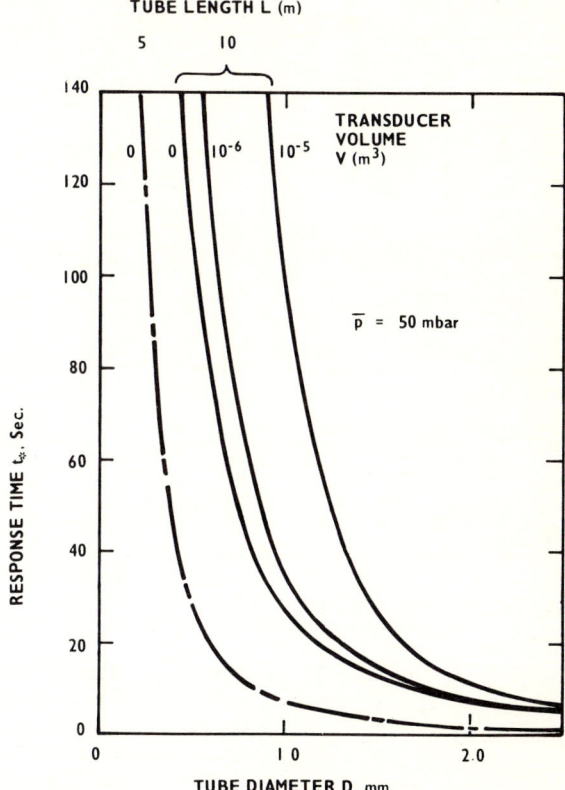

FIG. 4.1.5 Response time for air-purged pressure lines.

The frequency response for 3 mm dia. tubing for a range of tube lengths is shown in Fig. 4.1.6. Resonant frequencies are clearly seen, while for long tubes frictional damping predominates. From the previous section on purging requirements, the typical system with ~ 5 m length and ~ 0.003 m bore tubing can be seen to have a negligible response to fluctuating pressures at frequencies greater than 100 Hz for mean pressures of ~ 1 bar. However the gas density is an important variable, and amplitude ratio increases significantly at increased mean pressure.

Larger amplitude pressure fluctuations, particularly of total pressure, are considered in [4-8 to 4-10]. In [4-8] a system including a Pitot tube and a simple liquid manometer is shown to give negligible error in time-mean value of total pressure, except at frequencies near the resonant value (typically 1.0 Hz for a water manometer). Reference [4-9] describes the response of a highly damped system to a total pressure waveform similar to that experienced downstream of a turbine moving blade. The nonlinear effects are shown to produce errors of up to 2% in the time-mean value of pressure

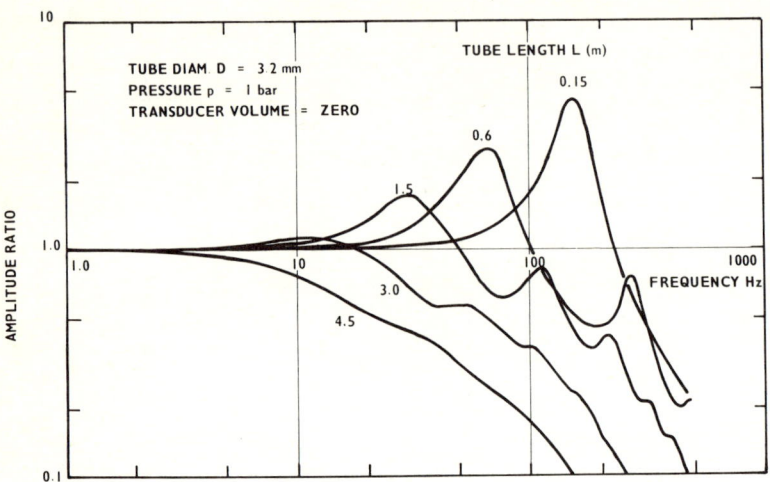

FIG. 4.1.6 Frequency response of air-filled tubing connecting transducer to measuring point.

recorded at fluctuation frequencies of 300 cps. The effect of Pitot-probe head geometry has also been investigated [4-10], a simple tube being shown to record time-mean pressure. Hooded Pitot tubes are shown in some cases to introduce significant errors in recorded mean pressure. The interpretation of recorded total pressure and its use with static pressure to calculate velocity is considered later.

4.1.1.5 Pressure Measurement in Turbines

The main problem of obtaining measurements of pressure in operating turbines is that of access. Permanently installed instrumentation is in practice often unreliable, and it may be preferable in many cases to use long probes which are inserted into the operating turbine via accurately positioned guide tubes. A typical installation for obtaining the radial distribution of steam conditions between the blade rows of an LP turbine is shown in Fig. 4.1.7.

Various types of sensing heads have been used to measure total and static pressure and flow direction in a three-dimensional field, some examples from [4-12 to 4-14] being shown in Fig. 4.1.8. For measurements in operating turbines these heads have certain disadvantages. Probe heads (*a*) and (*b*), Fig. 4.1.8, must be small to be incorporated into a probe body of typically 25 mm dia., and purging through small holes can give a long blow-down time t_*. Probe head (*c*) is very robust but requires extensive calibration in a wet-steam flow. Finally, head (*d*) requires the purging and recording of at least eight pressures, which can be very time consuming.

An alternative series of probe heads [4-11], which require minimum calibration and can be fitted with relatively large tappings and pressure lines

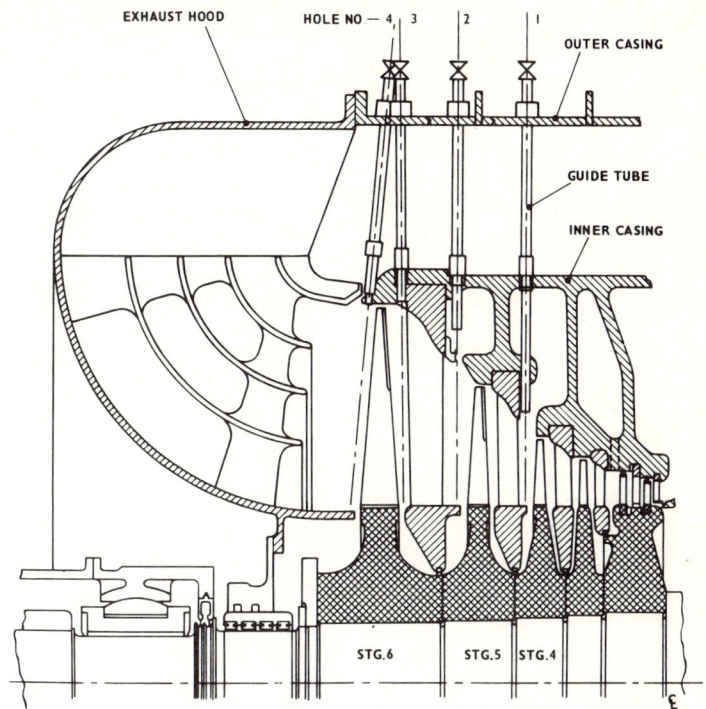

FIG. 4.1.7 Axial section through turbine showing position of access holes.

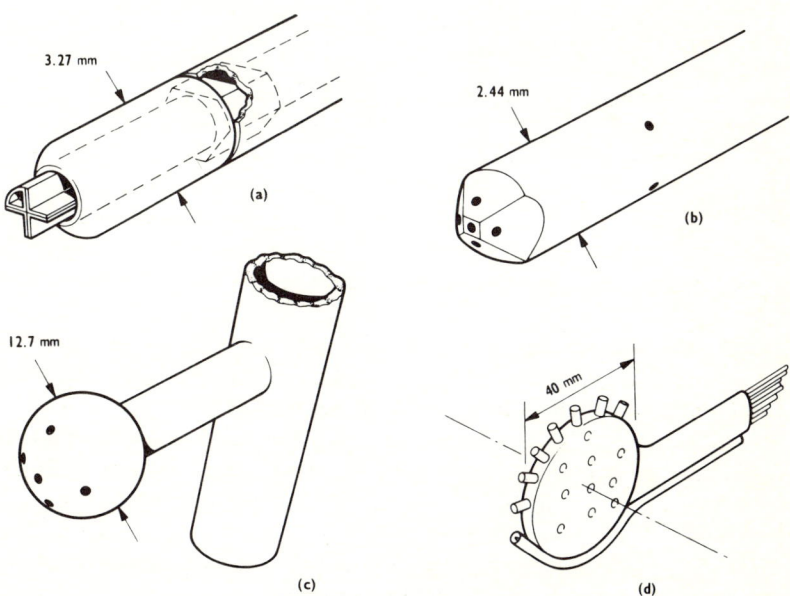

FIG. 4.1.8 Probe heads for measurements in three-dimensional flows: (*a*) Todd (1962); (*b*) Lindley and Whybrow (1969); (*c*) Finlayson and Roberts (1955); (*d*) Lagun, Symoyu and Elzarov (1966).

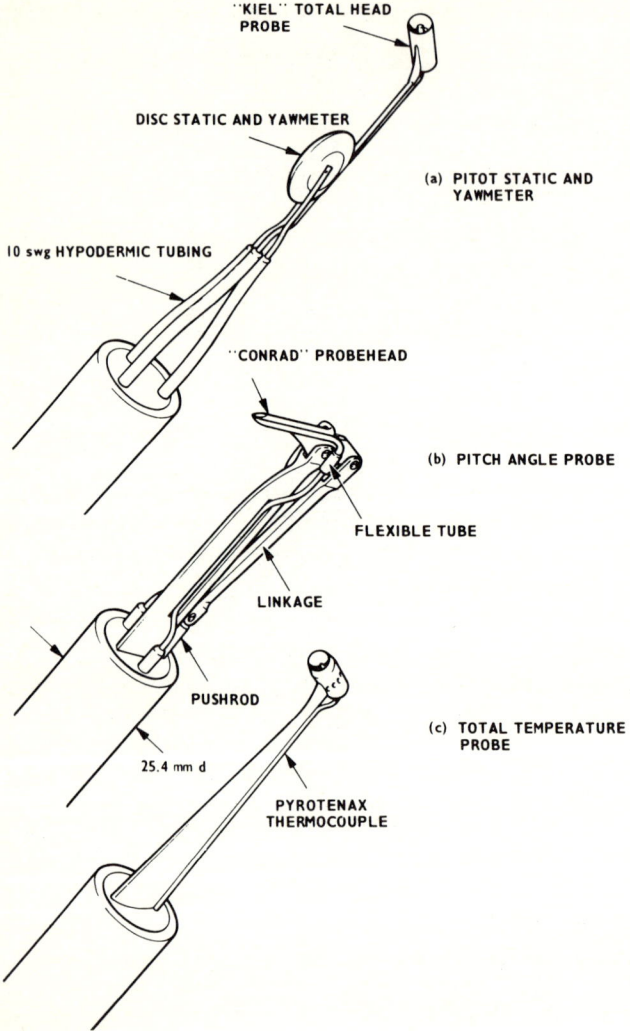

FIG. 4.1.9 CERL turbine probes.

for rapid purging are shown in Fig. 4.1.9. Probe (*a*) is rotated to obtain the flow direction in the circumferential plane (yaw angle) by nulling static pressures on each side of the disk. Total pressure is obtained from the Kiel probe. Disk static can be related to the flow static pressure via a calibration curve of pressure versus flow Mach number. The symmetry of the disk and the Kiel hood ensure the measurements are independent of the flow angle in the meridianal plane (pitch angle). A separate probe (*b*), Fig. 4.1.9 of the Conrad type, is set at the measured yaw angle and nulled to determine pitch angle. The inherent accuracy and purging speed of the probes outweigh the disadvantage of using more than one probe to measure the conditions required.

4.1.2 Temperature Measurement

Static temperature in an equilibrium flow of wet steam is, of course, the saturation value corresponding to the measured static pressure. The main purpose of measuring temperature in wet-steam turbines is therefore to detect regions of superheat or supersaturation. Conventional methods, such as the hooded thermocouple probe shown in Fig. 4.1.9c, have been used to record the total temperature of the flow, and the static temperature determined from the measured values of total temperature and total static pressure.

If the total temperature equals the saturation value corresponding to the total pressure, then the steam is saturated or wet, the wetness fraction of course not quantifiable from these measurements alone. On the other hand, if the stagnation condition is superheated, calculation of the isentropic values at the static pressure will determine whether the steam flow is superheated or wet.

In steam flows containing sprays of coarse water, the temperature recorded can be substantially less than the saturation value for the measured total pressure. This phenomenon has been noted where substantial quantities of coarse water are impinging on the probe, and it may be due to cooling by the water which is at the saturation temperature corresponding to the static pressure.

Of particular interest would be the recording of the temperature of supersaturated steam in a turbine, but the author knows of no published measurement.

4.1.3 Velocity Measurement

4.1.3.1 Interpretation of Pitot Pressure

Small size, simplicity, and inherent reliability make the Pitot static tube an attractive instrument for the measurement of velocity in flowing fluids. The device has been in use for many years, but the the detailed flow field in the mouth of a Pitot tube is extremely complex and little understood. However recorded pressures are found [4-15] to be sufficiently close to the true stagnation pressure to give little incentive to more detailed investigation. The use of a Pitot tube in a flow of wet steam requires a reexamination of its mode of operation, particularly where the liquid phase consists of a fine dispersion of droplets.

Experimental investigations in the literature are generally concerned with flows of gases containing large drops or particles (e.g. [4-16]), the probes used being more complicated than the simple Pitot tube in construction and operation. A theoretical approach is provided by [4-17, 4-18], the former being concerned with flows containing larger droplets (>2 μm dia.) than those anticipated in the condensation fog in a turbine. We shall consider briefly the method and conclusions of [4-18].

For a Pitot tube of typically 2 mm dia. in a wet-steam flow of velocity 200 m/s, the droplets in the flow will pass through the region near the Pitot tube mouth in ~ 20 μs. The relaxation times for fog droplets will be

$$\text{Inertial } \tau_I \cong 5 \text{ μs}$$

$$\text{Thermal } \tau_T \cong 60 \text{ μs}$$

as shown in Chap. 2. We may therefore expect significant deviation of droplet trajectories from the flow streamlines and almost total thermal inequilibrium (i.e., negligible phase change). On this basis the mathematical model used by Crane and Moore [4-18] considers the trajectories of constant diameter droplets and computes the total pressure of the wet-steam flow from the momentum loss of both phases.

To calculate the droplet trajectories a solution is required for the vapor phase-flow field in the vicinity of the probe mouth. The analysis was restricted for simplicity to incompressible subsonic flow, and it was assumed that the deviation in droplet trajectories had negligible effect on the velocity distribution of the vapor flow field. These limitations are not thought to introduce serious errors for wet-steam flows of low wetness fraction ($<15\%$) and at subsonic velocities.

A plane potential flow solution was therefore developed with a free streamline boundary condition as shown in Fig. 4.1.10. The tube width is defined as 2π and free-stream velocity as unity. It can be shown that the velocity potential ω is given by

$$\omega = -\frac{(1-C_h)^2(1+C)^2}{4(1-C)^2} - q \ln \frac{4(C-C_h)(1-CC_h)}{(1+C_h^2)^2(1-C)^2}$$

and the complex velocity C in the physical z plane is given implicitly by

$$z = -\frac{(1-C_h)^2}{4} \frac{(5-3C)(1+C)}{(1-C)^2} + (1+C_h^2)\ln\frac{(1+C_h)(1-C)}{2}$$

$$-\ln(C-C_h) - C_h^2 \ln(1-CC_h)$$

The solution is based on a free streamline velocity equal to free-stream value, implying that the pressure in the zone bounded by *CDEF* is also at the free-stream level. In practice, mass transfer will reduce this pressure, and reattachment of the free streamline will occur. However the solution is considered sufficiently accurate in the region of importance ahead of the tube mouth. To represent a purged Pitot tube, a source is included at C, of

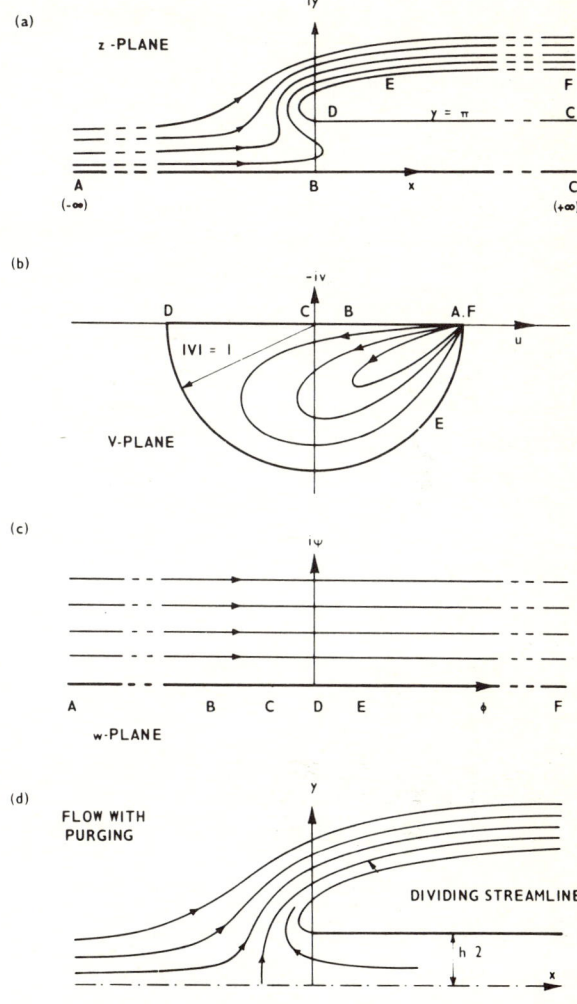

FIG. 4.1.10 Vapor phase flow model.

strength such that the velocity of the purging fluid far inside the tube is C_h (for purging outflow, C_h is negative). The flow field near the tube mouth for unpurged and purged cases are shown in Fig. 4.1.11.

The effects of differing densities in the main stream and the purge fluid may be incorporated, provided that both densities are constant. It is shown in [4-19] that the potential flow streamline pattern is as for fluids of equal density, provided the velocity of one fluid is modified as

$$|C|_a = |C|_{a,k} \left(\frac{\rho_b}{\rho_a} \right)^{1/2}$$

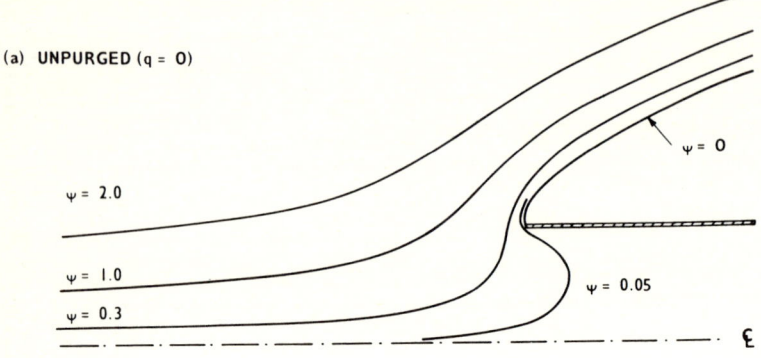

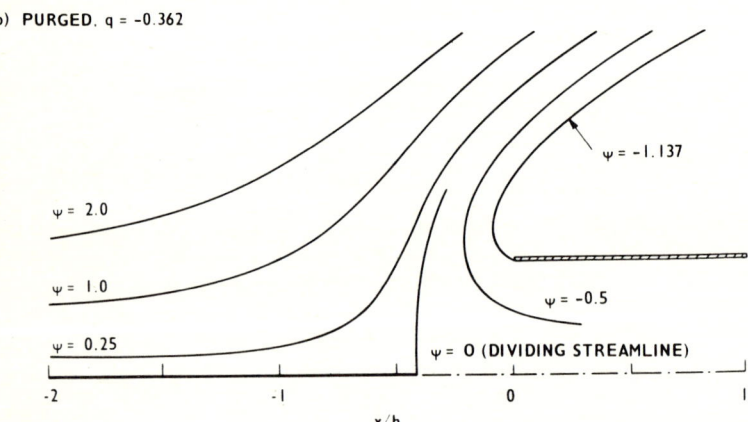

FIG. 4.1.11 Streamline patterns in potential flow round a two-dimensional Pitot tube.

where a and b denote the different fluids and k refers to the conventional kinematic solution.

Droplet trajectories in a known vapor velocity field can be calculated from the equations of motion as shown in Chap. 2. In dimensionless form the x, y components of droplet acceleration may be written

$$\frac{d\bar{u}}{dt}x = P_I(\bar{C}_x - \bar{u}_x)$$

$$\frac{d\bar{u}}{dt}y = P_I(\bar{C}_y - \bar{u}_y)$$

where droplet velocity u and time t are expressed in dimensionless form as

$$\bar{u} = \frac{u}{C_0} \qquad \bar{C} = \frac{C}{C_0} \qquad \bar{t} = \frac{C_0 t}{h}$$

Distances $\bar{x}$ and $\bar{y}$ are also dimensionless ratios of reference length h, which is incorporated in the inertia parameter

$$P_I = \frac{18\mu h}{C_0 d^2 \rho_f (1 + 2.53\,\text{Kn})}$$

The distance traveled by a droplet and its change in velocity over a small time interval $\Delta \bar{t}$ can be obtained by integrating the equations of motion, assuming $\bar{C}$ and P_I are constant over this small distance; the expressions for the x components are

$$\bar{u}_{x2} = \bar{C}_x - (\bar{C}_x - \bar{u}_{x1})\exp(-P_I \Delta \bar{t})$$

$$\bar{x}_2 = \bar{x}_1 + \bar{C}_x \Delta \bar{t} - (\bar{C}_x - \bar{u}_{x1})\frac{1 - \exp(-P_I \Delta \bar{t})}{P_I}$$

Drop trajectories calculated for different inertia parameters are shown in Fig. 4.1.12 for unpurged and purged Pitot tubes. The two inertia parameters correspond to the following conditions in low-pressure steam:

$$P_I = 5.71 \qquad h = 2.0 \text{ mm}$$
$$d = 0.1 \text{ μm}$$
$$C_0 = 340 \text{ m/s}$$
$$p_0 = 0.069 \text{ bar}$$

$$P_I = 0.186 \qquad h = 0.5 \text{ mm}$$
$$d = 0.6 \text{ μm}$$
$$C_0 = 340 \text{ m/s}$$
$$p_0 = 0.069 \text{ bar}$$

From the drop trajectories the contribution of the liquid phase to the recorded total pressure can be calculated, using the control volume shown in Fig. 4.1.13. This contribution, termed the overpressure Δp_t, is obtained from the reduction in momentum flux of the liquid phase

$$\Delta p_t = \frac{4Y}{1 - Y}\left(\frac{1}{2}\rho_g C_0^2\right)\int_0^{1/2}(1 - \bar{u}_{xe})\,d\bar{y}_{-\infty}$$

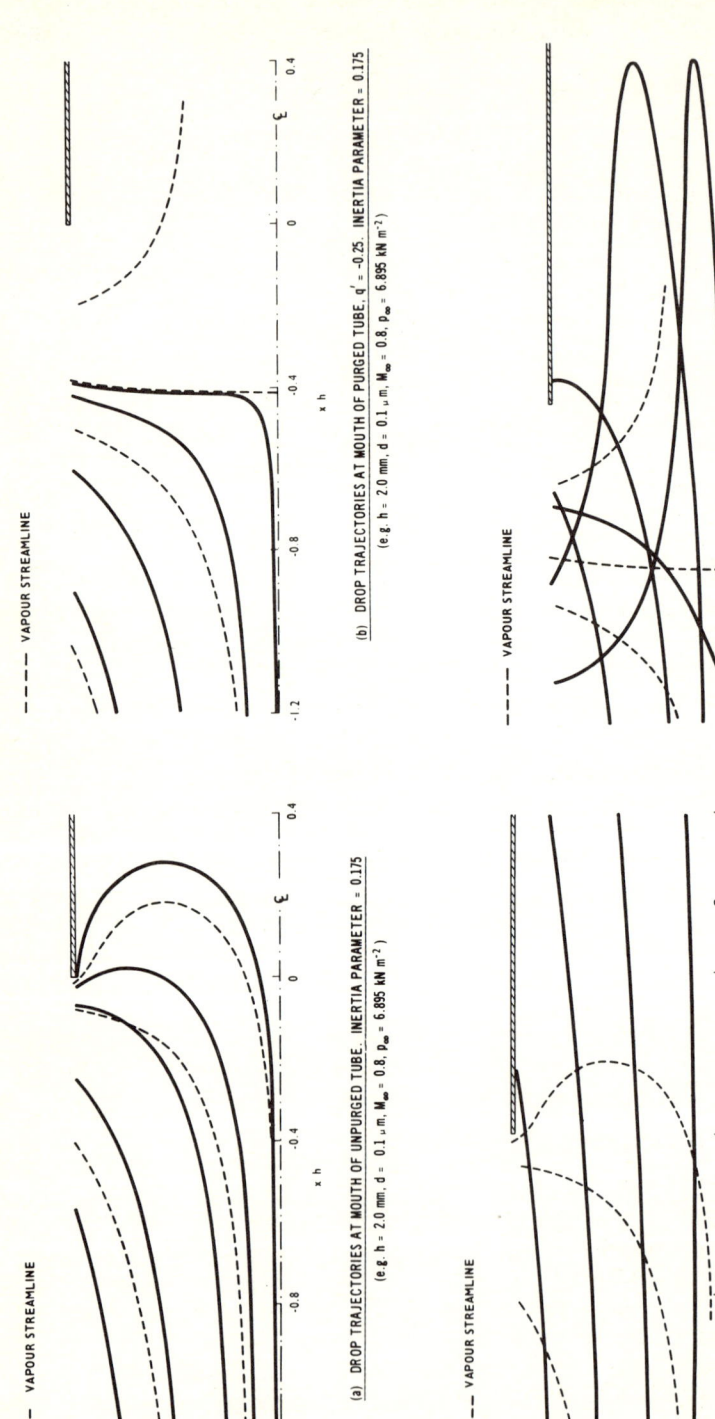

FIG. 4.1.12 Calculated trajectories of droplets at mouth of a Pitot tube.

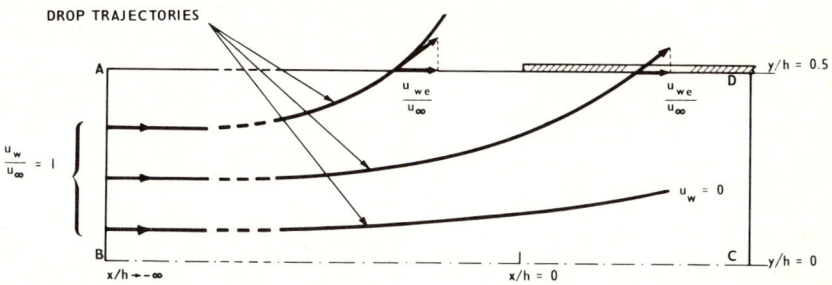

FIG. 4.1.13 Control volume for calculation of overpressure.

where $\bar{u}_{xe}$ is the x component of velocity of droplets leaving the control volume that entered on a streamline at $(-\infty, y)$. Rewriting this equation as

$$\frac{\Delta p_t}{\frac{1}{2}\rho_g C_0^2} = \frac{4Y}{1-Y} J$$

it was found that parameter J was substantially independent of P_I or purge velocity C_h. A constant value of 0.30 was therefore adopted. The velocity of the two-phase flow is therefore calculated by subtracting the overpressure from the recorded total pressure p_t to obtain the stagnation pressure of the gaseous phase. Then

$$C_0 = \left\{ \frac{2k}{k-1} \frac{p}{\rho_g} \left[\left(\frac{p_{tg}}{p} \right)^{(k-1)/k} - 1 \right] \right\}^{1/2} \quad (4.1.3)$$

where

$$p_{tg} = p_t - \frac{0.6 Y \rho_g C_0^2}{1-Y}$$

and isentropic index k for the vapor phase is taken as 1.3.

The above equations are solved by using an iterative process, a first approximation to C_0 being the value calculated for $p_{tg} = p_t$. An example of calculated velocities for specified total and static pressures is shown in Fig. 4.1.14 for a range of wetness fraction Y. Also shown is the velocity calculated by assuming the wet steam stagnates in equilibrium with isentropic index of 1.12, the so-called mixture method.

4.1.3.2 Laser-Doppler Anemometry

The problem of Pitot-pressure interpretation provides an incentive for the development of an absolute method of velocity measurement for wet-steam

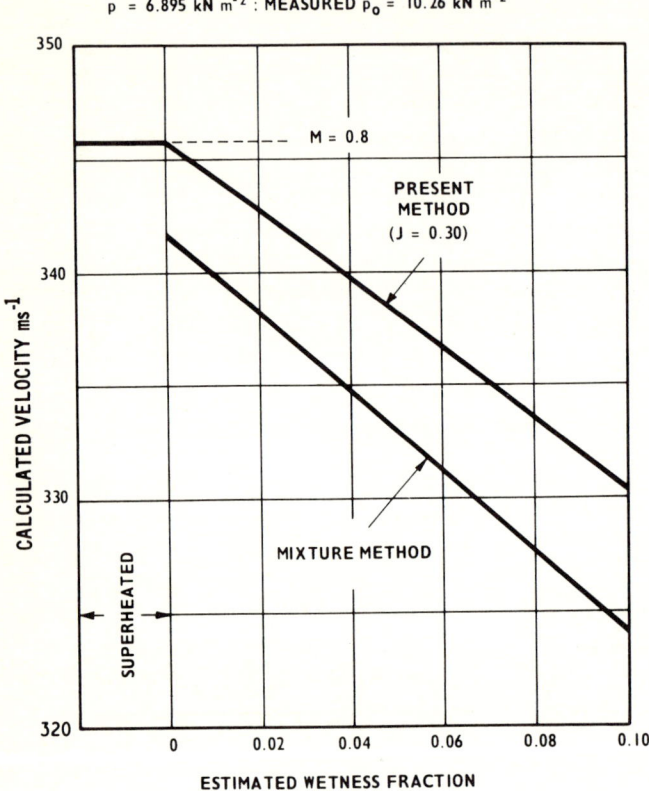

FIG. 4.1.14 Effect of estimated wetness fraction on velocity calculated from given static and total pressures.

flows. Recent advances in laser-Doppler techniques suggest that these optical methods may be a possible alternative. A basic problem in the use of such methods for measuring the velocity of a gas is to obtain an adequate scattered intensity from particles in the flow. The suspended particles (seeds) must be sufficiently small for slip to be negligible and, from Chap. 2, a condensation fog of 1 μm dia. drops may therefore be an ideal medium for laser anemometry.

A considerable amount of research has been done on the development and use of laser anemometers in recent years, a useful background to the subject being given in [4-20, 4-21]. Reliable, relatively simple systems are now commercially available, manufactured, for example, by DISA, BB.Co.–Goerz. A brief description of the basic systems follows, including the practical aspects of application to velocity measurement of wet steam.

The Doppler principle is well known, and the various anemometer arrangements are based on the simple linear relationship

$$f = \frac{\Delta\omega}{2\pi} = 2\,\frac{C}{\lambda}\,\sin\phi \qquad (4.1.4)$$

$\Delta\omega$ is the angular frequency shift of radiation scattered from a particle with velocity C. The incident radiation is of wavelength λ, angle ϕ being the half angle between incident and reflected rays.

The two commonly used optical systems are shown in Fig. 4.1.15, their characteristics being compared in detail in [4-22, 4-23]. The reference beam (or coherent) method, Fig. 4.1.15a, detects the Doppler frequency f by feeding mixed incident (reference) and reflected radiation to a photomultiplier. The corresponding output current i is proportional to the square of the electric field E, thus

$$i = \text{const}\,(E_{10}\cos\omega_1 t + E_{20}\cos\omega_2 t)^2$$

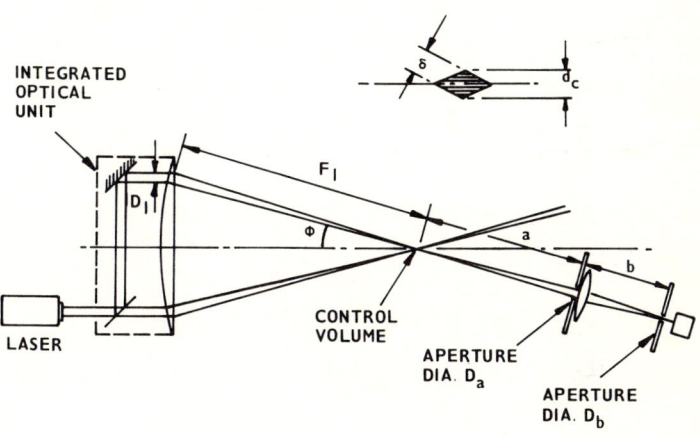

(a) REFERENCE BEAM SYSTEM

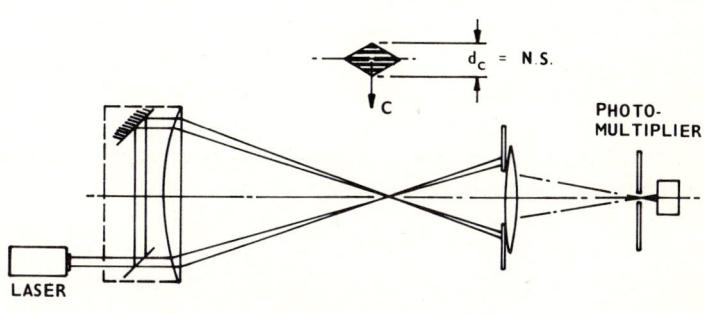

(b) FRINGE SYSTEM

FIG. 4.1.15 Optical systems for laser anemometers.

The photomultiplier response is limited to frequencies of typically $\not> 10^8$ Hz, and hence the device will only record time-mean values of field strength at typical radiation frequencies ω of 10^{14} Hz (visible light). Output will therefore be

$$i \cong \text{const} \frac{E_{10}^2 + E_{20}^2}{2} + E_{10}E_{20} \cos(\Delta\omega t) \qquad (4.1.5)$$

The ac output of the photomultiplier therefore oscillates at the Doppler (beat) frequency $\Delta\omega$, and various electronic processing methods are available for recording this signal.

In practice laser light is not quite coherent, and the distribution of intensity in the beam can be complex if the laser operates with multiple modes. The system should therefore be designed to operate on the simplest transverse mode (designated TEM_{00}) to give a Gaussian beam intensity distribution and coherence length l_{co} of several centimeters, where

$$l_{co} = n(\lambda + \Delta\lambda) = \left(n + \frac{1}{2}\right)\lambda \cong \frac{\lambda^2}{2\Delta\lambda}$$

for large numbers of waves n. To retain coherence the light path lengths for reference and reflected beams should be nearly equal and the control volume diameter d_c (Fig. 4.1.15) should be as small as possible. Focusing of the laser beam is therefore important; the minimum waist δ given by [4-24] being

$$\delta \cong \frac{F_1 \lambda}{D_1} \cong d_c \cos\phi$$

where F_1 is the focal length of the lens focusing a laser beam of diameter D_1. Both δ and D_1 are defined as the width of beam where the intensity is $1/e$ of the peak value of the Gaussian distribution. The aperture diameter D_a of the collecting lens, to maintain coherence, is shown in [4-22] to be

$$D_a < \frac{a\lambda}{d_c} \qquad (4.1.6)$$

where a is the distance from control volume to lens. Finally, a second aperture of diameter D_b is required in front of the photomultiplier to minimize the collection of stray light, where

$$D_b < \left(\frac{b}{a}\right)d_c \qquad (4.1.7)$$

and b is the distance from lens to photomultiplier. Thus the selection of lenses, half angle ϕ, and apertures is a compromise; if the control volume contains too few particles or the particle traverse time is short, the signal will contain intermittent blank or drop-out periods of random duration and a poor Doppler signal will be obtained. It is anticipated that the reference beam method is superior when the flow medium contains a high number-density of particles.

The second method, the fringe (or noncoherent) method, is shown in Fig. 4.1.15b and is applicable where the flow contains relatively few particles. Symmetrical crossed beams are used to produce interference fringes in the control volume of spacing s where

$$s = \frac{\lambda}{2 \sin \phi}$$

Particles crossing the fringes will produce a scattered intensity of frequency f, identical to the Doppler frequency

$$f = \frac{2C \sin \phi}{\lambda}$$

However there is now no coherence requirement and a large-aperture light collection system can be used. For optimum signal production, the fringe spacing should be approximately 4 times the droplet diameter, i.e.

$$\sin \phi \cong \frac{\lambda}{8d} \qquad (4.1.8)$$

and the control volume should contain an adequate number of fringes, e.g., 10-200.

The output signal from the photomultiplier can be processed electronically in several ways for display as droplet velocity. For instantaneous velocity a frequency tracking system can be used, available equipment being capable of handling Doppler frequencies of up to 40 MHz. Frequencies of up to 100 MHz can be detected on typical spectrum analyzers. For higher frequencies the photomultiplier detector can be replaced by a Fabry-Perot interferometer.

In terms of flow velocity, for typical values of half angle ϕ and laser wavelength λ, we may summarize:

Photomultiplier and frequency tracker	→ 120 m/s
Photomultiplier and spectrum analyzer	→ 300 m/s
Fabry-Perot interferometer	300 m/s →

An approximate expression for laser power requirement is given in [4-24]. Introducing typical values of photomultiplier efficiency and system light-collection efficiency, we obtain

$$P > (1.5)\,10^{-13}\,\frac{D_1^2 C}{I_\theta d^2 d_c} \tag{4.1.9}$$

where P is minimum laser power in watts. The scattering coefficient I_θ for small waterdrops is a strong function of particle size parameter α and scattering angle θ, where

$$\alpha = \frac{\pi d}{\lambda}$$

and θ is the angle between incident and scattered beams. Variation of I_θ is presented in Sec. 4.1.6.

In [4-25] a successful application of the fringe method for a wet-steam flow is described, the relevant details of the optical system (Fig. 4.1.15a) being

$\phi = 5°$ $\quad\quad\lambda = 514.5$ nm (argon laser)
$F_1 = 0.3$ m $\quad\quad d_c = 0.23$ mm
$D_a = 19$ mm $\quad\quad D_1 = 1.06$ mm
$D_b = 0.2$ mm $\quad\quad s = 3.0\ \mu$m

Velocity measurements up to 200 m/s were obtained to an estimated accuracy of ± 1% for steam of up to 0.05 wetness fraction. Fog-drop diameter was in the range 0.5–1.0 μm and, for a monodisperse distribution, up to 300 particles may have been in the control volume at any instant. A forward-scattering system was employed, a typical spectrum analyzer output being shown in Fig. 4.1.16. A satisfactory signal was obtained with a laser output power of 60 mW.

4.1.4 Measurement of Unsteady Flow

Recent experiments with hot-wire anemometers in wet steam have shown that this instrument may provide much information on the unsteady flow phenomena in turbines. Its main function is as a research tool, and as yet the interpretation of output signal is not fully understood. However the experiments of Wood [4-26] have already produced important data and these experiments are described below.

The use of hot-wire anemometers has been confined mainly to the measurement of low-velocity fields, the wires being found to have a short operating life in high-speed flows. Similarly, in flows containing water drops, a fine wire will have a limited life under intermittent bombardment. Tungsten wires of 5 μm dia. have been found to give an adequate life in LP turbine conditions, e.g., in wet-steam flows of Mach numbers up to ~ 1.3 and at 50

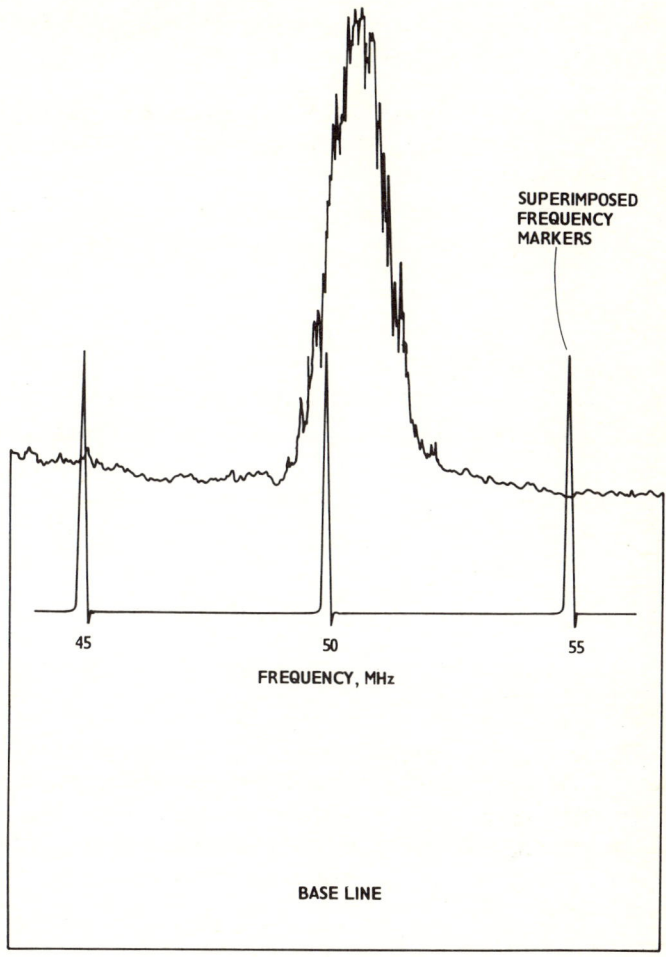

FIG. 4.1.16 Spectrum of typical Doppler signal (fringe mode) recorded on *x-y* plotter.

mbar dynamic pressure. Operating life is still relatively short and therefore the following procedures are adopted:

1. Batches of wires of similar properties are obtained so that, if a wire fails following calibration, measurements can be continued using a similar wire and the calibration not repeated.
2. The wire is shielded until data are required, the shield is withdrawn for a period of typically 10 s, and the output is recorded on magnetic tape for processing after the experiment.
3. The steady component of the recorded signal can be used as a calibration point, which minimizes the wire exposure time.

4 For increased stability and extended wire life, the wires are operated at a slightly reduced maximum overheat ratio (e.g., 1.6 instead of 1.8).

In the experiment of [4-26], standard DISA gold-plated tungsten wires were used in a DISA 55D01 constant-temperature anemometer system. Signals were recorded on a Sangamo 3500 tape recorder in the fm mode, which has a flat frequency response to 20 kHz. Measurements were made in the final stage of a 500 MW turbine and in the test section of the CERL steam tunnel. Correlation of results has been fairly successful on the basis described below.

4.1.4.1 *Steady Flow Sensitivity*

The steady response of a hot wire is determined by the balance of the heat input supplied by the anemometer and the convection of heat away from the wire by the surrounding fluid. Sensitivity can be expressed in terms of the variation of Nusselt number Nu for the wire with flow Reynolds number Re where

$$\mathrm{Nu} = \frac{i^2 R}{\pi \lambda_g l (T_w - T_e)}$$

$$\mathrm{Re} = \frac{CD\rho}{\mu}$$

Here wire current i and resistance R correspond to an effective mean temperature T_w for the wire. Fluid thermal conductivity λ and viscosity μ are evaluated at the total temperature T_t of the flow. Nusselt number is defined in terms of the equilibrium temperature T_e attained by the unheated wire in the flow. The recovery ratio $\eta(=T_e/T_t)$ has been found to be approximately 0.97 for wet-steam flows, with apparently no systematic dependence on Mach, Reynolds, or Knudsen numbers.

Generally correlations of Nu(Re) for a range of overheat ratios (i.e., range of T_w) do not collapse onto a single curve but show a "temperature loading." This loading was found to be enhanced in wet steam compared with air and is thought to be due to the evaporation of water deposited on the wire. The heat balance for the wire is therefore considered to consist of two components of heat output, corresponding to a heat flow rate q to evaporate water on the wire

$$q_1 = \rho C A h_{fg} Y [D + E(T_w - T_e)] \qquad (4.1.10)$$

and the usual convective heat transfer rate q_2 to the vapor flowing over the wire

$$q_2 = \pi \lambda_g l (T_w - T_e) \mathrm{Nu} \qquad (4.1.11)$$

Considering first the evaporative component, for the wet-steam dispersions anticipated in practice, it can be shown that collection efficiency for a 5 μm wire will be 100%. The mass flow rate m of water arriving on the wire is therefore

$$m = A\rho C Y$$

area A being the projected frontal area of the wire in the direction of the steam flow. Constant D in Eq. (4.1.10) therefore allows for possible splashing of drops on impact with the wire. The expression also includes a temperature-loading term, constant E, depending upon the rate of evaporation and the possibility of water being ejected from the wire by the boiling process or by flow forces. The constants D and E can be obtained from a wet-steam calibration for a range of overheat values. Results obtained so far suggest that their values depend to some extent on flow conditions.

The expression for q_2, the heat transfer rate to the vapor, can include the correlation Nu(Re) obtained from dry vapor tests, which is of the form

$$\text{Nu} = a + b\,\text{Re}^{1/2}$$

The final form of the heat balance is therefore

$$i^2 R = \rho C A h_{fg} Y [D + E(T_w - T_e)] + \pi \lambda_g l (T_w - T_e)(a + b\,\text{Re}^{1/2}) \quad (4.1.12)$$

and it can be seen that by testing over a range of overheat (or wire temperature T_w) we have the possibility of determining flow velocity C and wetness Y. Research is continuing on the significance of constants D and E, which may also supply information on fog-droplet size.

4.1.4.2 Fluctuation Sensitivity

The hot wire in wet steam responds to fluctuations in wetness in addition to fluctuations in mass flow and stagnation temperature. In terms of measured mean-square voltage, the fluctuation equation for the anemometer may be written

$$\frac{\overline{dV^2}}{V^2} = S_T^2 \frac{\overline{dT_t^2}}{T_t^2} + S_m^2 \frac{\overline{d(\rho C)^2}}{(\rho C)^2} + S_Y \frac{\overline{dY^2}}{Y^2}$$

$$+ 2R_{mT} S_T S_m \frac{\sqrt{\overline{dT_t^2}}}{T_t^2} \frac{\sqrt{\overline{d(\rho C)^2}}}{\rho C} + 2R_{mY} S_m S_Y \frac{\sqrt{\overline{d(\rho C)^2}}}{\rho C} \frac{\sqrt{\overline{dY^2}}}{Y}$$

$$+ 2R_{TY} S_T S_Y \frac{\sqrt{\overline{dT_t^2}}}{T_t^2} \frac{\sqrt{\overline{dY^2}}}{Y} \quad (4.1.13)$$

where S_T, S_m, S_Y are respectively the sensitivities to fluctuations in total temperature T_t, mass flow ρC, and wetness fraction Y. Correlation coefficients R are included for each combination of fluctuating variables.

The fluctuation sensitivities are obtained by logarithmic differentiation of the calibration equation to give

$$S_T = \frac{di/i}{dT_t/t_t}$$

$$= \frac{A_2(T_w - T_e)}{2i^2 R} \frac{nT_t}{k_t} - \frac{A_2 T_e}{2i^2 R} - \frac{A_1 EYT_e}{2i^2 R}$$

$$- \frac{A_2 m(T_w - T_e)}{4i^2 R} \frac{b\,\text{Re}^{1/2}}{a + b\,\text{Re}^{1/2}} \quad (4.1.14)$$

$$S_m = \frac{A_1 Y}{2i^2 R}[D + E(T_w - T_e)] + \frac{A_2(T_w - T_e)}{4i^2 R} \frac{b\,\text{Re}^{1/2}}{a + b\,\text{Re}^{1/2}}$$

$$S_Y = \frac{A_1 Y}{2i^2 R}[D + E(T_w - T_e)]$$

Note that the variation in conductivity and viscosity are included as

$$\lambda = \lambda_t + n(T - T_t)$$

$$\mu = \mu_t \frac{T}{T_t}$$

and it is assumed that the response of the anemometer control is such that we may consider wire resistance is constant, so that

$$\frac{di}{i} = \frac{dV}{V}$$

Fluctuations in flow direction about the normal to the wire cannot of course be deduced with a single wire and will appear in the mass flow component.

In certain cases it is possible to make simplifying assumptions in order to separate the several components of the fluctuation equation. For example in [4-27, 4-28], fluctuation modes in supersonic flows and boundary layers were deduced either by assuming modes of similar magnitude or by neglecting the modes of smallest magnitude. Similarly it is found that relatively little error is incurred in this analysis by assuming fluctuation of wetness Y is negligible. The main equation may then be written in the two forms

$$\frac{e'}{S_T} = \left[T'^2 + \left(\frac{S_m}{S_T}\right)^2 m'^2 + 2R_{mT}\left(\frac{S_m}{S_T}\right) T'm' \right]^{1/2}$$

or (4.1.15)

$$\frac{e'}{S_m} = \left[\left(\frac{S_T}{S_m}\right)^2 T'^2 + m'^2 + 2R_{mT}\left(\frac{S_T}{S_m}\right) T'm' \right]^{1/2}$$

where e' is the voltage fluctuation, and the primed variables represent the fluctuation amplitudes of the particular variable (e.g., $T'^2 = \overline{dT_t}^2/T_t^2$).

The results may be plotted as a fluctuation diagram, after [4-27, 4-28], as shown in Fig. 4.1.17, and hyperbolas fitted to the experimental points. The characteristics of the hyperbolas giving the best fit define T'_t, m', and R_{mT}. Some typical measurements of fluctuations obtained in a 500 MW turbine are shown in Fig. 4.1.18. From the results it is also possible to deduce the fluctuations in steam velocity u' and static temperature T'_s, and these are also shown.

4.1.5 Measurement of Wetness Fraction

Several methods of wetness fraction measurement have been devised of varying accuracy, complexity, and applicability. The selection of an appropriate method depends on the accessibility of the wet-steam flow and particularly on the form of the liquid phase, (i.e., coarse water or fog). The available methods are therefore grouped here according to their application and also on the basis of whether they are absolute methods or require calibration. Finally, the relative accuracies of some absolute methods are compared and some details given of a particular method being developed at CERL for measurement of wetness in turbines.

4.1.5.1 Coarse Water Measurement (Absolute Methods)

Mechanical separation Coarse water is by definition in the form of water films or rivulets on surfaces, or as large water drops (> 10 μm dia.) within the steam flow. Mechanical separation is therefore feasible and various designs of separating elements will be described in Chap. 7. Where it is necessary to extract a continuous sample of wet steam in, for example, laboratory tests or tests on bled-steam lines, simple separators can be successfully used provided that care is taken to avoid reentrainment of water in the steam flow.

Turbine interstage measurement For wet-steam research a useful device has been developed [4-29] for measuring the coarse water between turbine stages. A long cylindrical probe is inserted into the steam space through guide tubes as shown in Fig. 4.1.7. A sampling aperture at the inner end of the probe has area A and is fitted with a remotely operated shutter. The probe is rotated until the aperture faces the flow direction (as determined previously by

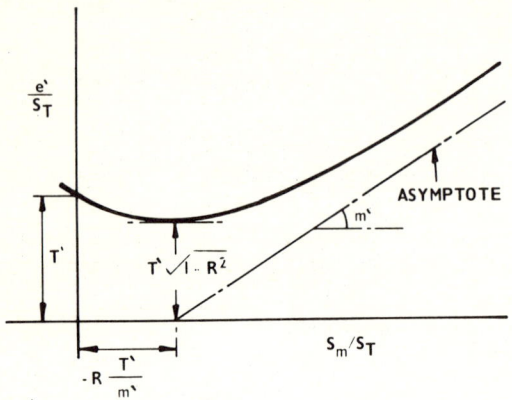

(a) CHARACTERISTICS OF FLUCTUATION HYPERBOLA

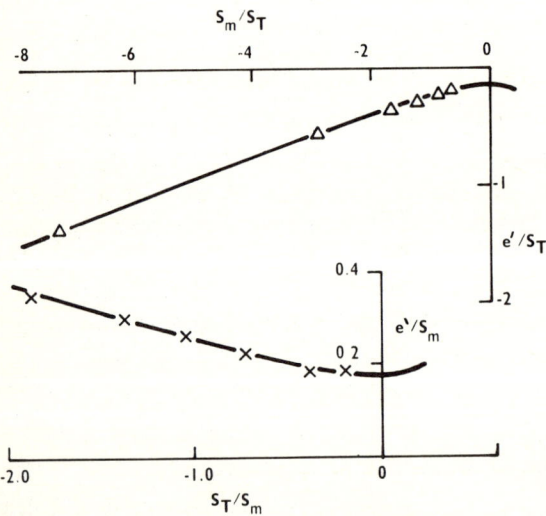

(b) TYPICAL DOUBLE FLUCTUATION DIAGRAM

FIG. 4.1.17 Unsteady-flow analysis.

aerodynamic probes), and the shutter is opened for a few seconds to expose a closed chamber containing an absorbent pad. As the chamber is closed, the main steam flow will pass around the probe, small fog droplets deviating very little from steam streamlines. Collection efficiency for coarse water droplets of > 10 μm will however be at least 95% as shown by calculations for cylinders in LP steam flows in Fig. 4.1.19 [4-30]. The mass of water δm absorbed by the pad during the exposure time δt is found by weighing, and the local coarse water wetness fraction is calculated from

Instrumentation for Wet Steam

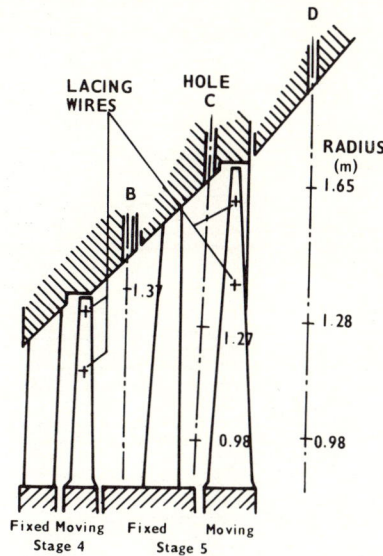

FINAL TWO STAGES OF THE LOW-PRESSURE TURBINE SHOWING PROBE POSITIONS

HOLE	RADIUS (m)	m'	T'	R_mT	u'	T_S'
B	1.39	0.17	0.045	0	0.17	0.045
C	1.28	0.065	0.072	0	0.088	0.068
C	1.00	0.040	0.035	0	0.048	0.033
D	1.77	0.18	0.080	0	0.17	0.079
D	1.40	0.071	0.046	0	0.081	0.045
D	1.10	0.080	0.029	0	0.084	0.029

FIG. 4.1.18 Measured flow fluctuations in a 500 MW turbine.

$$Y'' = \frac{\delta m (1 - Y)}{\delta t A \cos \phi C \rho_{sat}} \quad (4.1.16)$$

where ϕ is the steam pitch angle and C, ρ_{sat}, Y are estimated or measured steam velocity, saturated density, and (total) wetness fraction respectively.

It is assumed here that the coarse water is moving in the direction of the steam flow, although not necessarily at the same velocity; this has been found to be the case in LP turbines [4-31]. A typical measured distribution of coarse water between the penultimate and final stages of a 500 MW turbine is shown in Fig. 4.1.20.

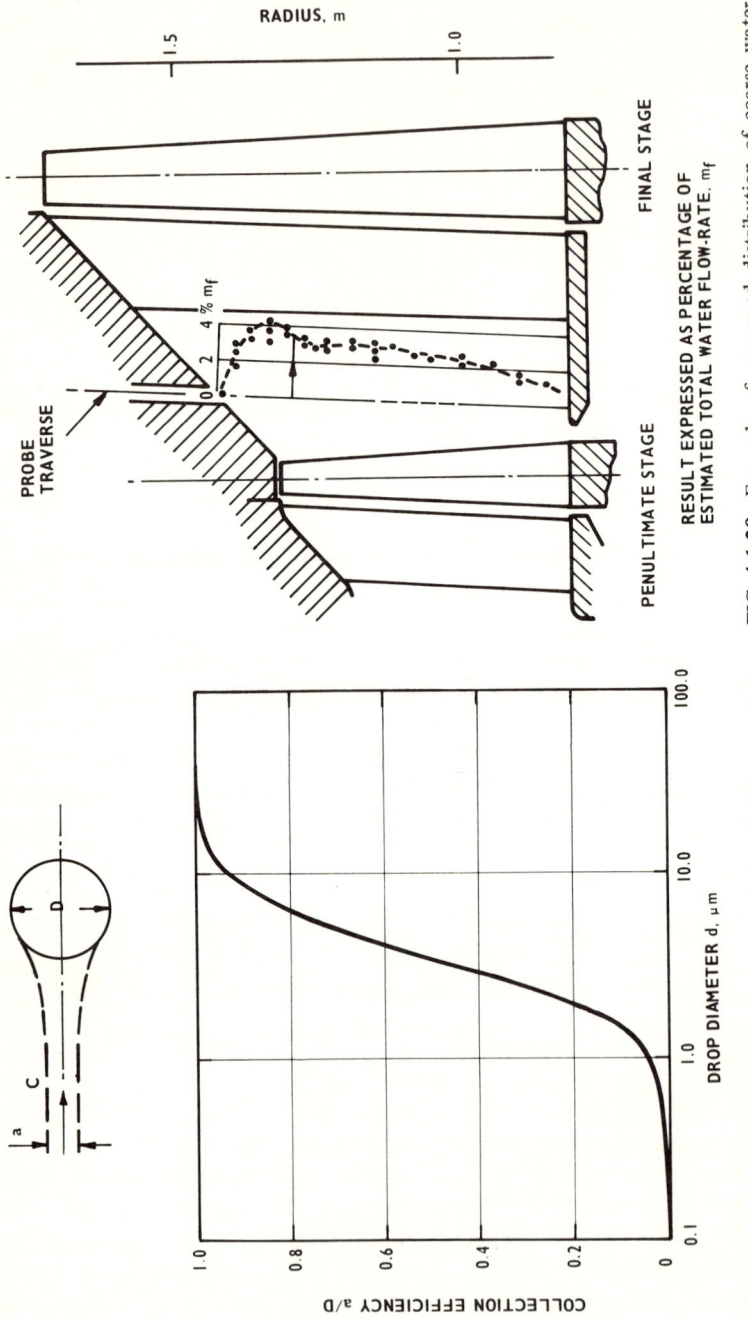

FIG. 4.1.19 Inertial deposition on a cylinder in LP wet steam ($C = 200$ m/s, $D = 25$ mm).

FIG. 4.1.20 Example of measured distribution of coarse water in a 500 MW turbine.

Tracer methods In addition to mechanical separation, tracer methods have also been used to measure coarse water flows. The basis of tracer methods is the addition to the wet steam, at a known mass flow rate m_t, of a solution of an easily detectable trace element. The concentration c_1 of the trace in the initial solution is known. Downstream of the injection point, where it is judged that the tracer solution has mixed thoroughly with the coarse water in the steam, a sample is extracted and concentration c_2 measured. The mass flow rate of the coarse water m_c then follows from

$$m_c = m_t\left(\frac{c_1}{c_2} - 1\right) \qquad (4.1.17)$$

It can be seen that adequate mixing is essential; in practice concentration c_2 may be very small. Both factors may reduce the accuracy of the method. Selection of the tracer is determined by the situation; typically sodium or lithium salts are used for chemical detection, sodium isotopes for radioactive tracing.

4.1.5.2 Fog Wetness Fraction (Methods Requiring Calibration)

Many devices will register an effect due to the presence of water in a steam flow that can possibly be exploited for wetness fraction measurement. However the process may depend upon several features of the flow such as droplet size, size spectrum and velocity, and elaborate calibration may be necessary. The production of wet-steam flows of particular compositions is extremely difficult, making calibration a major problem. However over a limited range of application a calibration may be possible, the method then having the advantages of simplicity, reliability, and ease of operation. Possible methods are here reviewed briefly.

Absorption methods Such methods are based on the reduction in intensity I of a radiation by absorption in the liquid phase during transmission through a fog. The typical absorption equation is

$$\frac{I}{I_0} = \exp(-Ax) = f(Y) \qquad (4.1.18)$$

where coefficient A must be related to wetness Y by calibration. In [4-32, 4-33] methods using β rays and infrared radiation are described. However, all methods must be considered to be at the laboratory stage.

Dielectric methods The dielectric constants for steam and water differ ($\epsilon_f/\epsilon_g \cong 80$) and therefore offer a means of determining the relative quantities of each component in a wet-steam mixture. The basic method involves the measurement of the dielectric constant of the wet-steam mixture flowing between parallel plates of area A, spacing x, for an applied voltage V at frequency ω. The governing equation is

$$\epsilon = \text{const}\ \frac{x}{V\omega e_0 A} = f(Y)$$

However as reported in [4-32] correction may be required for materials depositing on the plates.

Indirect methods involving the dielectric property of the mixture are also reported in [4-32], including the detection of the change in wavelength of a radiowave in passing through wet steam. In a similar manner the resonant frequencies of microwaves in cavities filled with samples of wet steam can also be shown to be a function of wetness fraction. As can be seen from the relative magnitudes of the dielectric constants, variation in the volume of water present will have the dominant effect, and all such methods will therefore be insensitive at wetness fraction of $0 \to 0.15$.

4.1.5.3 Fog Wetness Fraction (Absolute Methods)

Throttling method This method is well known and references may be found in many older texts on steam. The principle of the method is to transfer a sample by constant enthalpy expansion from a wet to a superheated state. The enthalpy H_2 can be determined from pressure and temperature measurements p_{t2} and T_{t2} of the superheated vapor. The wetness Y of the original sample can then be obtained from

$$Y_1 = \frac{H_{g1} - H_2 - q/m}{h_{fg1}} \qquad (4.1.19)$$

where H_{g1} is the total enthalpy of saturated steam at the sample pressure p_1, and q is the heat lost from the sample of mass m.

The problems of obtaining a representative sample are common to most of the thermodynamic methods and will be considered later. The particular problems of throttling are errors arising from heat loss and incomplete mixing following expansion. Adequate insulation and sufficient sample size can reduce heat-loss errors to acceptable proportions. The attainment of thermodynamic equilibrium is more difficult; as shown in Sec. 4.2, heat transfer processes between phases can require considerable time. The throttling orifices and expansion chambers must therefore be arranged with care to promote adequate mixing.

A practical limit to the application of the system occurs at low pressures as shown in Fig. 4.1.21. The pressure p_{t2} of the expanded sample is usually determined by the temperature of available cooling water for an auxiliary condenser, a typical minimum value being ~ 20 mbar. To promote heat transfer a minimum final superheat of at least $20°C$ is also advisable. From Fig. 4.1.21a the corresponding lowest pressure p_{t1} of a 10% wet sample will be ~ 10 bar.

Heating method Here a wet-steam sample is heated (usually electrically) until superheated; the heat required to dry the sample is detected by a rise in

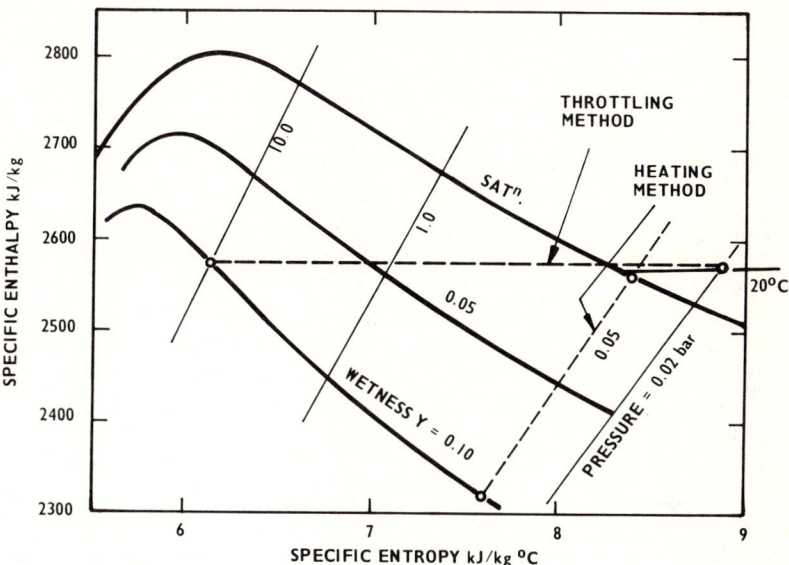

FIG. 4.1.21 Mollier chart for steam showing methods of wetness-fraction measurement.

steam temperature downstream of the heater. The equipment for a continuously sampling system is shown diagrammatically in Fig. 4.1.22a; detailed design and operation of this type of instrument will be considered later. From the energy equation for the flow the wetness fraction Y of the original sample may be written

$$Y_1 = \frac{(H_2 - H_1)_{\text{sat}}}{h_{fg1}} - \frac{Q_1}{mh_{fg1}} - \frac{q}{mh_{fg1}} \quad (4.1.20)$$

where $(Q_1 - q)$ is the heat input rate to make the sample mass flow rate m become dry saturated.

Mass flow rate m may be measured directly or may be deduced from the heat balance of a second heater placed in series such that

$$m = \frac{Q_2}{H_3 - H_2}$$

Condensing method If, instead of evaporating the water phase, the complete sample is condensed, the relative proportion of vapor phase may be deduced from the heat extracted by the condenser cooling water. The apparatus is shown diagrammatically in Fig. 4.1.22b and the sample wetness is obtained from

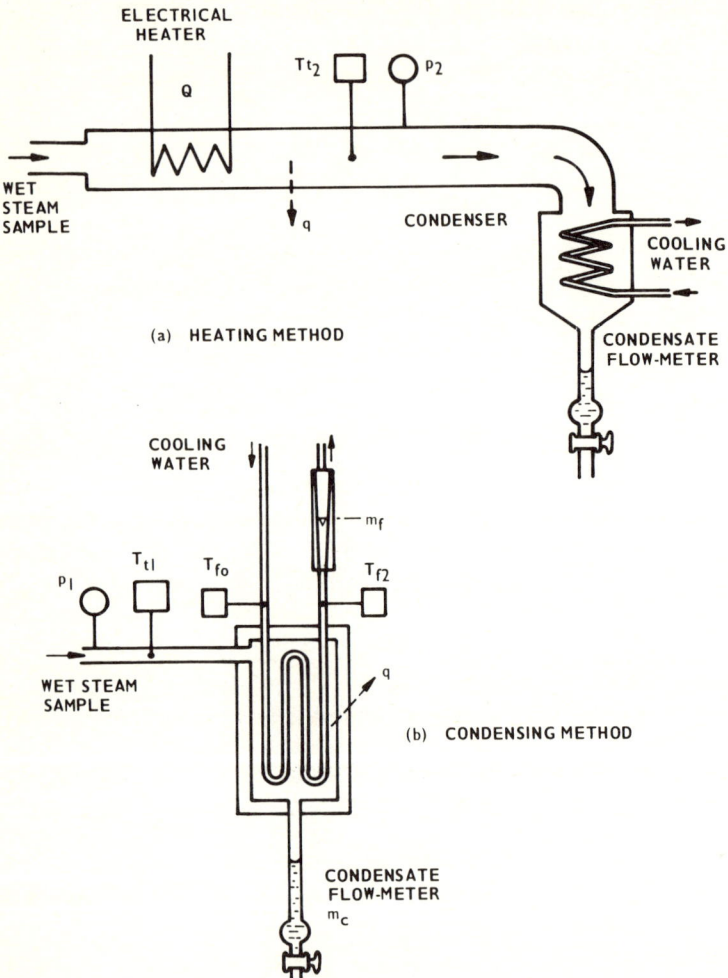

FIG. 4.1.22 Methods of wetness-fraction measurement.

$$Y_1 = \frac{H_{g1} - (m_f/m_c)c_{pf}(T_{f2} - T_{f0}) - h_{f2} - q/m_c}{h_{fg1}} \quad (4.1.21)$$

Accurate measurements are therefore required of condensate flow rate m_c and cooling-water flow m_f. Again heat loss q to the surroundings must be minimized, and as shown the sample is usually condensed within a vacuum flask.

Psychometric method In this method the wet-steam sample m is mixed with a flow of hot, dry air. Proportions are selected to ensure that the final mixture is not saturated. Measurements of temperature and pressure are made as shown in Fig. 4.1.23 [4-34], and the steam/air ratio x of the air before and after mixing is found psychometrically. By equating the heat lost by the air to the heat gained by the wet steam sample, an expression for the original steam wetness fraction Y can be obtained

$$Y_1 = \left(\frac{1}{h_{fg}}\right)\left[(T_S - T_2)c_{pg} + (T_1 - T_2)\left(\frac{c_{pa} + x_1 c_{pg}}{x_2 - x_1}\right) + \frac{q}{m}\right] \quad (4.1.22)$$

Again the method is susceptible to heat loss q but vacuum insulation can be incorporated. It can be seen that for negligible heat loss, mass flow measurements are not required, but the condition of $< 100\%$ humidity at exit necessitates a relatively large supply of hot air. Similarly for low-pressure wet steam a large extraction pump would be required.

4.1.5.4 Selection of an Absolute Method (Fog Wetness)

The thermodynamic methods described in the previous paragraphs involve the measurement of several variables, each measurement inevitably being subject to

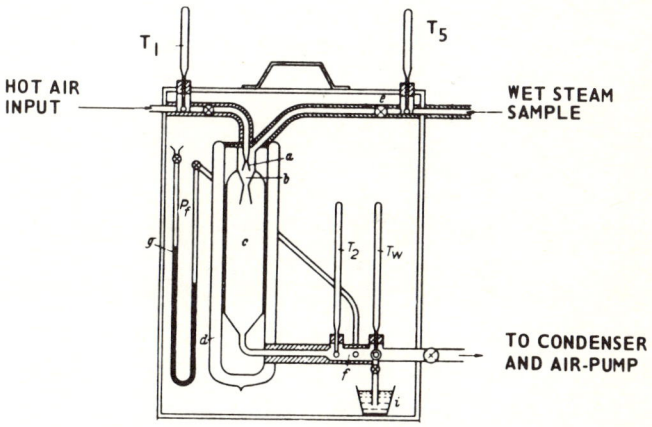

a	AIR NOZZLE	f	PSYCHROMETER SECTION
b	MIXING ZONE	g	MANOMETER
c	MAIN CALORIMETER	i	WATER SUPPLY FOR WET-BULB THERMOMETER
d	INSULATING VACUUM JACKET		
e	CONTROL VALVE		

FIG. 4.1.23 Calorimeter for psychometric method of wetness fraction measurement, from [4-35].

unknown systematic error. The probability of error will be difficult to assess, but we will define an accuracy of a general variable $V(\pm \delta V)$ as being the 99% confidence limit. Then error e_v is

$$e_v = \frac{\delta V}{V}$$

and where the variable is a function of several independent variables

$$V = V(a, b, c \ldots)$$

the error may be written

$$e_v = \left[\left(\frac{a}{V}\frac{\partial V}{\partial a}e_a\right)^2 + \left(\frac{b}{V}\frac{\partial V}{\partial b}e_b\right)^2 \ldots\right]^{1/2}$$

If all measurement errors e_a, e_b are within the same confidence limit, then overall error e_v is also within this limit.

Applying this analysis to the governing Eqs. (4.1.19)–(4.1.22) the four absolute methods, the error $e_{(\ )}$ in measured wetness fraction may be expressed as

(i) Throttling method

$$e_{(i)} = \frac{1}{Yh_{fg}}\left\{\left[p_{t1}\left(\frac{\partial H_{g1}}{\partial p_{t1}}\right)e_{p_{t1}}\right]^2 + \left[p_{t2}\left(\frac{\partial H_2}{\partial p_{t2}}\right)e_{p_{t2}}\right]^2 + \left[T_{t2}\left(\frac{\partial H_2}{\partial T_{t2}}\right)e_{T_{t2}}\right]^2 + \left(\frac{q}{m}e_q\right)^2\right\}^{1/2}$$

(ii) Heating method

$$e_{(ii)} = \frac{1}{Yh_{fg}}\left\{\left[p_{t2}\left(\frac{\partial H_2}{\partial p_{t2}}\right)e_{p_{t2}}\right]^2 + \left[p_{t1}\left(\frac{\partial H_1}{\partial p_{t1}}\right)e_{p_{t1}}\right]^2 + \left(\frac{q}{m}e_q\right)^2 + \left(\frac{Q_1}{m}e_Q\right)^2 + \left(\frac{Q_1}{m}e_m\right)^2\right\}^{1/2}$$

(iii) Condensing method

$$e_{(iii)} = \frac{1}{Yh_{fg}}\left\{\left[p_{t1}\left(\frac{\partial H_{g1}}{\partial p_{t1}}\right)e_{p_t}\right]^2 + \left[\left(\frac{m_f}{m_c}\right)c_p\Delta T_f e_{\Delta T}\right]^2 + \left[\left(\frac{m_f}{m_c}\right)c_p\Delta T_f e_{m_f}\right]^2 \right.$$
$$\left. + \left[\left(\frac{m_f}{m_c}\right)c_p\Delta T_f e_{mc}\right]^2 + \left[p_t\left(\frac{\partial h_{f1}}{\partial p_{t1}}\right)e_{p_{t1}}\right]^2 + \left(\frac{q}{m_e}e_q\right)^2\right\}^{1/2}$$

(iv) Psychometric method

$$e_{(iv)} = \frac{1}{Yh_{fg}}\left\{(c_{pg}\Delta T_s e_{\Delta Ts})^2 + (Z\Delta T_a e_{\Delta Ta})^2 + \left(\frac{Z\Delta T_a x_2}{x_2 - x_1}e_{x2}\right)^2 \right.$$
$$\left. + \left[\frac{Z\Delta T_a x_1(c_{pa} + x_2 c_{pg})}{c_{pa} + x_1 c_{pg}}e_{x1}\right]^2 + \left(\frac{q}{m}e_q\right)^2\right\}^{1/2}$$

where $\Delta T_s = T_s - T_2$
$\Delta T_a = T_1 - T_2$
$Z = (c_{pa} + x_1 c_{pg})/(x_2 - x_1)$

Errors in mass ratios x can be expressed in terms of the measured wet-bulb temperature T_w and pressure p as follows:

$$e_x = \frac{1}{p_f}\left\{\left[\frac{pT_w}{p - p_f}\left(\frac{\partial p_f}{\partial T}\right)e_{Tw}\right]^2 - \left(\frac{p_f p}{p - p_f}e_p\right)^2\right\}^{1/2}$$

where p_f is the partial (saturation) pressure of water vapor at temperature T_w.

The methods may now be compared by assigning the following error values to the various measured parameters:

Pressure	$e_p = 0.01$	typically to ±0.2 mbar
Temperature	$e_T = 0.005$	typically to ±0.5°C
Mass flow	$e_m = 0.005$	volume measurement of water
Electrical heat	$e_Q = 0.01$	typically to ±1.0 W
Heat loss	$e_q = 1.00$	for good insulation to ±2.0 W

Results are shown in Table 4.1.1 for the specific case of a wet-steam flow of fraction $Y_1 = 0.10$. With the exception of the throttling method, pressure was assumed to be 0.050 bar.

For application to LP turbine conditions, the heating method therefore appears appropriate; a description is now given of such an instrument being developed at CERL.

TABLE 4.1.1 Comparison of accuracy for different methods of wetness measurements

Method	Accuracy of wetness measurement (e_Y)	Main sources of inaccuracy
Throttling	0.010	Heat loss
Heating	0.016	Heat input, loss; sample mass flow
Condensing	0.087	Cooling-water temperatures, mass flow; sample mass flow
Psychometric	0.023	Outlet pressure measurement

4.1.5.5 Details of a Probe Using the Heating Method

A wetness probe of this type is described in [4-35], the general arrangement being reproduced in Fig. 4.1.24. Isokinetic sampling is achieved by orientation of the probe in the flow direction and adjustment of suction until the sample is induced at the flow velocity. For this purpose a long cylindrical intake nozzle is constructed with internal and external static pressure tappings. Downstream of the intake the sample flows over an electrical heating coil, through a series of baffles, and finally over a coiled resistance thermometer before being extracted and condensed for mass-flow measurement.

As can be seen, the probe is large and can be used only in a turbine exhaust hood. A new arrangement has therefore been developed at CERL [4-36] as shown in Fig. 4.1.25. The system is now incorporated into a 25 mm dia. probe that can be inserted between turbine stages in the same manner as the aerodynamic probes described earlier. The arrangement also enables the basic operation of the probe to be verified, which is essential for a system that does not rely on calibration.

Thermodynamics of probe From the theory of wet-steam flow (Chap. 2) we may expect the thermal relaxation time for water collecting on the probe's internal surfaces to be relatively long. The establishment of thermodynamic equilibrium downstream of the heater will therefore be the main design problem for the probe. This can be shown by assuming, for simplicity, (*a*) all the heat from the electrical heater is transferred to the vapor phase, and (*b*) the liquid phase is deposited on the probe walls of the intake bend and is evaporated downstream of the heater by heat transferred from the superheated vapor.

At outlet to the heater, as shown diagrammatically in Fig. 4.1.25, there will therefore be a steam flow rate m_g at temperature T_2, and a water film flow rate m_f at temperature T_1, where

$$m_g = (1 - Y)m$$
$$m_f = Ym$$
$$T_2 = T_1 + \frac{Q}{(1 - Y)mc_{pg}}$$

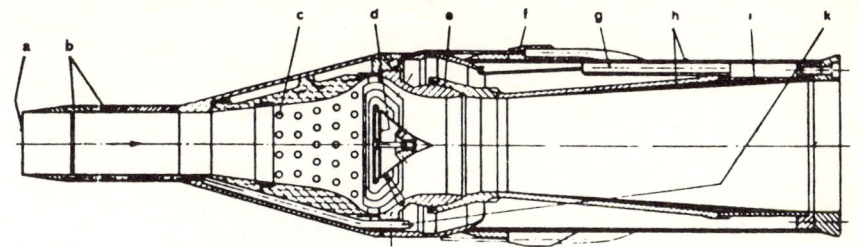

a Tip of measuring head.
b Static pressure taps.
c Heating.
d Stagnation cone.
e Ball and socket joint.
f Cable line.
g Sliding resistance for pitch angle measurement.
h Double casing with
i Araldite insulation.
k Measuring leads.

Longitudinal section through the measuring head

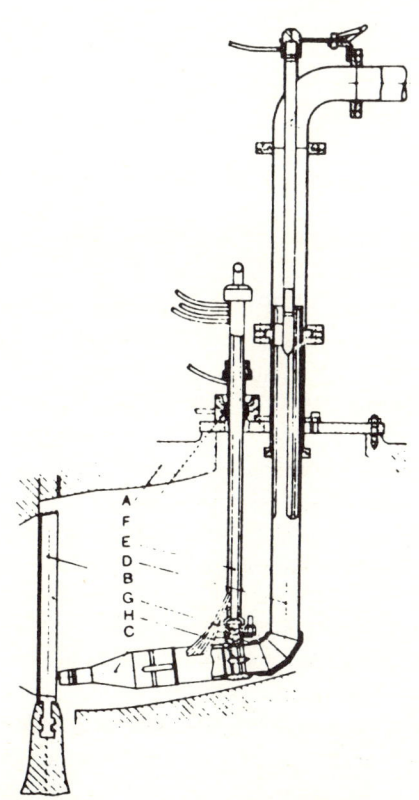

A Exhaust steam branch.
B Runner blade.
C Measuring head.
D Telescopic tube.
E Actuator tube.
F, G Ball and socket joint.
H Spider-web type thermometer.

Cross-section through the measuring unit

FIG. 4.1.24 Early version of a sampling probe for wetness measurement, from [4-36].

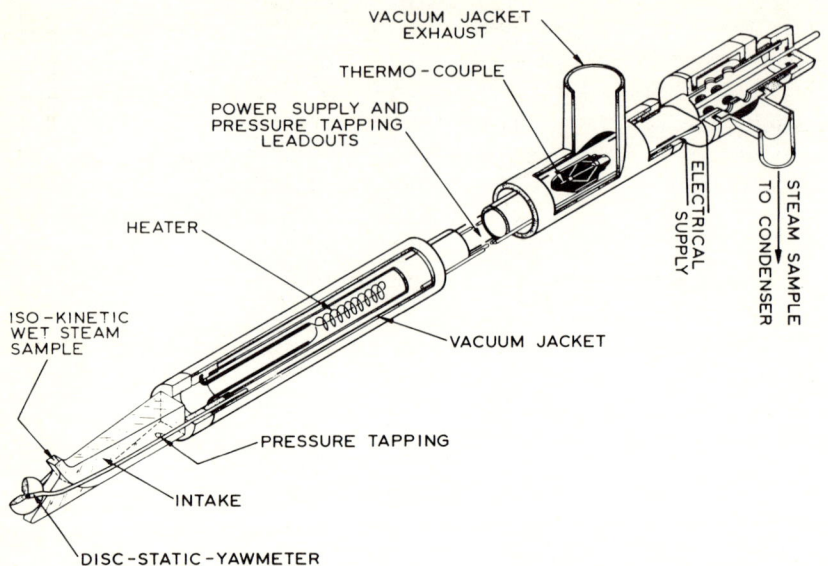

FIG. 4.1.25 CERL wetness probe.

where m is the total sample flow and Q is the heat input from the electrical heater.

For a small element some distance s downstream of the heater we may also write a heat balance

$$m_g c_{pg} d(T - T_1) = \alpha(T - T_1)\pi D ds = h_{fg}\, dm_f$$

on the assumption that there is negligible pressure drop along the tube and the water remains at the saturation temperature T_1. The heat transfer coefficient α is given by the usual correlation for flow inside a tube on the assumption that the water flows as a uniform film

$$\alpha = 0.023 \left(\frac{\mu}{DC\rho_g}\right)^{0.2} \left(\frac{\lambda_g}{c_p \mu}\right)^{0.67} (c_p \rho_g C)$$

For small initial wetness Y_1 (i.e., <0.10) we can assume for present purposes that changes in m_g, α, and ρ_g along the tube are negligible, and integrate the heat balance equation to give

$$T = T_1 + (T_2 - T_1)e^{-as}$$

where $a = \pi D\alpha/m_g c_p$. Substituting this expression into the equation for water flow rate gives

$$\frac{dm_f}{ds} = \frac{\pi D\alpha}{h_{fg}}(T_2 - T_1)e^{-as}$$

Integrating this equation we can define a critical mixing length s_c at which the water mass flow rate becomes zero

$$s_c = \frac{1}{a}\ln\left(\frac{Q/Q_0}{Q/Q_0 - 1}\right)$$

where $Q_0 = m\,Y_1\,h_{fg}$, the heat required to evaporate the liquid phase. It can be seen that if $Q = Q_0$, then the critical length $s_c \to \infty$. For a practical solution some residual superheat at the dry-out position is inevitable, the temperature elevation being

$$T_c - T_1 = (T_2 - T_1)e^{-as_c} = \left(\frac{Y_1}{1 - Y_1}\right)\frac{h_{fg}}{c_{pg}}\left(\frac{Q}{Q_0} - 1\right)$$

This simple representation demonstrates clearly the characteristics of a probe based on the heating method. Some typical quantities have been calculated for the particular example

$$p = 0.050 \text{ bar}$$
$$C = 50 \text{ m/s}$$
$$D = 15 \text{ mm}$$
$$Y_1 = 0.10$$

and are shown in Fig. 4.1.26. Critical lengths are on the order of $0.5 \to 2.0$ m for practical designs. For short lengths the steam temperature T_2 increases rapidly, increasing the risk of significant heat loss. For low heat loss and low residual superheat, mixing lengths of as much as 2 m are necessary. The process of detecting the dry-out heat requirement is shown in Fig. 4.1.26b. The classical concept of constant steam temperature until dry out is reached is not attained unless an extremely long mixing length is specified. For $s = 0.5$, for example, residual superheat is significant, and a change in slope of the T-Q line does not occur until $Q/Q_0 = 2$. The required dry-out value Q_0 is obtained by extrapolating the superheated section of the characteristic back to the Q axis.

Isokinetic sampling Isokinetic sampling is obviously preferable, but as can be seen from Fig. 4.1.24, complicated arrangements may be required to obtain universal orientation of the probe intake and minimum flow disturbance ahead of the probe. In addition the internal pressure of the probe must be carefully controlled to induce the sample at the free stream velocity. In practice sampling may fall short of this ideal, particularly where, for use in operating

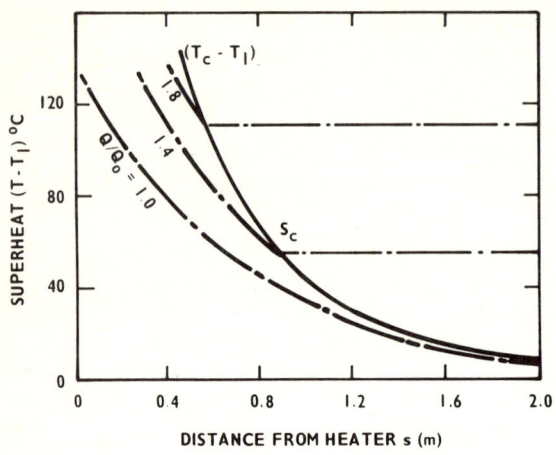

(a) CRITICAL LENGTH OF MIXING ZONE

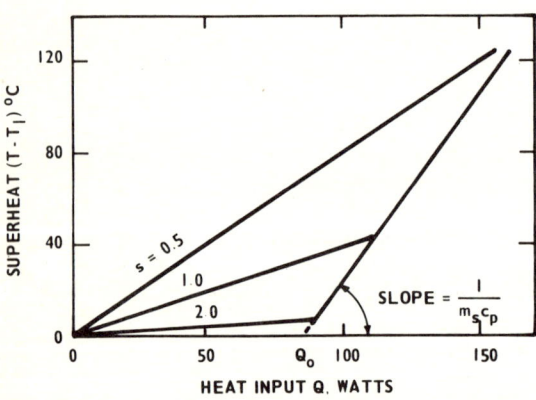

(b) PROBE OPERATING CHARACTERISTICS

FIG. 4.1.26 Wetness probe thermodynamics.

turbines, a simple robust intake design is incorporated (e.g., Fig. 4.1.25). To determine the likely error in measured wetness fraction resulting from nonisokinetic sampling, a two-dimensional potential flow calculation has been made of the flow in the inlet region for a range of sampling rates.

In the potential flow field, the outline of an arbitrary probe intake was represented by a series of j line sources of unspecified strength $\sigma_{1 \to j}$. The intake was made to form a flow boundary by specifying zero normal velocity at the midpoints of each line source. Sampling rate was specified by the velocity C_h normal to an internal surface of the probe head, as shown in Fig. 4.1.27a. The velocity at any point in the flow field is calculated from a linear equation incorporating source strengths $\sigma_{1 \to j}$ with coefficients determined by intake geometry. Hence the specification of velocities normal to the flow boundary enables a set of j simultaneous linear equations to be formed that can be solved for $\sigma_{1 \to j}$ (the Neumann problem). The velocity at any point in the flow field may now be calculated and droplet trajectories obtained in the manner described earlier for the Pitot-tube investigation (Sec. 4.1.3).

The error e_Y in the proportion of water to vapor in the sample arising from nonisokinetic sampling is, from Fig. 4.1.27b

$$e_Y = \frac{Y - Y_T}{Y_T} = \left(\frac{e}{a} - 1\right)(1 - Y)$$

As may be expected e_Y varies with suction ratio a/b and the inertia parameter P_I of the droplets (based on the free stream velocity C_∞ and intake width b). Results for a particular intake shape are shown in Fig. 4.1.28. For typical LP turbine conditions of

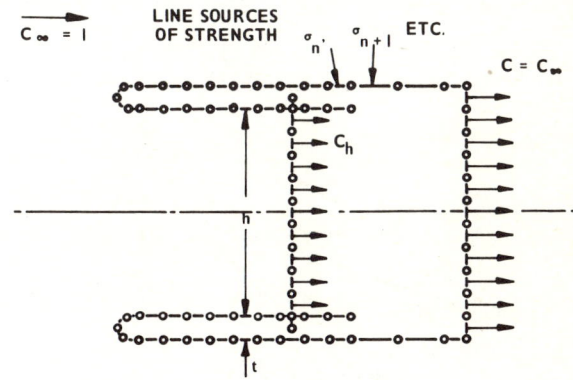

(a) INTAKE POTENTIAL FLOW SCHEME

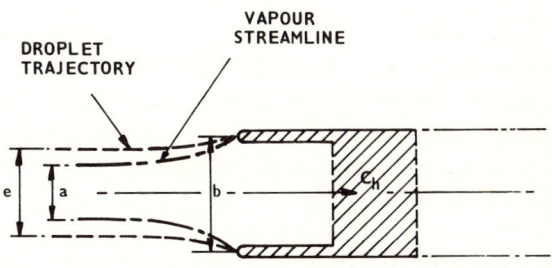

(b) SAMPLING NOMENCLATURE - INSUFFICIENT SUCTION

FIG. 4.1.27 Nonisokinetic sampling theory.

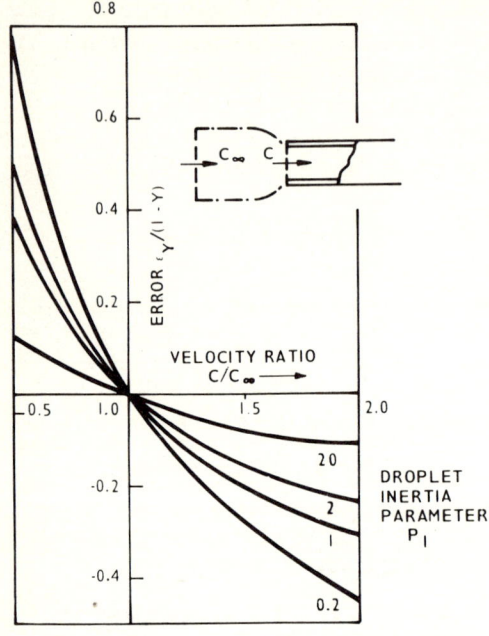

FIG. 4.1.28 Error in wetness fraction from nonisokinetic sampling.

$$P = 0.050 \text{ bar}$$
$$C_\infty = 200 \text{ m/s}$$
$$b = 10 \text{ mm}$$
$$d = 2.0 \text{ } \mu m$$

the results indicate that suction rate can be some 20% nonisokinetic (i.e., $0.8 < a/b < 1.2$) before the error in sample wetness fraction exceeds ± 0.05.

4.1.6 Measurement of Droplet Size

4.1.6.1 Large Droplet Size Measurement

For droplets of coarse water, typically greater than 10 μm dia., several methods are available; their suitability depends mainly on the convenience of the particular application. The following are a few available methods; no doubt several more methods can be devised.

Photographic method Droplets of >10 μm dia. can be resolved on film using a suitable optical magnification system. To avoid misleading reflections, back lighting is preferable; the system shown in Fig. 4.1.29a is known as focused direct shadowgraph. With a telephoto camera lens, the depth of focus of the recording system is small, which eliminates blurred images arising from water on windows, etc. The photograph of droplets in the narrow plane of focus is usually suitable for automatic analysis on modern particle-sizing equipment. This method is probably the simplest, most reliable technique for obtaining droplet-size spectra, but can be time consuming.

Needle probe A simple probe has been developed [4-37] based on the conduction through a water drop that bridges needles spaced a small distance s apart. Careful construction of the probe is required to ensure adequate insulation of the input and outlet electrical connections and also to control the positioning of the needle points to within a few microns. In practice the method is probably limited to droplet diameters $> 50\ \mu m$.

Of course if $s > d_{max}$, the largest droplet in the spectrum, no conduction counts will be recorded. For smaller spacings, the count rate obtained will depend on the probability P of a drop of diameter d spanning the gap s, and on the droplet cumulative number distribution $N(d)$. From Fig. 4.1.29b, probability P is proportional to the crosshatched area, i.e.

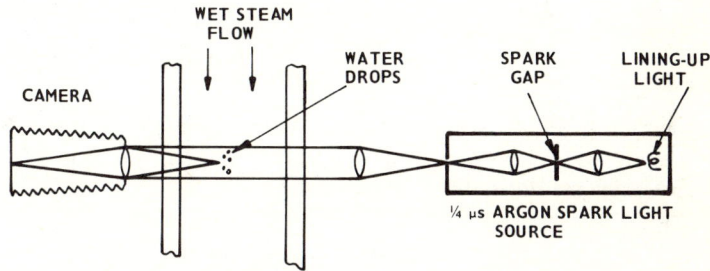

(a) FOCUSSED DIRECT-SHADOGRAPH SYSTEM FOR DROPLET PHOTOGRAPHY

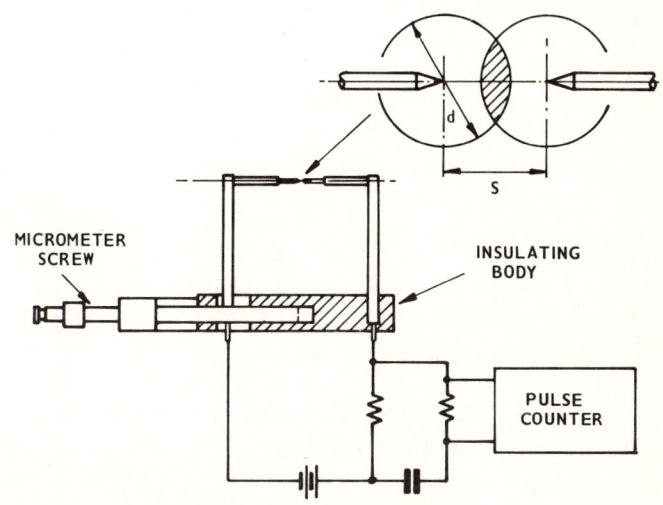

(b) NEEDLE PROBE FOR DROPLET SIZING

FIG. 4.1.29 Devices for large droplet size measurement.

$$P(s,d) = \text{const}\left[d^2 \cos^{-1}\left(\frac{s}{d}\right) - s(d^2 - s^2)^{1/2}\right]$$

The total count rate at spacing s will be

$$C_s = \text{const} \int_s^{d\max} P(s,d) \frac{dN(d)}{dd}\, dd$$

To obtain the droplet distribution $N(d)$ from the measurement of $C(s)$ two methods are available. A series of values of spacing s are used, and the droplet number probability distribution obtained from

$$\left(\frac{d\bar{N}}{dd}\right)_n = \frac{C(s)_n - \sum_{i=n-1}^{1} \frac{d\bar{N}}{dd} i \int_{s_i}^{s_{i-1}} P(s_n,d)\, dd}{\int_{s_n}^{s_{n-1}} P(s_n,d)\, dd}$$

where the integrals are computed from the expression for P and n refers to the interval $d_n \to d_n + \Delta d$. A simpler method involves the assumption of a particular distribution

$$\frac{dN}{dd} = f(d, a_1, a_2, \ldots a_j)$$

where a_1 to a_j are assignable constants. Inserting this expression in the basic integral for count rate, we obtain j simultaneous equations using the count rates C from j values of spacing s. The constants a can then be calculated, giving the number distribution.

Magnesium oxide slide method For droplets at relatively low velocities, size distribution can be obtained from the impressions made in a soft surface of magnesium oxide [4-38]. The soft layer can be formed by exposing a microscope slide to burning magnesium. A suitable mechanism is required to control the exposure time of the slide in the wet-steam flow.

The oxide layer must be thicker than the largest drop in the sample. May [4-38] calibrated the system to relate crater diameter to droplet diameter, and found the crater size independent of droplet velocity. The method was considered accurate down to droplet diameters of 10 μm.

4.1.6.2 Fog Droplet Size Measurement ($d < 3.0$ μm)

Well-established optical techniques are available for measuring the size of droplets when drop diameter is of the same order as the wavelength of light. Several basic texts are available [4-39, 4-40] that describe the phenomenon of light scattering; a brief summary follows.

The secondary radiation from a spherical drop produced by an incident plane electromagnetic wave has an intensity I_θ, which is a function of angular displacement θ from the incident direction. For unit incident intensity

$$I_\theta = \frac{F(\theta)}{k^2 s^2}$$

where s is distance from the droplet center, and wave number $k = 2\pi/\lambda$. The detailed angular distribution function $F(\theta)$ is a complex function of droplet size parameter $\alpha = \pi d/\lambda$ and refractive index m. The detailed distribution is given by the Mie theory, which shows that the scattered field is composed of two groups of partial waves I_1 and I_2 that are polarized perpendicular and parallel to the plane containing incident and scattered rays. The variation in the angular distributions I_1 and I_2 with size parameter α is particularly relevant to size measurement. As shown in Fig. 4.1.30, the scattering pattern progresses from symmetrical to increasingly forward directed as size parameter increases. The Mie theory has been verified by careful experiment and so can form the basis of absolute methods of size measurement.

From Fig. 4.1.30, the scattering pattern becomes symmetrical (Rayleigh scattering) and independent of drop size for diameters of < 0.02–0.03 μm. At the upper end of the range, the Mie solution converges progressively more slowly, and for $\alpha > 30$ solutions are not available. Within the range $0.02 < d < 2.5$ μm, several methods of deducing drop size from scattering pattern have been used.

Angular scattering The basic approach is to record the angular variation of intensity using the system shown in Fig. 4.1.31a. Droplet size is then deduced by comparing the measured angular position of the distribution minima with Mie theory values for the wavelength λ of light used. A version of this approach employs white light illumination when the scattered distribution is a color spectrum (higher-order Tyndall spectrum). Particle sizing is obtained from the angular position of the red lines that coincide with the theoretical minima for the green light distribution.

Dissymmetry A simplification to complete angular scanning is known as the measurement of dissymmetry. Intensities are measured at two angles θ_1 and θ_2 equally disposed from the normal to the incident direction. Dissymmetry is defined as

$$\frac{I_{\theta 1}}{I_{\theta 2}}$$

where $\theta_1 + \theta_2 = \pi$. At $\theta_2 = 45°$, it is found to increase monotonically from unity in the Rayleigh region through an order of magnitude to $\alpha = 5$.

Polarization ratio With a polarizing filter, the ratio of the intensities of the partial waves I_1 and I_2 can be measured and droplet size deduced from a

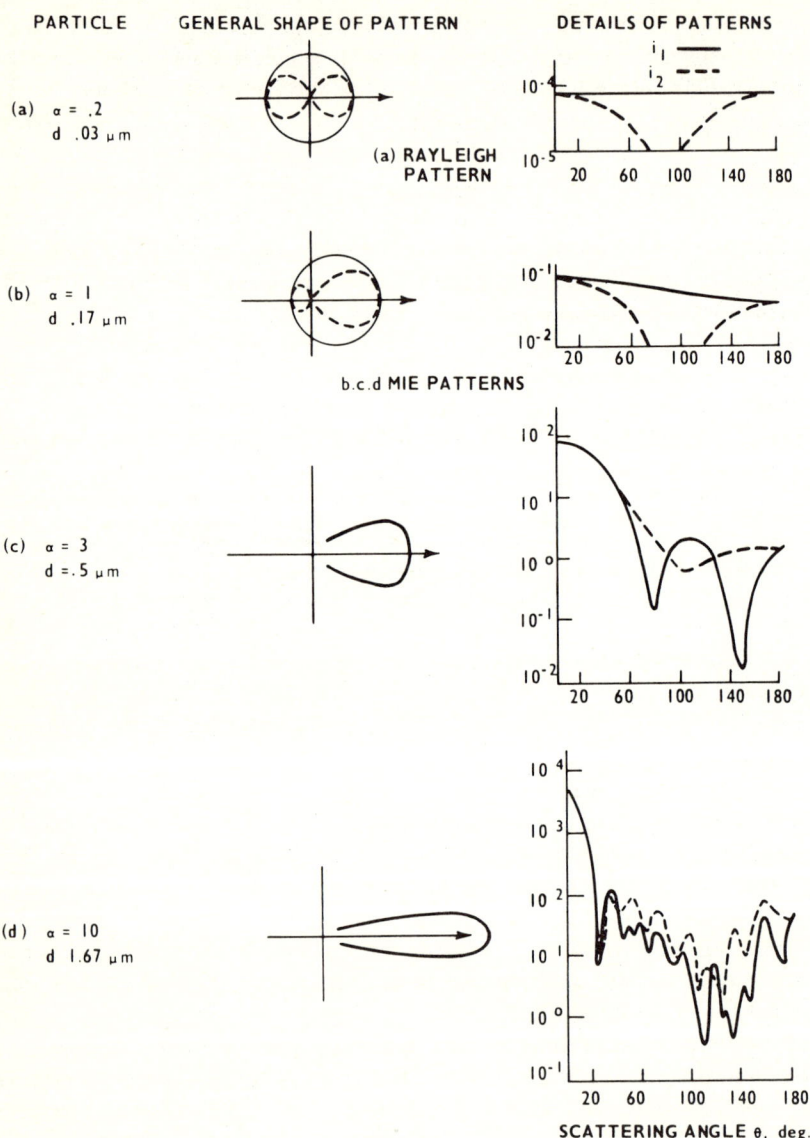

FIG. 4.1.30 Mie theory scattering of monochromatic light by single spheres of refractive index 1.5.

single angle of observation. Polarization ratio at $\theta = 90°$ increases from zero to unity over the droplet size range 0.05–2.0 μm.

Extinction The attenuation of a parallel beam of light due to scattering from a droplet will also be a function of particle size parameter α, and is expressed in terms of extinction coefficient E, where

$$E = \frac{\text{total flux scattered by the droplet}}{\text{flux geometrically incident on the particle}}$$

The extinction coefficient is also calculable from the Mie theory, and tables of $E(\alpha)$ are given in [4-41] for water drops where, over most of spectral range, refractive index $m = 1.33$. The function is shown in Fig. 4.1.32, and a diagram of the required optical system for extinction measurement is shown in Fig. 4.1.21b. For a suspension containing n droplets per unit volume, each of frontal area A, the flux δf removed from the incident direction over a small distance δs is therefore

$$\delta f = -fnAE\delta s$$

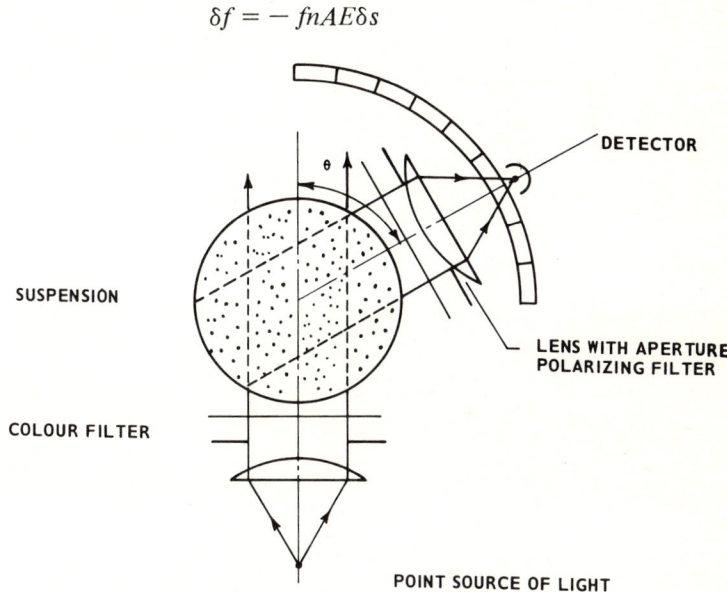

(a) ESSENTIALS OF SCATTERING APPARATUS

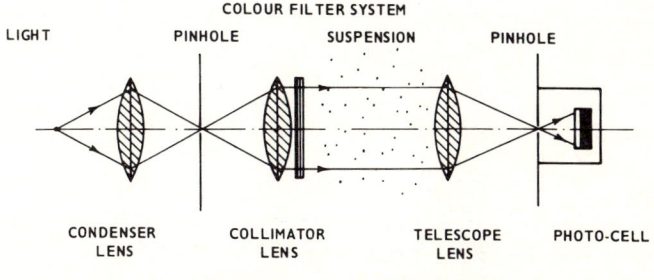

(b) THE ESSENTIALS OF EXTINCTION APPARATUS

FIG. 4.1.31 Methods of deducing drop size.

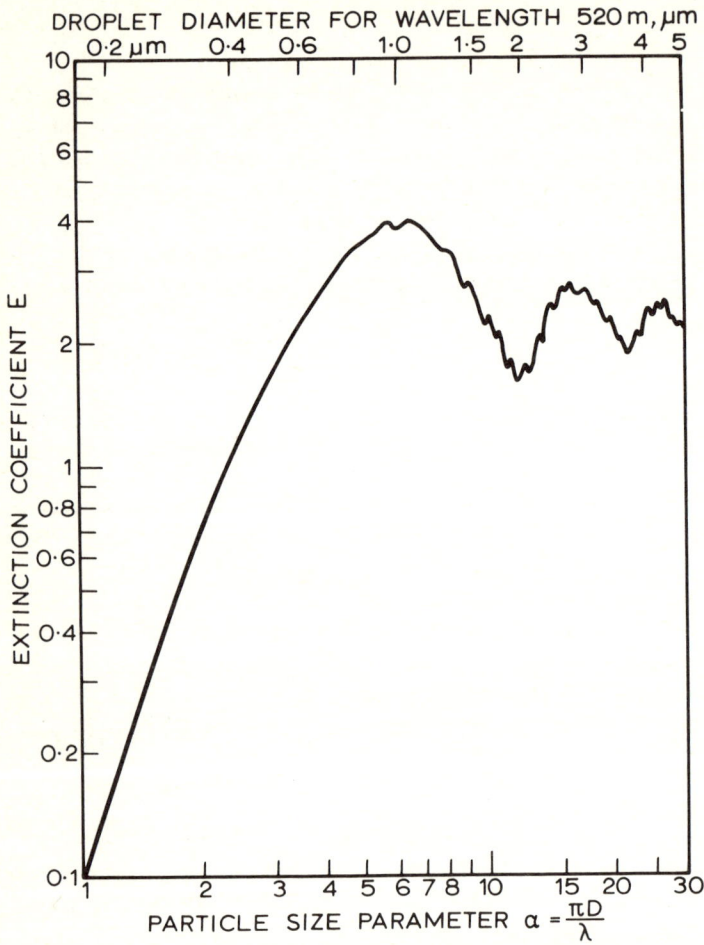

FIG. 4.1.32 The Mie theory extinction curve for water droplets refractive index $m = 1.33$.

where f is the incident flux per unit area. Integrating for a finite thickness suspension we obtain the Bouguer law

$$\frac{f}{f_0} = \exp(-nAEs)$$

Flux ratio is known as the transmittance, and the exponent as the turbidity of the suspension. In principle therefore, for known values of f/f_0, n, s, this equation can be solved for A (hence drop diameter d) from the known $E(\alpha)$ distribution. The detailed procedure is described in the next section.

4.1.6.3 Details of the Extinction Method

The extinction method is usually more suitable than angular scattering for the measurement of fog-drop size. The method is inherently less affected by multiple scattering and therefore requires a less sensitive detector. The geometry of the optical system is also more suitable for application to measurements in turbines. The experimental procedure follows.

Monodispersion, known-wetness fraction The Bouguer law can be expressed in terms of wetness fraction Y, assuming no slip between phases, as

$$\frac{f}{f_0} = \exp\left[-\frac{3s}{2}\frac{E(\lambda d)}{d}\left(\frac{Y}{1-Y}\right)\frac{\rho_g}{\rho_f}\right]$$

Transmittance f/f_0 can be measured using a single monochromatic source of known wavelength λ. In [4-42, 4-43], for example, a continuous laser was used. For known wetness Y and densities ρ, the above expression then provides a value of ratio E/d. From the Mie curve of $E(\alpha)$, for given λ, the value of d corresponding to the calculated value of E/d can be found.

Monodispersion, unknown wetness Further information is now required and, as in [4-44, 4-45], a series of transmittance values is obtained for different light wavelengths λ by using a range of interference filters. Results are plotted as $\ln(f_0/f)$ versus (π/λ) on log-log axes. Writing the Bouguer law in log form

$$\ln \cdot \ln\left(\frac{f_0}{f}\right) = \ln E + \ln \phi$$

and

$$\ln\left(\frac{\pi}{\lambda}\right) = \ln \alpha - \psi$$

where $\phi = \frac{3}{2}(s/d)[Y/(1-Y)]\rho_g/\rho_f$
$\psi = d$

It can be seen that the experimental curve will be of the same form as the basic $E(\alpha)$ curve, but displaced from it by $\phi - \psi$ as shown in Fig. 4.1.33. By determining these displacements, both d and Y can be calculated. Obviously the accuracy of the process improves with increased numbers of experimental points (light wavelengths).

Polydispersion, unknown wetness The transmittance of a polydispersion may be written

$$\frac{f}{f_0} = \exp\left[-\frac{\pi s}{4} \int_0^\alpha E(\lambda, d) N(d) d^2 \, dd\right]$$

where $N(d)$ is the cumulative number distribution of droplets per unit volume. This integral reduces to the equivalent monodisperse form if the following effective mean values are used:

$$\bar{E} = \frac{\int E N d^2 \, dd}{\int N d^2 \, dd}$$

and

$$d_{32} = \frac{\int N d^3 \, dd}{\int N d^2 \, dd}$$

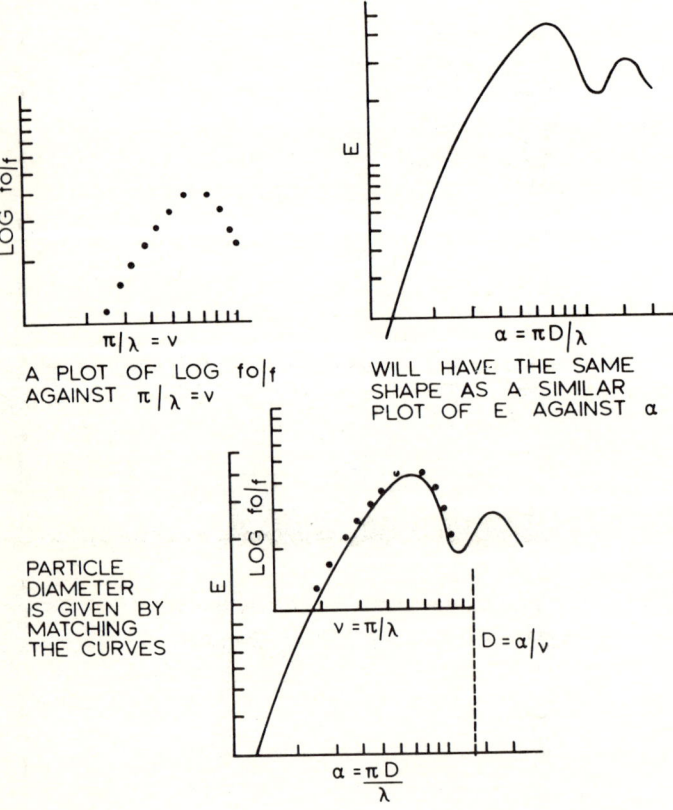

FIG. 4.1.33 The determination of particle size from extinction measurements.

the Sauter mean diameter. The above processes for monodispersions can then be applied to determine diameter d_{32} and wetness Y provided the form of the $\bar{E}(\alpha_{32})$ curve is known.

The particular procedure adopted will depend to some extent upon the measured transmittance variation, and the size range and type of droplet spectrum thought to be present. For example in [4-46] it is shown that, for drops < 0.3 μm dia., the $\bar{E}(\alpha_{32})$ curve is almost independent of the width or skew of the droplet distribution and corresponds closely to the standard curve for a monodispersion over the range $0 < \alpha < 4$, as shown in Fig. 4.1.34. Hence no information on distribution can be extracted from the measurements but,

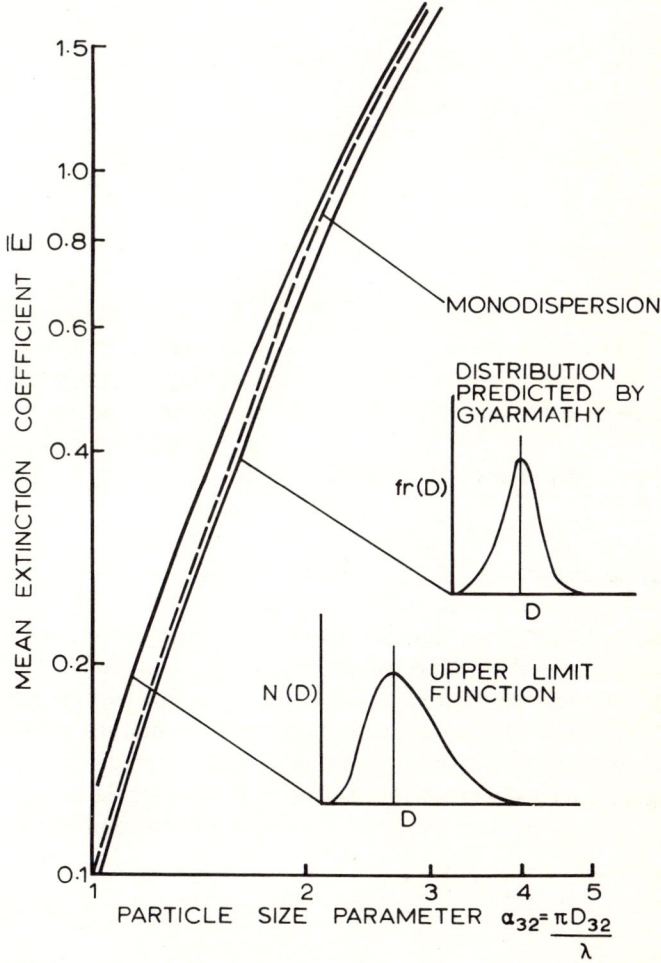

FIG. 4.1.34 The mean extinction coefficient for two size distributions of small particles.

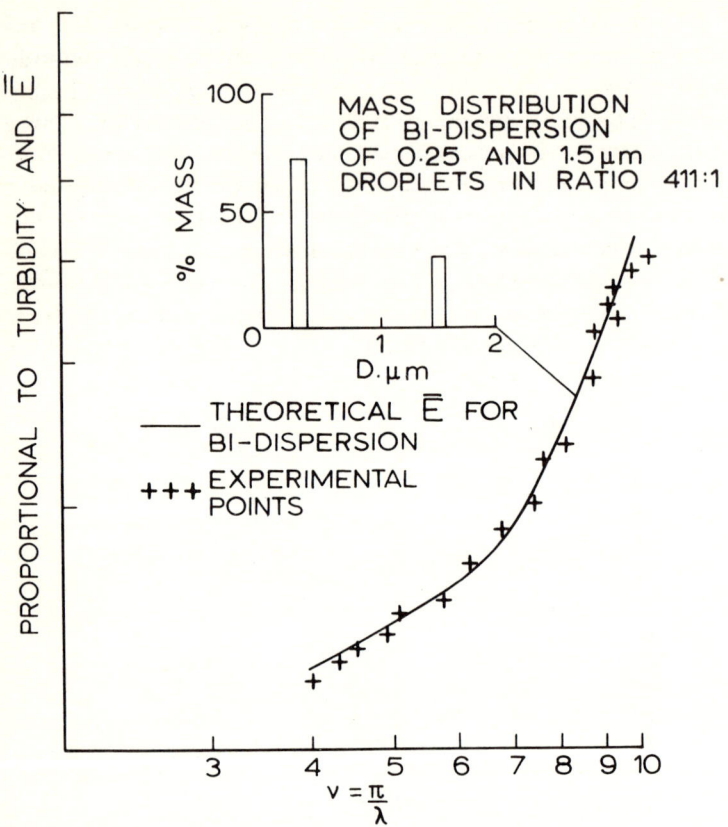

FIG. 4.1.35 Comparison of theoretical turbidity ($\bar{E}$) for bidispersion with experiment.

as shown in [4-44], accurate measurements can be made of mean drop diameter d_{32} and wetness fraction Y.

In some circumstances, for example where secondary nucleation occurs, the drop-size distribution may be distinctly bidispersed. An effective $\bar{E}(\alpha_{32})$ curve can then be synthesized [4-44] using values for two monodispersions (Fig. 4.1.35).

Several methods have been used to interpret results for the more general case of a relatively wide size distribution, but of single modal size. The complete inversion of the turbidity integral equation is attempted by Shifrin and Perelman [4-47], using an approximate expression for the $E(\alpha)$ distribution

$$E = 2\left(1 - \frac{\sin \delta}{\delta} + \frac{1 - \cos 2\delta}{2\delta^2}\right)$$

where $\delta = \alpha(m-1)$. Using the Mellins transform, the droplet number distribution $N(d)$ is calculated from the measured distribution of transmittances $f/f_0(\lambda)$. However the accuracy of the method depends on the provision of a wide range of experimental points which may, for small droplets, necessitate measurements at impractical light wavelengths. Recent work of Walters [4-48] also shows the method to be ill conditioned for size distributions of high negative skew, which often occur in wet steam.

A basically stable approach is to assume that the size distribution conforms to a particular function that is characterized by a few key constants. The constants are varied systematically until an $\bar{E}(\alpha_{32})$ curve is obtained that fits the experimental results. In [4-49] an error function of variable width is used; Walters [4-48] finds the upper limit distribution of Mugele and Evans [4-50] a more realistic function, where both width and skew can be specified. A typical result is shown in Fig. 4.1.36.

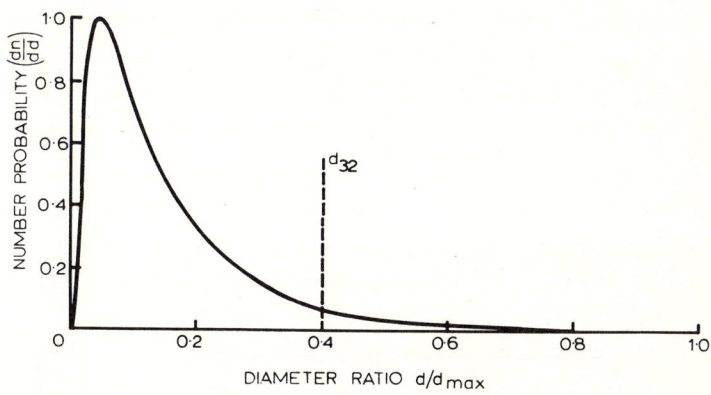

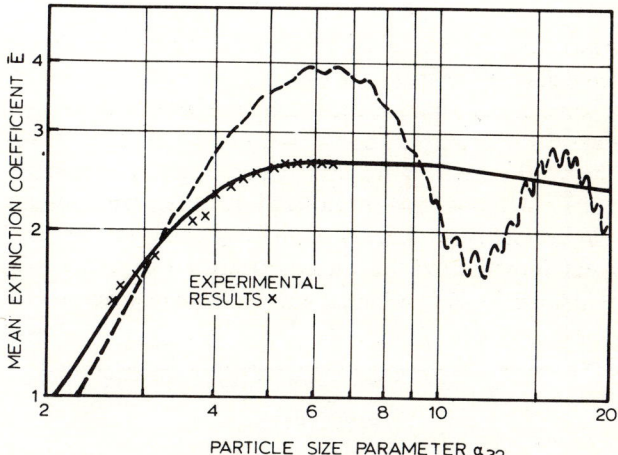

FIG. 4.1.36 The spectral turbidity of a polydispersion.

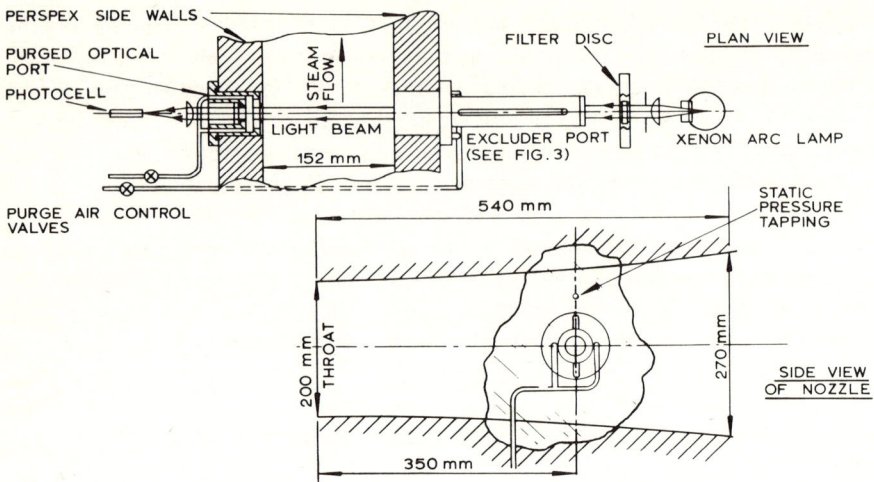

FIG. 4.1.37 Experimental arrangement and position of optical axis in working section nozzle profile.

4.1.6.4 Equipment for Extinction Measurement

The analysis methods of the previous subsection require experimental measurements of high accuracy with the minimum of spurious scatter. The method therefore depends upon the equipment used to obtain the transmittance measurements. The elements of the system follow.

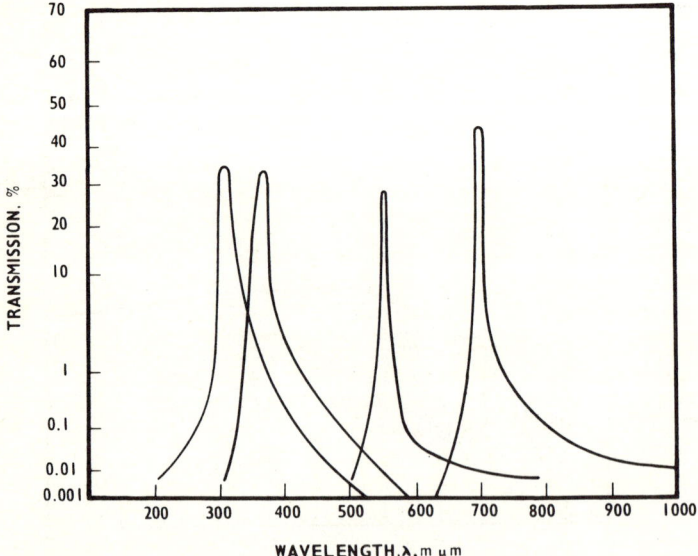

FIG. 4.1.38 Typical transmission curves for interference filters (courtesy of Balzer's Ltd.).

Light source A collimated beam of monochromatic light is required at several wavelengths. As shown in Fig. 4.1.37, this is conveniently achieved by using a xenon-arc lamp (typically 75 W, 0.3 mm square arc size), which has a spectral output from 300 to 800 nm. Collimation can be achieved to within 1 min of arc. For use in the field, ranges of light wavelength are conveniently obtained by typical interference filters as shown in Fig. 4.1.38.

Optical ports The means of passing the light beam into and out of the wet steam is critical to the success of the method. In [4-44] purged quartz windows were successfully employed, the flow of warm air keeping the window surface free from water, grease, etc. A special feature, shown in Fig. 4.1.39, was also incorporated to enable the system transmittance f_0 to be measured immediately following the measurement of the transmitted intensity f of system plus wet steam. The excluder tube in Fig. 4.1.39 is introduced and purged with air to exclude all wet steam between the optical ports.

Receiving system The optical requirement of the receiving system is considered in detail in [4-40]. The main requirement is to obtain sufficient angular resolution to substantially eliminate scattered light from the beam entering the photocell. The angle subtended from focusing lens to pin hole at the entry to the photocell should be less than $(3.84/\alpha)$ rad. Typically, for a 10 cm focal length lens, a 5 mm aperture will be adequate for resolving the scattering of 2 μm droplets. Finally, a miniature photocell of wide spectral response is required, such as the Cadmium Blue cell, Hayakawa type 5BC-102.

This system was tested using monodispersions of sulfur droplets in a clear solution of sodium thiosulfate. By adjusting the production of the hydrosol, the droplet size could be varied and the resulting confirmation of the system

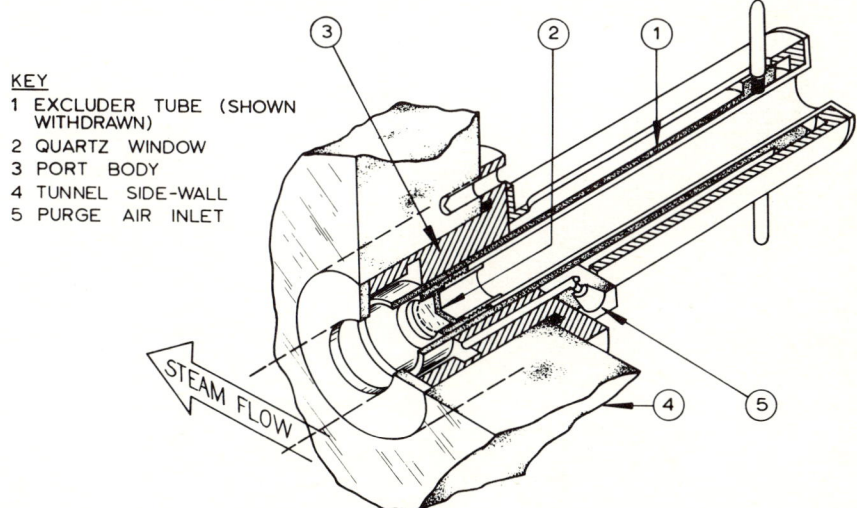

FIG. 4.1.39 Fog-excluder port.

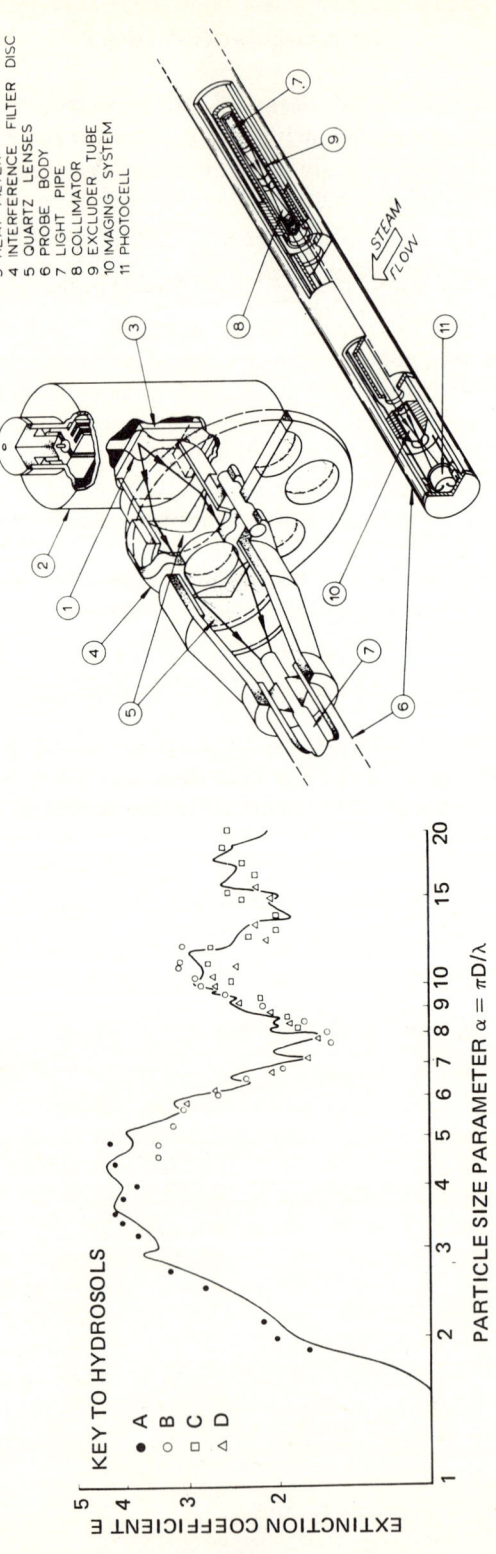

FIG. 4.1.41 Proposed turbine optical probe.

FIG. 4.1.40 Comparison of Mie theory with experiment.

4.2 A LIGHT-SCATTERING PROBE FOR DROPLET SIZE AND WETNESS FRACTION MEASUREMENT IN TWO-PHASE FLOWS

A. Ederhof

For measuring wetness, droplet size distributions, and numbers of droplets the so-called light scattering probe was developed at the Institut für Dampf- und Gasturbinen der RWTH Aachen [4-51]. With this probe one can also estimate the approximate velocities of single droplets. The principle of measurement of the probe, suitable types for different applications, and results are given and discussed.

4.2.1 The Principle of Measurement of the Light-Scattering Probe

Figure 4.2.1 shows the principle of measurement of the probe. The arc of a strong light source L is imaged in the plane at V. The optical transmission path consists of a lens, an aperture B_1, and an objective O_1. If there is one particle of a particle flow within this light beam, it will cause light scattering and will appear to an observer to be an independent light source. The intensity of the scattered light detected at a fixed angle θ gives information

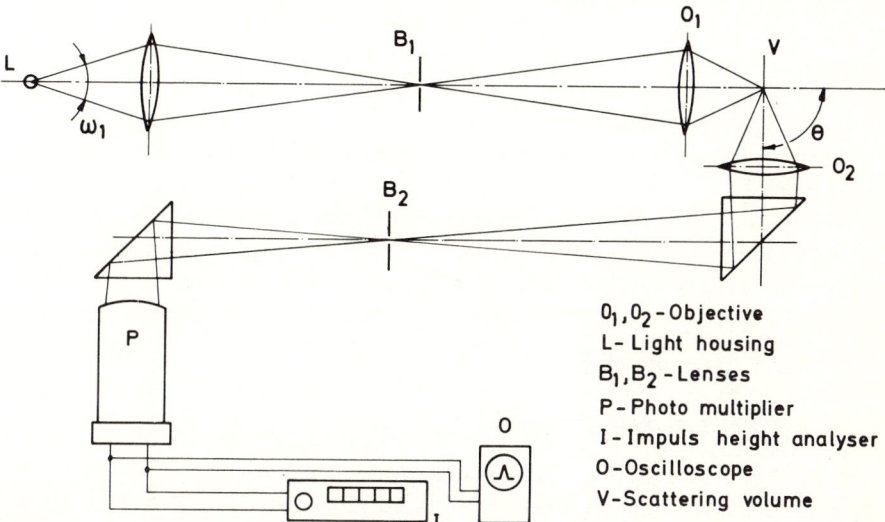

FIG. 4.2.1 Essentials of light-scattering probe.

O_1, O_2 - Objective
L - Light housing
B_1, B_2 - Lenses
P - Photo multiplier
I - Impuls height analyser
O - Oscilloscope
V - Scattering volume

on the size of the scattering particle. The mean scattering angle of the probe is 90°. The scattering volume is the volume enclosed by the transmitted light beam and the scattered one. It can be reduced or magnified over a wide range.

After passing along a second light path consisting of an objective O_2, an aperture B_2, and a prism, the scattered light is intercepted by a photomultiplier, where light pulses are changed into voltage pulses. These pulses are recorded on an oscilloscope and a pulse-height analyzer. The pulse widths, which are the durations of the droplets in the scattering volume, give the approximate droplet velocity values.

For interpreting the results some corrections for different errors have to be made. These are

1. Coincidence errors, caused by two or more droplets flowing simultaneously through the scattering volume
2. Fringe errors caused by droplets touching the scattering volume
3. Multiple scattering errors
4. Errors caused by variable light intensity
5. Interference errors

These corrections are part of a computer program analyzing the records obtained. In the special case of wetness fraction calculation, all droplet masses multiplied by the droplet numbers are summed. This liquid flow compared with the steam or air flow gives the value of wetness fraction.

4.2.2 Calibration of the Light-Scattering Probe

It is possible to calculate fairly accurately the approximate calibration curve of the probe, including the optical particularities. But this curve describes the relative intensity versus droplet diameter; i.e., the absolute intensity of the scattered light depends on the intensity of the lighting beam, which is different for the different probe versions. So we have to calibrate each version and to correct the theoretical curve, if there are deviations between calculations and calibrations.

There are two different methods for calibration: (*a*) within the range 0.1–5 μm dia. there is the possibility of atomizing latex hydrosols; (*b*) for bigger diameters we have developed the atomization of water by ultrasonic waves.

4.2.3 Example of Applications

4.2.3.1 Low-Pressure Probe

Figure 4.2.2 shows the low-pressure probe, which consists of the lamp housing, the casing with the casing head, the two fingers containing the objective lenses, and the photomultiplier with the cooling chamber for

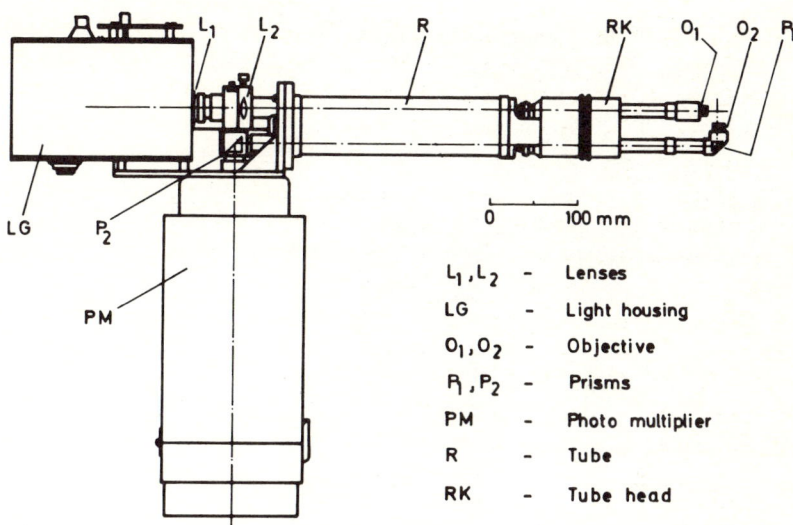

FIG. 4.2.2 Light-scattering probe for measuring wetness and droplet distribution.

reducing multiplier background noise. This version was constructed for measurement behind the last moving-blade plane of the low-pressure part of condensing turbines. It can be adapted to different dimensions by lengthening the casing of the probe.

This application causes no thermodynamic problems (temperatures and pressures are moderate) but introduces problems in measuring technique. Normally droplets in steam turbines are so small (in the range .1–100 μm) that they need a small scattering volume to keep coincidence errors low. Additionally, because of the high flow speeds, the durations of the droplets in the scattering volume are rather small (within ns range). This implies a fast-pulse height analyzer. As such instruments reach their limits of capacity in this range, the application of the probe is limited by electronics. Nevertheless, it is possible to extend this limit by using special apertures, which are much longer in the direction of the flow than in the transverse direction. Droplets then remain for longer times in the scattering volume and reduce the required resolving time of the pulse-height analyzer.

Figure 4.2.3 shows the installation of the probe in the low-pressure part of a combined steam/gas-turbine power plant in Alsdorf, Germany. The probe was flanged directly to the turbine casing; it can be rotated axially and traversed radially. At the rear end of the probe are the lamp housing and the photomultiplier cooling chamber.

In Fig. 4.2.4 are plotted two measured droplet-size distributions. One histogram refers to the blade root, the other one to the blade tip. The mean droplet size is 4 μm at the blade root and about 17 μm at the blade tip. The value of wetness at the root is larger than at the tip.

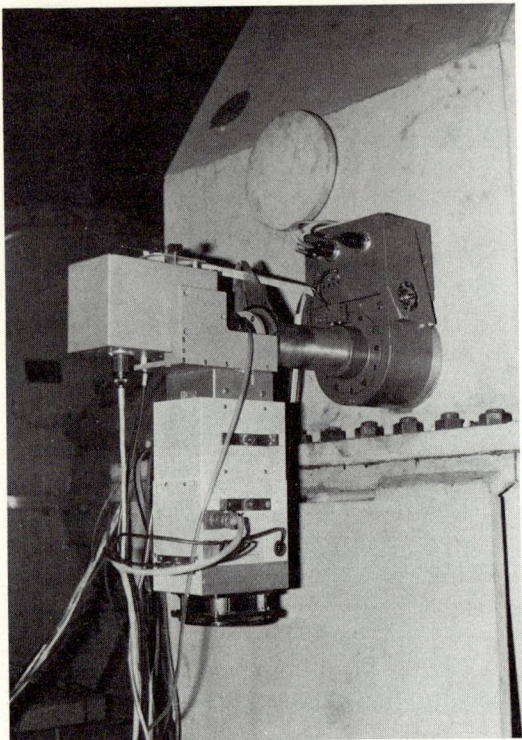

FIG. 4.2.3 Probe attached to LP turbine casing.

4.2.3.2 *High-Pressure Probe*

Light-water nuclear power plants supply wet steam in the live-steam lines to steam turbines. The small amount of wetness at the turbine inlet is not only important for the determination of the turbine efficiency, but can also cause erosion of the steam pipe and the high-pressure blading in some cases. A high-pressure probe was therefore built, because little is known about droplet sizes and because steam wetness cannot be measured accurately. The difficulties encountered resulted essentially from the thermodynamic properties of the steam: pressures up to 80 bars, and temperatures up to 300°C. As the diameter of the probe casing cannot fall below a certain value (about 70 mm), there has to be an adjusting device that can traverse the probe against a force of approximately 5000 kp.

The probe has also to be traversed through a stuffing box to form a seal against the live-steam pressure. It is necessary to mount two windows in the probe casing head because the scattering volume is outside the probe. These windows must be transparent and resistant to steam pressure and temperature. They also have the purpose of keeping high-pressure steam out of the optical system. The probe casing and the stuffing box also have to be resistant to

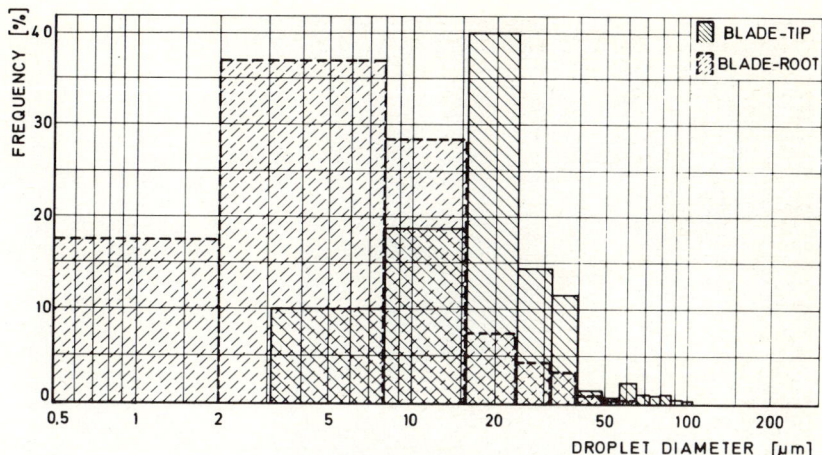

FIG. 4.2.4 Droplet distribution.

temperature and corrosion. Further, several injunctions have been issued by the competent technical control board (TÜV) guaranteeing the safety of the probe.

Figure 4.2.5a shows the finished probe. The polished and chrome-plated probe casing is traversed by a yoke and two screwed spindles using a geared motor. The probe slides in a stuffing box connected with a gate valve and the live-steam line (Fig. 4.2.5b). A contact manometer ensures that the probe casing cannot be pulled out of the stuffing box with the gate valve opened nor be moved into the steam line with the gate valve closed.

The specification of the windows at the casing head seemed to be temporarily unrealizable, because all the tested glasses lost their surface qualities very rapidly. Finally, sapphires were found to be suitable in tests carried out for weeks. The sapphires were cemented with a special adhesive. The probe is being tested. First measurements in the live-steam line of the VAK-Kahl were successful. Further experiments are planned.

4.2.3.3 Probe for Drift Measurements at Cooling Towers

The requirements for a probe for drift measurements at cooling towers are different from those previously described. Here the droplet sizes are considered to be in the range 1–300 μm, sometimes even greater. The droplet numbers and the droplet velocities are smaller than in turbines. On the other hand, there are some operational problems:

1 The probe has to be traversed at the chimney level of the cooling tower
2 It has to work in heavy sprinkling
3 Registration and analysis of the test data must be made outside the cooling tower, because of its bad facilities for instrumentation and small accommodation for the operating crew

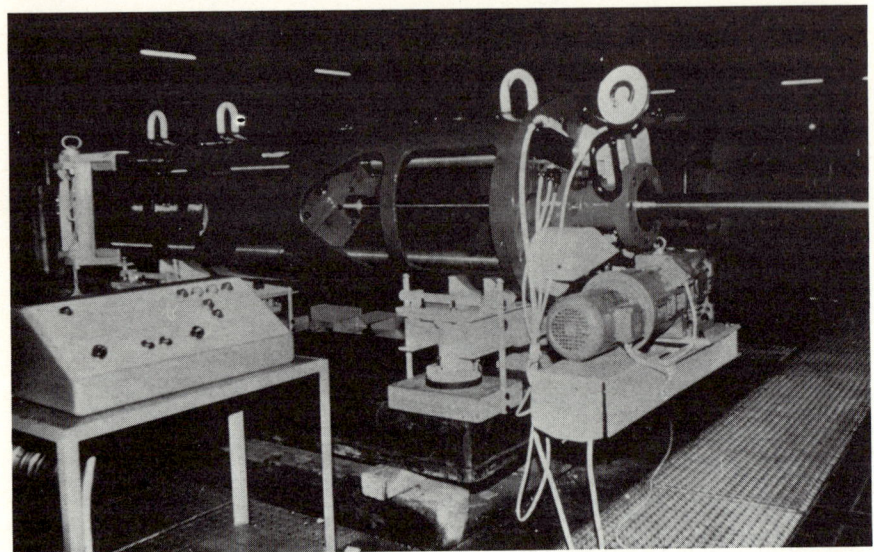

(a)

(b)

FIG. 4.2.5 Arrangement for measuring with a light-scattering probe in live-steam lines.

4   The probe has to be so sturdy that the inevitable rough handling during the transport to the top and during the mounting at the top are withstood without damage.

Figure 4.2.6a shows the version adapted for this purpose. Lamp housing, cooling chamber, and a part of the optics are installed within a stainless-steel

(a) View on probe

(b) Probe traversing cooling tower

FIG. 4.2.6 Light-scattering probe for drift measurements in cooling towers.

frame. To protect against moisture, this frame is covered with a waterproof stainless-steel casing. The probe casing is sealed by rubber bellows. The heat of the lamp and multiplier is carried away by compressed air, introduced and exhausted through two funnels. This compressed air also gives a small amount of overpressure in the cover and prevents water entering in case of leakage. The measuring cables are conducted through a waterproof shield down the

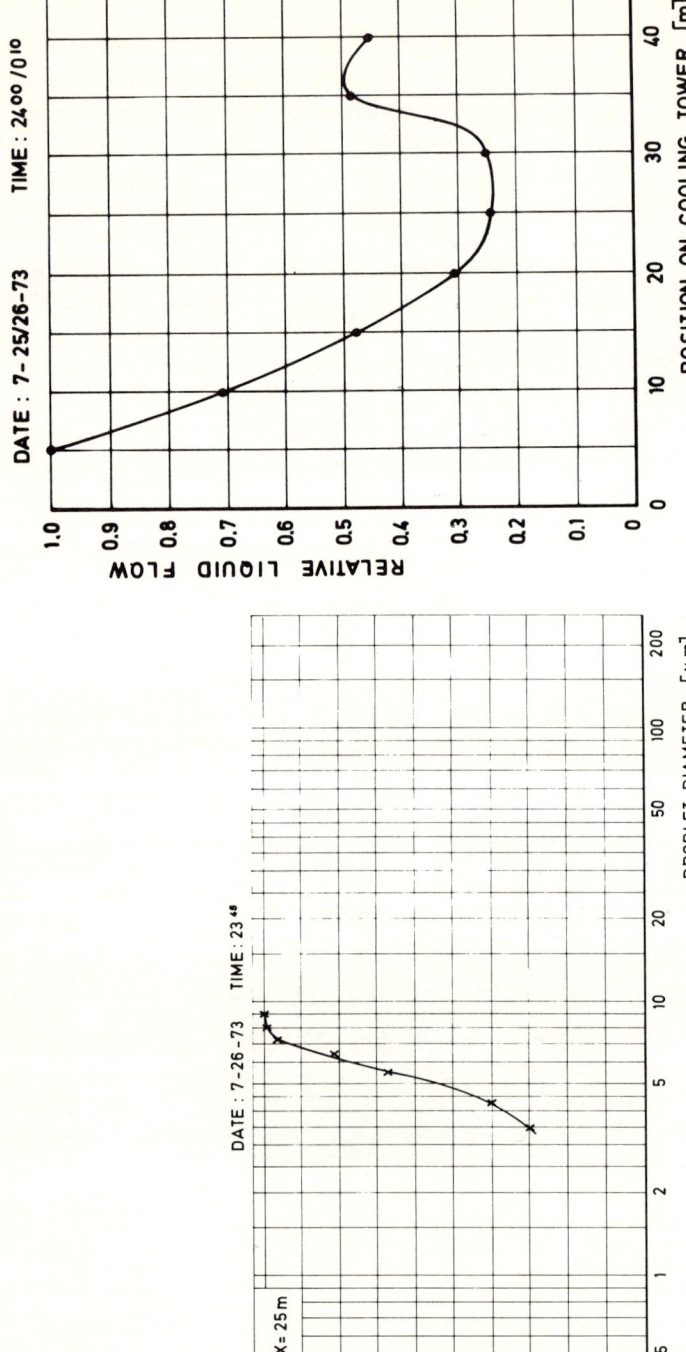

FIG. 4.2.8 Liquid flow distribution across cooling tower.

FIG. 4.2.7 Cumulative frequency curve.

cooling tower shell to the recording chamber at the bottom. A bracket holding a bearing pulley was attached to the probe casing. This bearing pulley rests on a supporting cable spanning the diameter of the chimney (Fig. 4.2.6b). With the help of a traction rope, the probe can be traversed between two opposite points on the cooling tower rim. The measuring cables are in sliding suspension from the supporting strand. The probe casing was surrounded initially by a sponge-rubber mat; this was later omitted.

Figures 4.2.7 and 4.2.8 show some results of measurements taken at the cooling tower of the power plant Neurath, RWE. Figure 4.2.7 shows a droplet-size distribution as a cumulative-frequency curve for a fixed measuring point; the number of droplets larger than 10 μm was found to be small. Figure 4.2.8 shows the variation of liquid flow over a cooling tower traverse; there are significant differences across a diameter, which can be attributed in part to the influence of wind, in part to other unexplored causes. The absolute values of liquid flow in a cooling tower drift are 0.1–4 g/kg dry air.

NOMENCLATURE

a	acoustic velocity
A	area
c	concentration of tracer
c_p	specific heat at constant pressure
C	vapor or gas velocity or count rate
d	drop diameter
d_c	control volume diameter for laser-Doppler anemometer
D	internal diameter of connecting tubes or aperture diameter
D_1	diameter of laser beam
e	measurement error
e'	voltage fluctuation
E	extinction coefficient
f	frequency or light flux
h	Pitot-tube diameter or specific enthalpy (static)
H	specific enthalpy, stagnation value
i	electrical current
I	intensity of radiation
k	index of isentropic expansion or compression
Kn	Knudsen number
L	length of pressure lines
m	mass flow rate or refractive index
n	number of droplets per unit volume
N	cumulative number distribution
p	pressure
Δp	amplitude of pressure oscillation
P	dimensionless relaxation parameter for wet steam

Pr	Prandtl number
Q	electrical heat input rate
R	electrical resistance or correlation coefficient
s	distance or optical fringe spacing or needle spacing
S	hot-wire system voltage sensitivity
t	time
t_*	response time of an intermittently purged system
T	temperature
u	droplet velocity
V	volume (of transducer)
x	steam/air ratio
Y	wetness fraction (by mass flow rate)
α	particle size parameter ($= \pi d/\lambda$) or heat transfer coefficient
γ	relaxation times of wet-steam mixtures
θ	scattering angle from the incident direction
λ	wavelength of electromagnetic radiation or thermal conductivity
μ	dynamic viscosity
ρ	density
ϕ	half angle between incident and reflected rays or flow pitch angle
χ	ratio of specific heats
ω	angular frequency or complex potential function

Subscripts

x, y	Cartesian coordinates
i	initial value
I	inertial
T	thermal
t	stagnation value
$*$	value at time $t = t_*$
f	liquid phase
g	vapor phase
w	hot wire
e	equilibrium, unheated value

REFERENCES

4-1 Moore, M. J., Langford, R. W. and Tipping, J. C.: *Proc. Inst. Mech. Eng.*, vol. 182, pt. 3H, 1967-1968.
4-2 Smith, A.: private communication.
4-3 Schuder, C. B., and Binder, R. C.: *Trans. ASME, J. Basic Eng.*, Dec., 1959.
4-4 Iberall, A. S.: *J. Res., Nat. Bur. Stand.* vol. 45, RP 2115, July, 1950.
4-5 Brown, F. T.: *Trans. ASME, J. Basic Eng.*, Dec., 1962.
4-6 Anderson, B. W.: "The Analysis and Design of Pneumatic Systems," Wiley, New York, 1967.
4-7 Bergh, H., and Tijdemann, H.: *NRL-TR F238, Nat. Aero. and Astro. Res. Inst.*, Amsterdam, 1965.

4-8 Horlock, J. H., and Daneshyar, H.: *J. Mech. Eng. Sci.*, vol. 15, no. 2, 1973.
4-9 Samoilovich, G. S., and Yablokov, L. D.: *Teploenergetika*, vol. 17, no. 9, pp. 70–73, 1970.
4-10 Weyer, H., and Schodl, R.: *Trans. ASME, J. Basic Eng.*, Dec., 1971.
4-11 Moore, M. J., and Sculpher, P.: *Proc. Inst. Mech. Eng.*, vol. 184, pt. 3G(111), 1969–1970.
4-12 Finlayson, P. C., and Roberts, A. G.: *Doc. No C/4954*, Brit. Coal Res. Assoc., 1955.
4-13 Todd, K. W.: Paper to the Conf. on Reheat Steam Turbines of Great Output, Inst. Fluid Flow Mach., Polish Acad. Sci., Gdansk, 1962.
4-14 Lagun, V. M., and Simoyu, L. L.: *Teploenergetika*, vol. 13, no. 6, 1966.
4-15 Bryer, D. W., and Pankhurst, R. C.: "Pressure Probe Methods for Determining Wind Speed and Direction," H.M. Stationery Office, London, 1971.
4-16 Anderson, G. H., and Mantzouranis, B. G.: *Chem. Eng. Sci.*, vol. 12, pp. 233–242, 1960.
4-17 Dussourd, J. L., and Shapiro, A. H.: *Jet Prop.*, vol. 28, no. 24, 1958.
4-18 Crane, R. I., and Moore, M. J.: *J. Mech. Eng. Sci.*, vol. 14, no. 2, 1972.
4-19 Hopkins, D. F., and Robertson, J. M.: *J. Fluid Mech.*, vol. 29, p. 273, 1967.
4-20 Angus, J. C., Morrow, D. L., Dunning, J. W., and French, M. J.: *Ind. Eng. Chem.*, vol. 61, no. 2, Feb., 1969.
4-21 Durst, F., Melling, A., and Whitelaw, J. H.: *Comb. Flame*, vol. 18, pp. 197–201, 1972.
4-22 Drain, L. E.: *J. Phys. D.*, vol. 5, pp. 481–495, 1972.
4-23 Wang, C. P.: *J. Phys. E. (Sci. Instrum.)*, vol. 5, p. 763, 1972.
4-24 Durst, F., and Whitelaw, J. H.: *Opto-electronics*, vol. 5, pp. 137–151, 1973.
4-25 Crane, R. I., and Melling, A.: to be published.
4-26 Wood, N. B.: *Inst. Mech. Eng., Conf. publ.* 3, 1973.
4-27 Morkovin, M. V.: AGARDograph 24, NATO Paris, 1956.
4-28 Kistler, A. L.: *Ballistic Res. Lab. Rep.* 1052, 1958.
4-29 Williams, G., and Lord, M.: *Third Conf. on Steam Turbines*, Gdansk, Sept., 1974.
4-30 Crane, R. I.: *Int. J. Mech. Sci.*, vol. 15, pp. 613–631, 1973.
4-31 Thorpe, A. D., and Wood, M. R.: *Proc. Inst. Mech. Eng.*, vol. 182, pt. 3H, p. 130, 1967–1968.
4-32 Rutz, J.: *Kernenergie 14 Jahrgang Heft* 6/1971.
4-33 Lück, W.: *VDJ Bericht*, no. 97, p. 117, 1966.
4-34 Kasprzyk, S. von: *Brennst-Wärme-Kraft*, vol. 16, no. 7, July, 1964.
4-35 Christ, A., and Wulff, W.: *Escher Wyss News*, vol. 33, 1960.
4-36 Langford, R. W., and Moore, M. J.: to be published.
4-37 Wicks, M., and Dukler, A. E.: *Heat Transfer Conf. V*, 39, 1966.
4-38 May, K. R.: *J. Sci. Inst.*, vol. 17, Sept., 1940.
4-39 Vander Hulst, H. C.: "Light Scattering by Small Particles," Wiley, New York, 1957.
4-40 Hodkinson, J. R.: in C. N. Davies (ed.), "Aerosol Science," Academic, New York, 1966.
4-41 Penndorf, R. B.: *J. Opt. Soc. Amer.*, vol. 47, no. 11, p. 1010, 1951.
4-42 Petr, V.: *Inst. Mech. Eng., Conf. Publ.* 2, 1970.
4-43 Gyarmathy, G., and Lesch, F.: *Inst. Mech. Eng., Conf. Publ.* 2, 1970.
4-44 Walters, P. T.: *Inst. Mech. Eng., Conf. Publ.* 3, 1973.
4-45 Chareyre, R., and Manas, B.: *Inst. Mech. Eng. Conf. Publ.* 3, 1973.
4-46 Dobbins, R. A., and Jizmagian, G. S.: *J. Opt. Soc. Amer.*, vol. 56, p. 1345, 1966.
4-47 Shifrin, K. S., and Perelman, A. Ya.: "Proceedings Second Interdisciplinary Conference on Electromagnetic Scattering," Gordon and Breach, New York, 1965.

4-48 Walters, P. T.: to be published.
4-49 Wallach, M. L., and Heller, W.: *J. Phys. Chem.*, vol. 68, p. 924, 1964.
4-50 Mugele, R. A., and Evans, H. D.: *Ind. Eng. Chem.*, vol. 43, no. 6, pp. 1317–1324, 1951.
4-51 Marx, P. P.: Dissertation, RWTH, Aachen, 1970.

CHAPTER 5

Experimental Development of Wet-Steam Turbines

A. Smith

5.1 INFLUENCE OF MOISTURE ON BLADING EFFICIENCY

In conventional fossil-fired power stations, the gain in overall efficiency obtained from the adoption of higher superheat and reheat temperatures is predominantly thermodynamic; improvements from the reduction in moisture content are small since this is usually restricted to the last two or three stages of the LP turbine, which accounts for only a small proportion of the overall heat drop. Economic and technical considerations in water-moderated reactor systems, however, have favored a reduction in initial steam temperatures, so that the expansion is now predominantly wet. Attention has consequently been refocused on the additional losses suffered by blading when operating in wet steam.

The first published reference to wet-steam losses was made by Karl Baumann [5-1], who proposed that a good approximation to the wet isentropic efficiency of blading could be obtained by multiplying the dry efficiency by the mean-equilibrium dryness fraction, the isentropic heat drop being based on the value read from a Mollier diagram. It is worthy of note, however, that Professor A. Stodola did not agree with his pupil, and in his treatise [5-2] specifically mentions that no additional wetness loss was evident from his experiments on an eight-stage reaction turbine.

Today few would support this finding, but Professor Stodola's article does serve to clarify that the steam tables are based on unit mass of mixture, and that the enthalpies quoted below the saturation line are debited with the reduction in latent heat caused by the presence of water. The Baumann factor therefore accounts for a supplementary wetness loss in addition to that resulting from the reduction in isentropic heat release caused by the presence of water.

In 1931, Soderberg [5-3] made an important contribution to wetness-loss studies by subdividing the sources of loss into three effective areas before quantifying them. Initially he calculated the momentum required from the steam to accelerate the moisture component from rest at the blade trailing edges up to a given proportion of the mean steam discharge velocity, on both the stationary and moving rows. He then considered the individual contributions of the vapor and water phases to the change in absolute tangential momentum across the moving-blade rows, from which a braking loss was deduced. Finally, he estimated the losses associated with centrifuging the water deposited on the moving-blade surfaces to the cylinder wall.

The result of these combined wetness losses on 50%-reaction blading, set at 0.4 opening coefficient o/p, indicate that the loss factor α, defined by $\eta_w = \eta_d(1 - \alpha Y_m)$, decreases with the ratio of blade speed to steam efflux

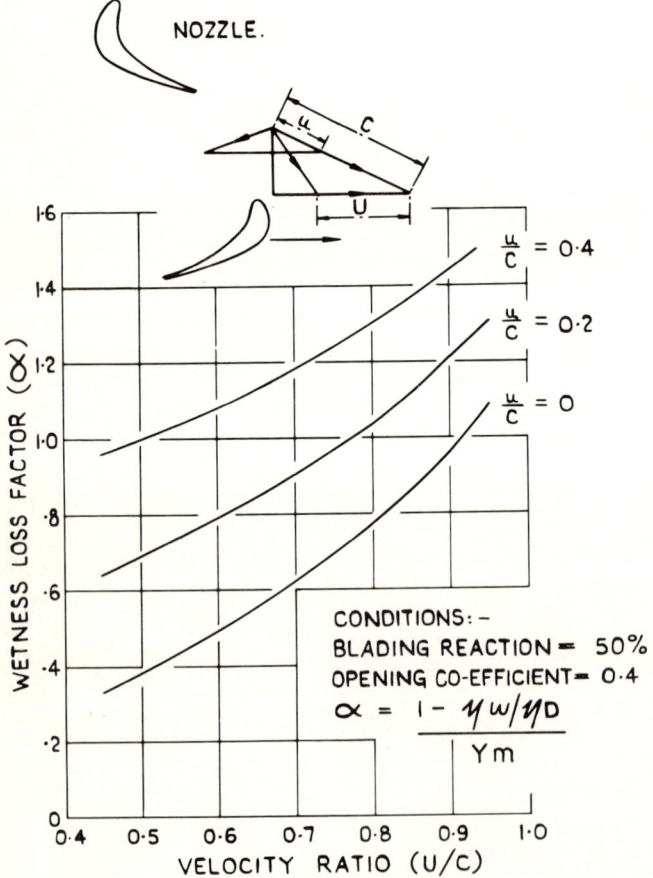

FIG. 5.1.1 Soderberg prediction of variation in wetness-loss factor with velocity ratio.

velocity U/C (Fig. 5.1.1). The loss factor is also shown to be sensitive to the ratio of droplet to vapor velocity u/C leaving the blade rows; this tends to fall as the u/C ratio approaches zero, because of the overriding influence of Soderberg's moisture acceleration loss over the blade-braking term.

Soderberg had effectively restricted his analysis to wetness losses associated with the second and subsequent generations of droplets, neglecting those losses associated with the formation of smaller first-generation droplets. Consequently, reversion shock losses were omitted that, if located near blade suction surfaces and of sufficient strength, could cause additional boundary-layer separation losses. Drag losses associated with the resulting first generation of droplets were also omitted. Analytical difficulties associated with reversion, the effective blade tip leakage, and possible stirring losses close to the turbine cylinder walls, however, have tended to favor an experimental approach.

Because of the high heat drop necessary to produce wet steam of representative quality from a dry and saturated initial steam condition, most turbine wet-loss experiments have been made with water injection into either initially superheated steam or air [5-4]. The results have consequently been more applicable to the later stages in a wet expansion, where the majority of the water droplets present are likely to be of the coarse second-generation variety. An example of this technique obtained from a seven-stage 50%-reaction set in the "Kate" experimental HP turbine at Heaton Works, indicates that the wetness-loss factor α falls with the isentropic velocity ratio of the blading U/C_0 (Fig. 5.1.2), following the trend predicted by Soderberg. Reliable results however are difficult to obtain by this method, because no effective heat-balance check can be made to ensure that the vapor and injected water are in thermal equilibrium. Consequently, there might have been more water present to cause losses than would have been produced by work extraction.

More recently, a two-stage wetting procedure has been adopted in which the Works boiler supply has been initially desuperheated to within 5°C of the saturation temperature level at the prevailing inlet turbine pressure (2.3-3.7 bars). When the thermal mixing is satisfactory by heat balance, the steam is wetted from a ring of Watson pressure-jet atomizers before the blading, the water temperature having been raised to within 3°C of the saturation level at inlet to the turbine. By altering the size of the jets and the pressure difference across them, some control of the droplet spectrum size is also possible, down to about 50 μm Sauter mean diameter ($d_{32} = \Sigma d^3 / \Sigma d^2$).

Results of this technique showed that the wetness-loss factor tends to fall with droplet size and increase in steam pressure because of the greater proportion of the droplets remaining steamborne (Fig. 5.1.3), a trend that was subsequently confirmed by introducing a steam-blast atomizing technique to produce droplets of less than 10 μm dia. (lower curve). This advantage can be lost, however, at low-velocity ratios where droplet collection on the blade

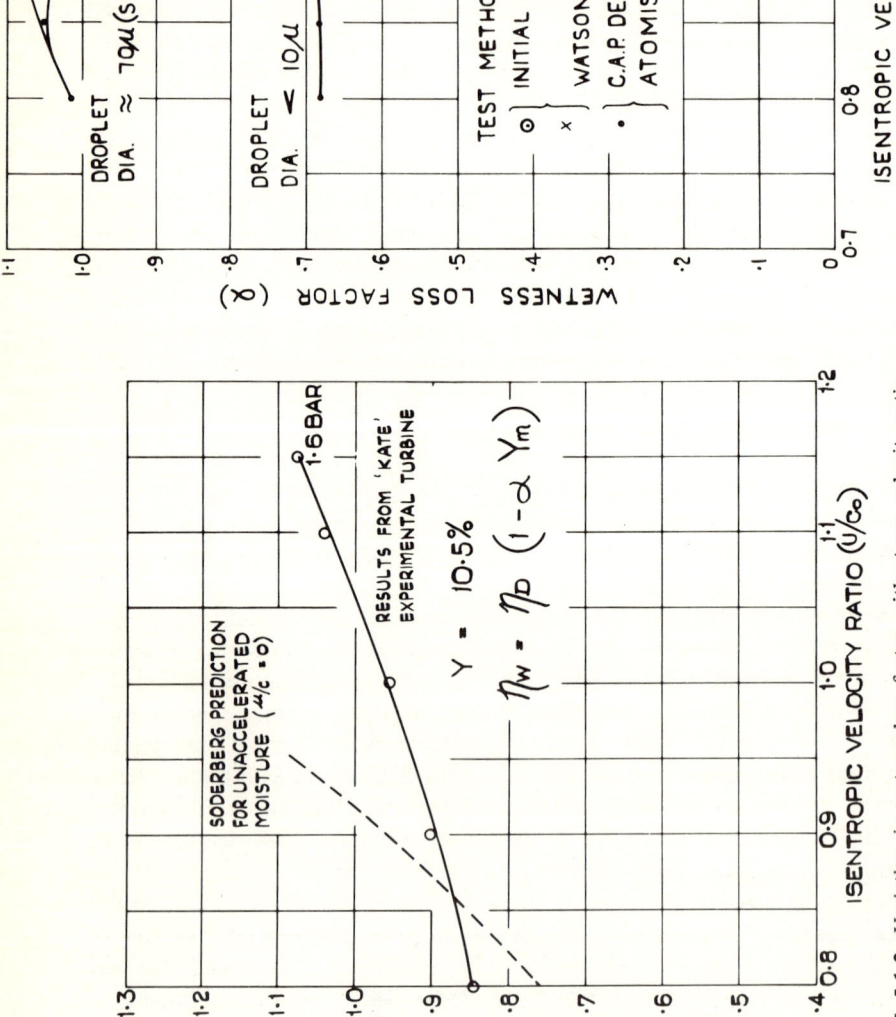

FIG. 5.1.3 Change in wetness loss with pressure and droplet size (results by water injection into seven-stage 50%-reaction experimental turbine).

FIG. 5.1.2 Variation in wetness-loss factor with steam velocity ratio (early results by droplet injection into seven-stage 50%-reaction experimental turbine).

concave surfaces is improved. It is equally important to note that, for a given blading configuration, wet efficiencies would peak closer to the dry optimum velocity ratio with smaller droplets than with larger ones, which suggests that for the best overall turbine efficiency, early stages of the wet expansion should be designed for a higher velocity ratio than those that follow, to compensate for the smaller mean droplet size.

To illustrate the influence of blade-tip geometry on wetness losses, wet-to-dry efficiency ratios are plotted against percentage wetness for two sets of blading of identical profile, aspect ratio (height/chord), and velocity ratio (Fig. 5.1.4). These show that, at equivalent clearances, unshrouded radial-clearance blading tends to suffer less than the shrouded multisealing contact type of blading, the reason probably lying in the absence of circumferential recesses in the cylinder wall to accommodate the shrouds where water can collect and brake the blade tips.

Miller and Schofield [5-5] are to be thanked for having published wetness-loss results in the reversion zone of a turbine. It would appear that their five-stage HP turbine was supplied with steam at different levels of superheat, the wetness being produced naturally by extracting work from the shaft. Difficulties arose, however, when attempting to isolate the additional losses encountered in the wet region of the expansion. The presentation giving the greatest consistency was obtained by plotting the ratio of the wet-to-dry isentropic efficiencies against a weighted average moisture content, this being

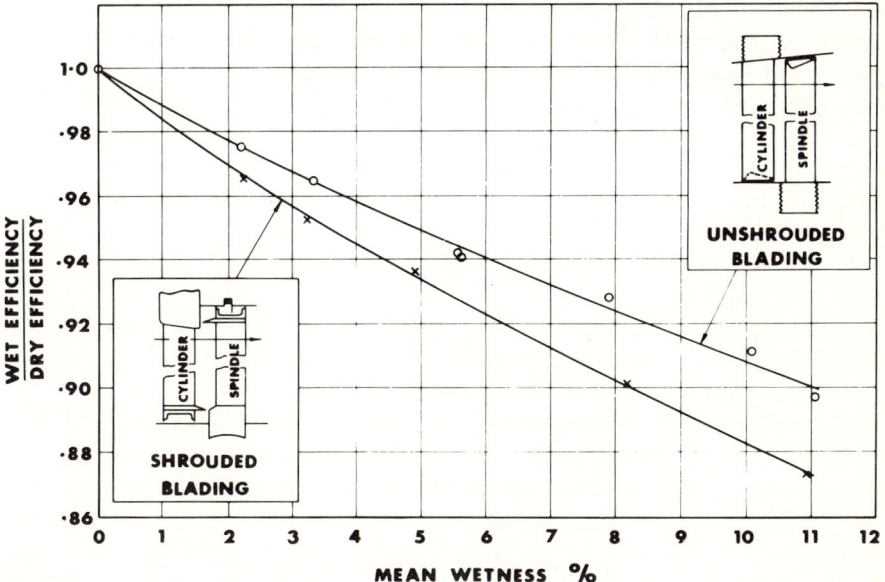

FIG. 5.1.4 Fall off in efficiency with wetness for shrouded and unshrouded 50%-reaction blading at similar clearance, velocity ratio, and pressure level.

found by multiplying the arithmetic mean wetness by the ratio of heat to work in the wet zone to the overall heat to work.

This procedure, which tends to minimize experimental scatter, differs from the more usual presentation, in that wetness instead of efficiency is corrected for that part of the expansion in the dry zone. Miller and Schofield's results, however, can be transposed into the more usual wetness-loss factor α using the expression

$$\alpha = \left(1 - \frac{\eta_0}{\eta_d}\right) \bar{Y}_m$$

where

$$\bar{Y}_m = \frac{\Delta hw}{\Delta hw + \Delta hd} Y_m$$

Wetness-loss factors at elevated pressures were found to be high between 0 and 4% wetness, the loss factor at 1% mean wetness reaching 2.3; at wetnesses of 8%, however, the factor had fallen to 0.9 or somewhat below the Baumann value of unity (Fig. 5.1.5). The reason for the high loss at low wetnesses was quoted as being a reversion influence.

Experiments have also been made at Heaton Works on our 14-stage 50%-reaction experimental HP wet-steam turbine "Mary," in which the wetness was again produced naturally from an initially superheated condition by extracting shaft work (Fig. 5.1.5). Test procedure was complicated by the need to change only one variable at a time, this being impossible to fulfill completely in a fixed-turbine geometry because of the change in thermodynamic conditions caused by varying degrees of wetness. An acceptable compromise was achieved, however, by running tests at a selected exhaust-pressure level, keeping the overall steam-density ratio constant to maintain a uniform velocity ratio across each stage. The influence of wetness on blade efficiency was subsequently established by varying the initial superheat. Before each set of readings, however, it was necessary to correct the velocity ratio by trimming the rotational speed to allow for the change in heat drop in the respective dry and wet zones; moreover, simultaneous inlet-pressure adjustments were also necessary to correct the density ratio.

The influence of change in velocity ratio on the wetness losses at the prescribed back pressure was subsequently established by repeating this procedure at different rotational speeds. Having completed one cycle of the test program in this way, the procedure was repeated for each back pressure.

Initial results from "Mary" (Fig. 5.1.5) appear to confirm that high wetness-loss factors exist in the low-wetness zone. The measurement of small changes in efficiency at low theoretical wetness levels, however, is prone to experimental error, and as Miller and Schofield intimated, these tend to be

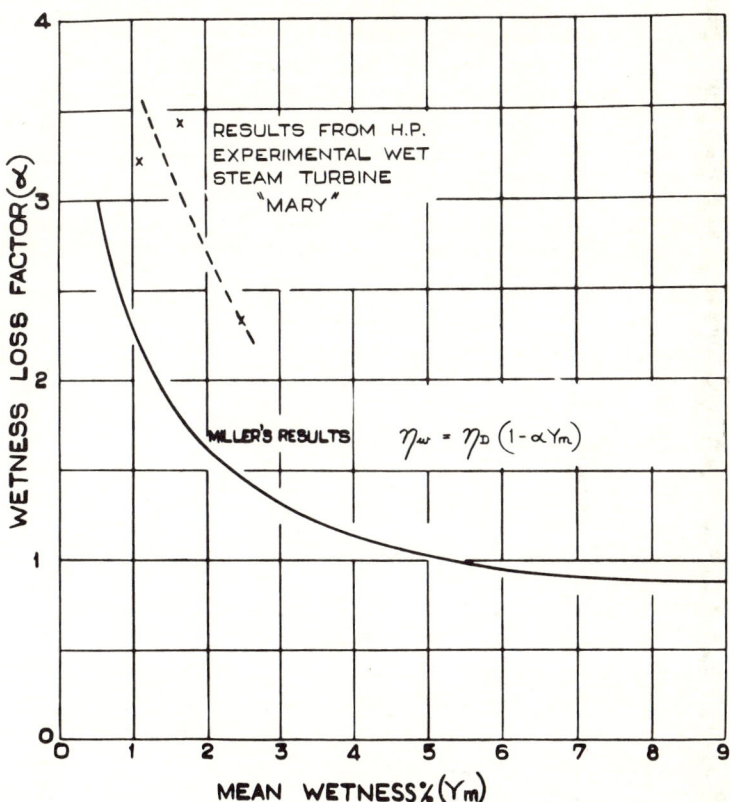

FIG. 5.1.5 High-pressure wetness-loss factors (wetness produced by work extraction).

amplified if the weighting process is applied to efficiency change. Careful check tests are therefore required, but at the moment evidence would support Miller's finding that wetness-loss factors are considerably higher than average at low-wetness levels.

Droplet-size measurements employing light-extinction techniques have recently been made in the Mary turbine through sight glasses located between blade rows. These show 1–2 μm droplets to be present after the fourth stage at an inlet pressure and superheat condition of 6.3 bar and 5°C, the equilibrium saturation line having been crossed in the second stage. Moreover, these measurements indicate that the proportion of water represented by the droplets is only 0.1%, the equilibrium moisture content being 1%. Instead of the steam becoming suddenly opaque, however, as observed in HP nozzles when the Wilson limit is reached [5-6], progressive diminution of the light intensity took place until the fog completely extinguished light transmission at the fifth stage. It would therefore appear that instead of reversion being

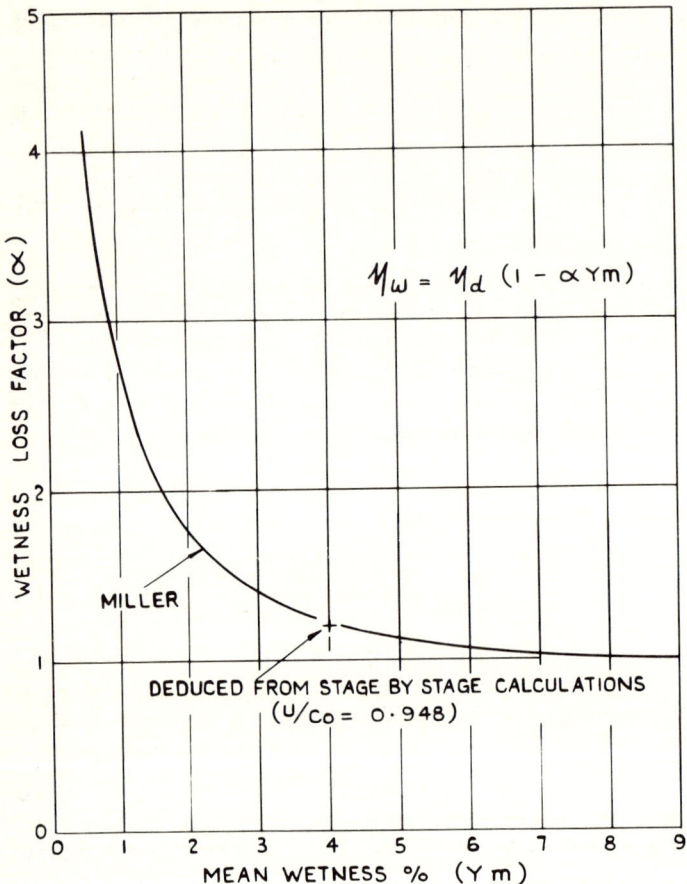

FIG. 5.1.6 Low-pressure wetness-loss factors (wetness produced by work extraction).

triggered at some specific point in the expansion, progressive nucleation takes place over a number of blade rows.

These observations would agree with a suggestion by Moore [5-7] that the vapor close to the convex blade surfaces, which is subjected to a considerably higher initial expansion rate than average, may reach its Wilson limit earlier. The vapor in the remainder of the blade passage, however, remains supersaturated to different degrees until it passes sufficiently close to the suction surface of a subsequent blade row to suffer reversion.

A process is consequently envisaged where successive reversions take place as the expansion proceeds, until supersaturation has been effectively relieved throughout the steam flow. The pressure discontinuity associated with reversion may therefore be superimposed on the adverse pressure gradient on blades created by the suction-surface deceleration, the interaction being sufficient to induce boundary-layer separation. Should such separation extend over several blade rows, this may explain the high wetness losses measured by Miller and ourselves.

An experimental LP turbine, discussed by Miller and Schofield [5-5], would appear to suffer even higher wetness-loss factors than their HP turbine at low mean wetnesses (Fig. 5.1.6), the tendency being for the factor to steady out at the Baumann value of unity at mean wetnesses above 6%. This compares with a value of approximately 0.9 for their HP turbine at the same wetness.

An overall wetness-loss factor of 1.2 was also estimated from model LP turbine results at Heaton Works [5-8] at a design exhaust wetness of 9%, which is in close agreement with the Miller and Schofield results. Since this mean value exceeds the Baumann value of unity, there would appear to be some justification for their claim of high losses at low wetnesses.

5.2 MOISTURE-REMOVAL DEVICES

Water-removal devices have been employed in wet-steam turbines for over half a century [5-9], their original function being to reduce erosion damage rather than to improve blading efficiency.

Extraction devices can be subdivided into two classes: those designed to remove potentially damaging free water from the blade path, and more sophisticated separators intended to remove droplets as well as free water from the steam. Because of the low steam and water velocities necessary to achieve effective droplet separation, the latter type are usually externally mounted or incorporated in reheaters between the HP and LP turbine cylinders, where their bulk can be accommodated. An exception to this was the fitting of relatively long internal separators to GEC turbines at Rolphton (Ontario) and Winfrith (steam-generating heavy-water reactor) [5-10].

In conventional fossil-fired plants, wetness problems are normally confined to the LP turbine, where erosion potential tends to rise with blade speed and increase in unit rating. Water extraction is consequently restricted to this region, cylinder belts usually being located downstream from moving blade rows so that advantage can be taken of their centrifuging action on surface-water deposits to drive water radially outward towards the belt.

Low-pressure blade erosion control returned to prominence during 1962, when last-row erosion shield lives of only three months were predicted for the new 500 MW, 3000 rpm machines ordered by the Central Electricity Generating Board, the last row tip speeds of which were 545 m/s. This resulted in steps being taken to establish the effectiveness of LP cylinder-water extraction before the last stage, which was coupled with a search for new materials to withstand the erosive influence of droplet impact at high speed.

Because of the proximity of the "Stella" Southpower station to Heaton Works, measurements were made there by Parsons on a 23-year-old 60-MW turbine to determine the effectiveness of its LP water-extraction belts. These were located after the penultimate moving-blade rows of the double-flow LP turbine cylinder (Fig. 5.2.1) with their trailing edges overlapping the inlet edges of the belts by 29% of the blade axial width, a technique subsequently recommended by Deich et al. [5-11]. No steam bleed was employed from the belt in this instance, although bleeds of 1 or 2% of the total flow were frequently employed to improve water extraction.

A metering orifice was installed at the base of the water-filled sealing loop between the extraction belt and the condenser, to prevent detrained water

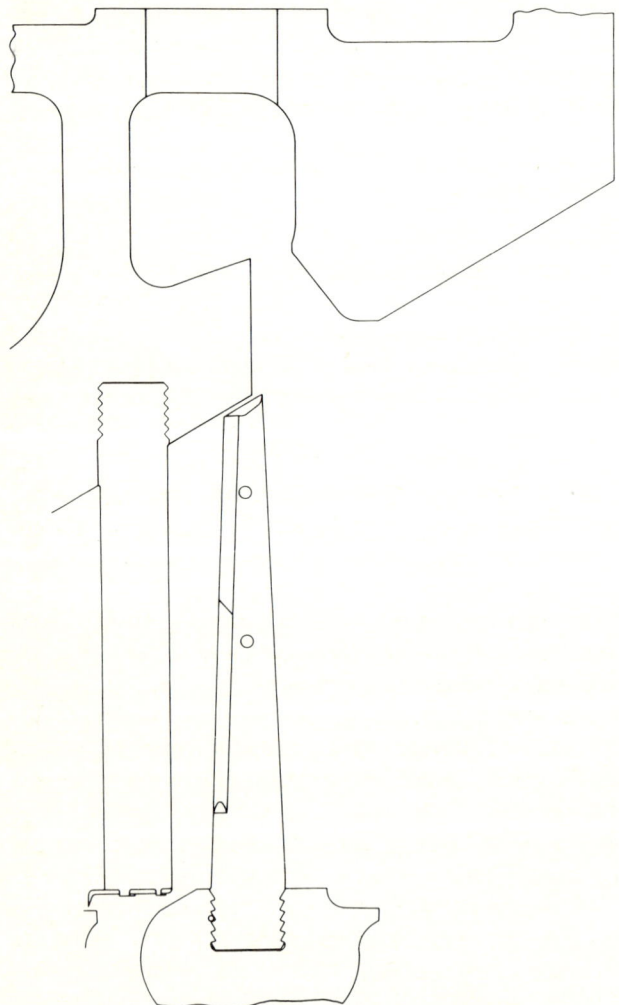

FIG. 5.2.1 Water extraction belt behind penultimate row in 60-MW turbine at Stella power station.

flashing into steam as the static pressure was reduced in flowing through the orifice. It was found that only 3% of the theoretical water available could be extracted, the load and local equilibrium moisture content being 66 MW and 11% respectively. Choking of the flow was thought to be responsible for the poor performance, but an increase in the drainage pipe and metering orifice sizes did not improve the extraction rate. It was suspected that a feed-heating belt located two stages upstream might be removing most of the free water with the bled steam, but the valve to this heater was ultimately closed without improving the water-removal rate from the extraction belt.

Subsequently, an experimental program was initiated on a third-scale four-stage model of a 500 MW, 3.46 m last-row tip diameter, LP turbine "Bess," which had been built for aerodynamic studies prior to construction of the Ferrybridge machines. The rotational speed of 9000 rpm was also three times the full-scale value, to duplicate blade speed and Mach numbers. Provision for cylinder-water extraction experiments had been built into the model before the last stage as at Stella, in the form of extraction belts whose inlet shape could be varied by fitting suitable circumferential rings. The influence of steam bleed on water-extraction rates could also be assessed by passing the extracted mixture through a Weber-type centrifugal separator before condensing the vapor phase for Rotameter measurement (Fig. 5.2.2).

Extraction rates from an open Stella-type belt were again disappointing, being only 1% of the equilibrium water available at the design wetness of 7.7% and with 1% steam bleed; moreover, the best of the inlet-fairing arrangements succeeded in raising this to only 1.8%. The reason for these low extraction rates was thought to lie in the low proportion of the total droplet population reaching the concave blade surfaces, their diameter being calculated to be 0.18 μm at entry to the last-stage nozzles on a simple one-dimensional basis, with a deposition rate of rather less than 5% of the total.

With erosion being identified with the stripping action of the steam of this surface water into large slow-moving drops from the nozzle trailing edges [5-12], it could not be assumed that the proportion being deposited was insufficient to cause an erosion hazard to the following moving blades. Direct extraction of this potentially dangerous water through slots in hollow last-stage nozzles was subsequently considered to offer advantages over the extraction-belt arrangement [5-13 to 5-15].

From preliminary cascade tests [5-15], a slot positioned at the nozzle trailing edges was found to be the most effective, a bleed of 1% being necessary in the Parsons arrangement to maintain the necessary depression within the blade for drainage purposes. The bleed rate for optimum extraction, moreover, is closely linked with the position, size, and shape of the slot; no mention of bleed, however, is made by Kirillov et al. [5-14]. Both Kirillov and we found, however, that a slot placed close to the trailing edge on the pressure surface of the nozzles was very effective and possessed

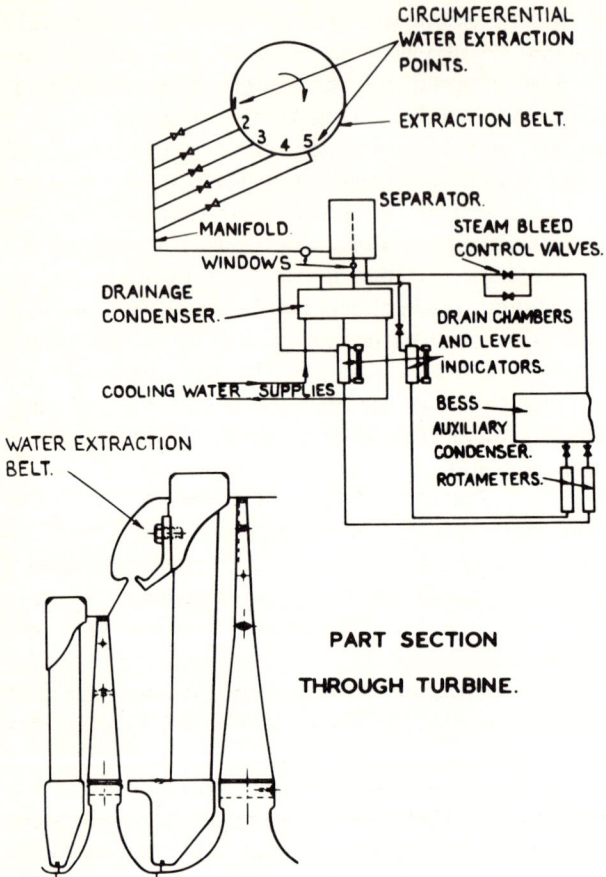

FIG. 5.2.2 Experimental arrangement of cylinder-water extraction experiments on third-scale model LP turbine.

certain constructional advantages over the trailing-edge position. Kirillov also advocated additional slots around the leading edge for maximum drainage effectiveness.

Hollow last-stage diaphragm nozzles incorporating pressure face slots were subsequently tested in the Bess third-scale model turbine, the slots extending over the outer 30% of the blade height where erosion potential was likely to be at a maximum because of the high peripheral speed of the following blade row. Extraction rates, however, were again so disappointing (less than 1% of the available moisture) that an introscope was fitted behind the last blade row so that the nozzle trailing edges could be viewed through the moving blades. By raising the inlet temperature to the model above the design value, it was possible to reduce the wetness level at exhaust to about 4%, where the mist density was sufficiently low to enable two or three of the nozzle convex surfaces to be examined in the vicinity of their trailing edges. With no suction applied to the hollow nozzles, small quantities of water appeared to flow around the trailing edges from the pressure to suction surfaces, where it formed narrow radial films in what was thought to be zones of flow separation close to the trailing edges. No stripping action of these films could be observed, however, either by eye or by high-speed photography. A steam bleed of $\frac{1}{2}\%$ was found to be sufficient to remove these films entirely, the actual quantity of water removed being found to be only 0.33% of the theoretical water present in the steam.

Although the model turbine experiment did not offer direct evidence that blade erosion could be reduced by direct nozzle drainage, more conclusive experiments were made at Plymouth on a GEC 60-MW turbine fitted with trailing-edge slots in the hollow nozzles of the last-stage diaphragm, indicating that this technique could be highly effective in controlling erosion in a machine that had been prone to erosive attack.

Concurrently, a hollow-bladed last-stage diaphragm was fitted to one of the six exhaust flows of a Ferrybridge 500-MW machine, supplementing the cylinder extraction belt fitted before the last stage. As in the Bess model turbine arrangement, slots were provided in the pressure surfaces near the trailing edges, steam and any surface water exhausting between the inner and outer cylinder casings to a back pressure closely corresponding to the condenser vacuum. Heat-balance checks on the upstream cylinder-extraction condenser indicated that, as in the model, little water was being removed from the circumferential belts and, after 1000 hr of high-load running, the cylinder covers were lifted for the erosion shields to be inspected. Erosion damage was found to be slight in all six LP turbine flows; additionally, no significant difference in erosion damage to the blade shields behind the slotted and solid nozzles could be detected. The situation was found to be much the same after a further 35,000 hr of high-load running, and now there would appear to be little doubt that the erosion danger on these 3000-rpm 500-MW LP turbines had been exaggerated, shield life expectancy being at least 10 yr.

Interest in HP turbine water extraction dates from the design of large machines for boiling water (BWR) and pressurized water (PWR) reactors, although the problem is equally relevant to the Canadian Candu system, this being effectively a PWR with heavy instead of light water in the primary cooling circuit. In common with the steam-generating heavy-water reactor (SGHWR), these systems produce dry saturated steam, either in the reactor or in external heat exchangers, at pressure levels of 40–70 bars, so that the respective inlet enthalpies to the turbines fall below the peak saturation level. Consequently, the steam is throttled to a slightly wet condition before it enters the HP turbine and, if no measures were subsequently taken to extract moisture and reheat the steam, a terminal wetness in excess of 20% would result.

Because of the lower overall heat drop of the nuclear cycle, as compared with a modern fossil-fired reheat system, the turbine expansion is usually divided into HP and LP cylinders only, no intermediate-pressure turbine being necessary. The crossover pressure in the case of Pickering power station [5-17] is 5.25 bars, after which the steam is passed through a water separator and is superheated in a live-steam reheater before entering three double-flow LP turbines (Fig. 5.2.3). The equilibrium moisture fraction at exhaust from the HP turbine can consequently reach 12%, which not only introduces an erosion risk to the exhaust blading but also can reduce the overall HP turbine efficiency by some 6 points.

Since interstage cylinder-extraction belts tend to increase HP turbine-shaft lengths, experiments were initiated at Parsons to remove moisture through narrow circumferential slots in the cylinder wall above the tips of unshrouded moving-blade rows. A seven-stage experimental 50%-reaction turbine was employed with a 4-mm slot located close to the axial center line of the fourth-stage moving row. Because of the small heat drop over the turbine, however, it was necessary to wet the steam artificially by a ring of nozzles before the first stage. At a slot-pressure level of 1.5 bars, it was found that 18% of the available water would be removed, the equilibrium wetness being 7.3% after having made allowances for the additional shaft work extracted.

The validity of this result was questioned on the grounds that little time was available for the superheated steam and injected water to reach thermal equilibrium, so that should a proportion of the injected water remain unevaporated, the extraction effectiveness would tend to be exaggerated; moreover, the coarseness of the water produced by this particular injection technique was not considered to be representative. It was subsequently decided to conduct a much more comprehensive development program on an old paper mill turbine "Inveresk," using steam naturally wetted from a superheated condition by work extraction.

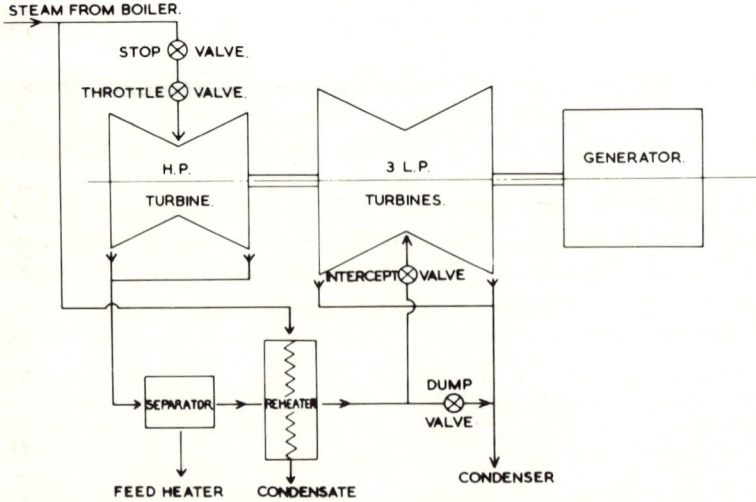

FIG. 5.2.3 Simplified diagram of steam path for Pickering turbine.

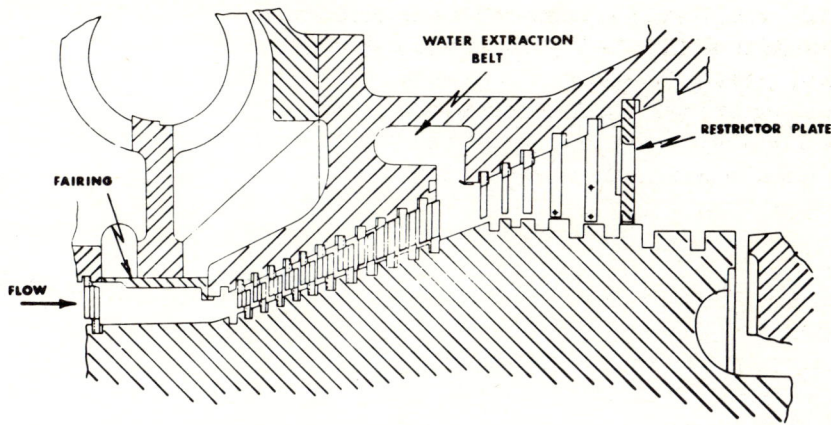

FIG. 5.2.4 Part section through experimental turbine.

A circumferential cylinder belt, originally used for axial thrust balancing, was machined to receive circumferential rings to model different extraction arrangements. To enable the equilibrium moisture content to be calculated before extraction, however, all but the first stationary-blade row downstream from the belt were removed, so that by knowing the inlet enthalpy, brake output, and condensed flow, the total enthalpy drop per unit mass could be calculated. The moisture fraction before the belt could subsequently be established at the measured extraction pressure from steam tables. The reason for including the one stationary downstream row was to preserve the aerodynamic flow condition in the vicinity of the belt.

In all, 29 of the 50%-reaction stages remained, the last four of which were replacements, representing the proposed HP turbine-blade geometry at Pickering (Fig. 5.2.4). Provision was also made to establish the influence of steam bleed on the water-extraction rate by piping the mixture initially through a centrifugal Weber separator, which was followed in series by a 1-μm cut-size wire-mesh separator, before condensing the dried steam for metering (Fig. 5.2.5).

Rotational speed and pressure ratio were normally kept constant during tests so that the heat drop, and therefore the mean velocity ratio across the blading, remained nominally unchanged. Thus by keeping the inlet superheat constant at 8°C, a sensibly constant moisture fraction of 6% could be maintained before the belt over the test range of extraction pressures.

The maximum extraction belt pressure was limited to 3 bars at a wetness of 6%, which raised the turbine inlet pressure to its maximum permissible value, the residual heat drop downstream from the belt being dissipated by a restrictor to protect the turbine exhaust casing and condenser. A lower extraction pressure limit of 1 bar was also imposed by inaccuracy of the torque measurements of the hydraulic dynamometer at low powers. The

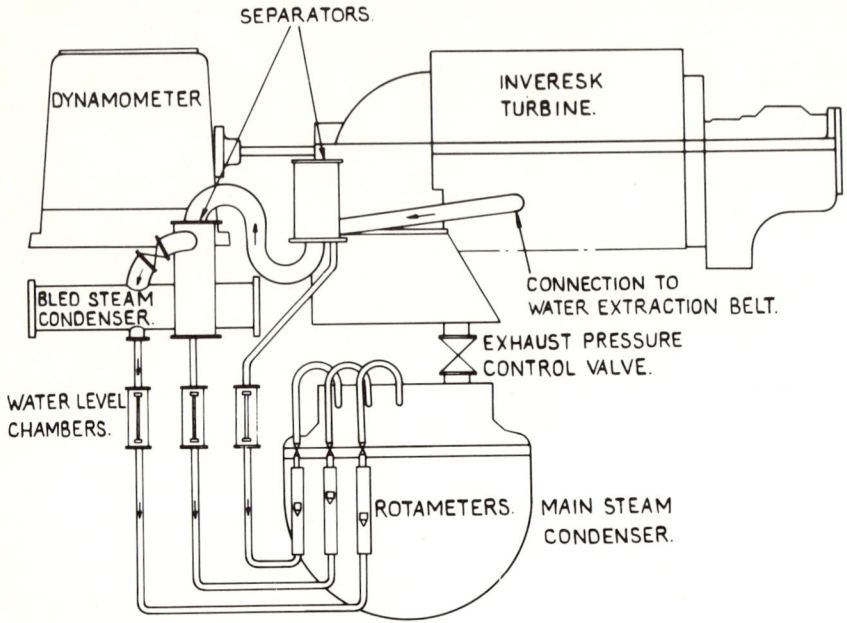

FIG. 5.2.5 Cylinder-water extraction arrangement on Inveresk turbine.

efficiency of each belt was expressed as the percentage of water extracted, compared to the theoretical equilibrium water content of the steam before the belt, and for test convenience the bleed quoted was the condensed-steam quantity after external water separation, expressed as a percentage of total inlet turbine flow.

The performance of a conventional open belt was improved from 23 to 30% at 1.2 bars by a 5% steam bleed, but fell at 2.7 bars to 16 and 23% respectively (Fig. 5.2.6). Attempts to improve efficiency by reducing the width of the port opening from 50 to 20% of the blade height were unsuccessful, as indeed were hooked arrangements designed to divert water into the belt (Fig. 5.2.7).

A system employing a louvred entry ultimately proved to be the most effective, the belt opening being faired flush to the cylinder wall by equally spaced plates around the circumference with small axially disposed drainage slots between them. Without bleed, the efficiency of this system was rather worse than the open belt, the respective values of 1.2 bar being 20 and 23%, but with steam bleeds in excess of 1%, a substantial improvement in efficiency resulted, a value of 46% being recorded with 5% bleed at 1.2 bar, which compares with 30% for the open belt (Fig. 5.2.6).

At higher pressure levels, efficiencies again fell, as in the case of the open belt, the respective zero and 5% bleed values at 1.2 bar falling from 20 and 46 to 12 and 38% at 2.7 bars (Fig. 5.2.6). However, performance was found to

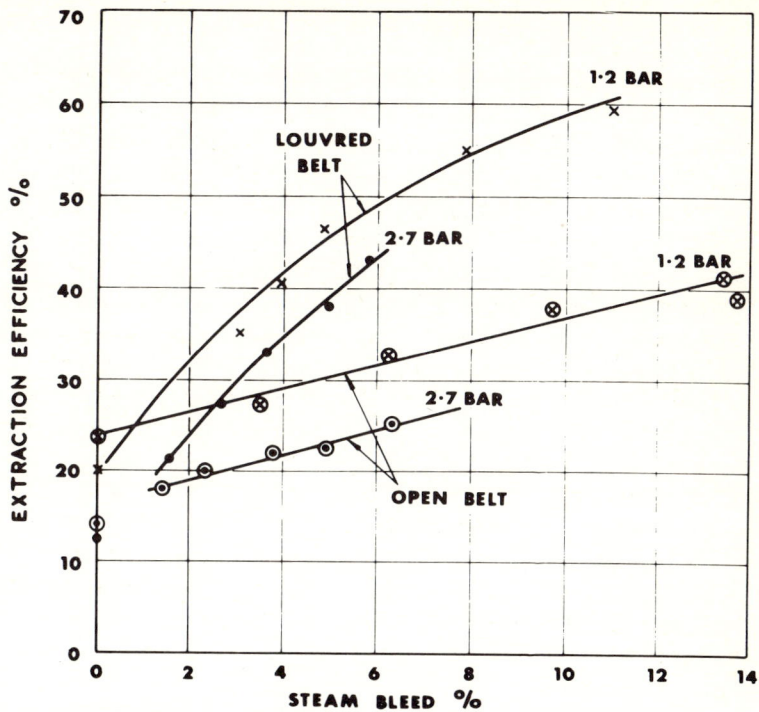

FIG. 5.2.6 Influence of steam bleed on extraction efficiency at two pressure levels on open and louvred belt arrangements in Inveresk.

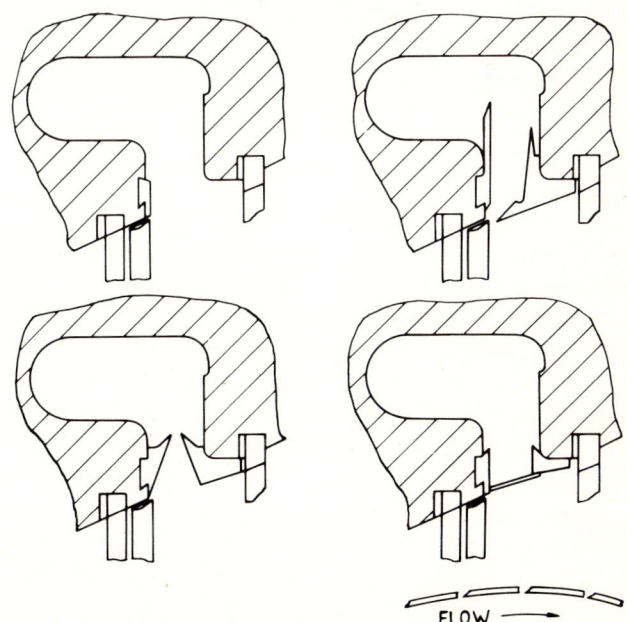

FIG. 5.2.7 Inveresk cylinder-water extraction belt arrangements.

be relatively insensitive to detailed changes in the louvre chord of the drainage-gap width.

The relative inferiority of the open belt over the louvred entry with bleed was thought to lie in the reatomization of the free water flowing along the cylinder wall at the sharp "corner" at entry to the belt. The louvred entry tended to avoid this by the bleed encouraging water to drain through the slots between the louvres into the quiescent zone beneath, where it was not subjected to the reatomizing influences of the main steam flow. Visual attempts to check this theory were impossible, however, the opacity of the fog being such that light transmission was stopped in 25 mm.

The severity of the falloff in extraction performance with pressure, coupled with large extrapolation necessary to estimate extraction rates in full-scale HP turbines from Inveresk results, resulted in the construction of a new high-pressure wet-steam experimental turbine "Mary" (Fig. 5.2.8).

The fourteen 50%-reaction stages of this machine were subdivided into three sections, the first five stages being designed to have the same mean $\dot{P}$ [5-18] as the Pickering HP turbine, in an attempt to reproduce the nucleation conditions of the full-scale machine. The following five stages constituted a transitional section that fed the remaining four stages, modeled to have the same opening coefficients and aspect ratios as the full-scale machine. The stationary blades of this final section were mounted in a separate cylinder liner so that blading geometries could be more easily changed.

The turbine was designed for a maximum inlet pressure of 20 bars, which was as high as the Works boilers would allow, the inlet superheat being controlled by an elaborate desuperheating arrangement down to about 6°C. The steam was subsequently made wet by extracting work from a directly-coupled hydraulic dynamometer to a condition of 7 bars and 6% wetness before the belt. Although the resulting turbine was only 1.2 m long

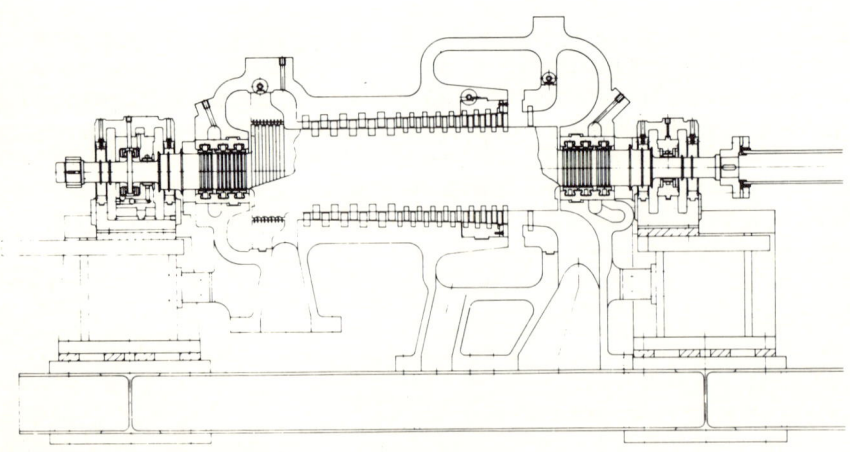

FIG. 5.2.8 Cross section through HP experimental turbine Mary.

between bearing centers, it produced 900 kW at a maximum rotational speed of 10,000 rpm.

Provision was made to change the extraction geometry after the last moving row, the belt being followed by a further stationary row of blading to maintain a representative flow condition past the water extraction zone. Two sets of sight glasses were also provided along each side of the cylinder cover so that steam conditions could be viewed across the top of the shaft between each of the first six stages.

Test procedure closely followed the earlier Inveresk program, rotational speed and pressure ratio being held constant to keep overall blade-velocity ratio and exhaust wetness sensibly unchanged. Although the centrifugal force imposed on blade-surface water deposits was approximately 5.5 times that of the full-scale machine, it was considered that the reduction in scale over Pickering by a factor of 5.5 would result in the same proportion of moisture being centrifuged to the cylinder walls by the moving blades as in the full-scale machine.

Where the lower test pressure levels of Mary overlapped the higher levels of Inveresk (2 bars), it was found that extraction efficiencies of the open belt were only about 2 points lower (17%) with no bleed, and 2 points higher (28%) with 5% bleed (Fig. 5.2.9). From this double logarithmic plot of extraction efficiency against pressure, it was found that the results from the new machine tended to follow a linear extrapolation of the old, their rates of falloff in efficiency with pressure agreeing closely.

Unlike the open belt, however, efficiencies of the louvred arrangement remained unaffected by pressure (Fig. 5.2.10), but fell below the Inveresk levels of 41 and 15% at 2 bars either with or without bleed, their respective values being 17 and 5%.

The reason for the generally poorer extraction performance of the open belt at higher pressures had been given as a smaller proportion of the blade-water deposits being centrifuged to the cylinder walls for extraction, the higher steam densities tending to strip and reentrain a greater proportion of the total rate of surface water deposition from the moving-blade trailing edges than at lower pressures. Such an explanation proved more difficult to reconcile with the louvred results, however.

The distribution of the water at the extraction point in Mary was subsequently located from a radial traverse by an isokinetic sampling probe. This involved removing a series of steam samples through a total head probe that had been oriented into the direction of flow. The rate of bleed was controlled to remove steam from the flow at a rate corresponding to the velocity of flow, so that entrained water would remain aerodynamically undeflected at the entrance to the probe. The water-to-vapor proportions of the samples were subsequently determined by passing the sample first through a water separator where the water phase was measured by Rotameter and condensing the remainder to measure the vapor phase. Contrary to

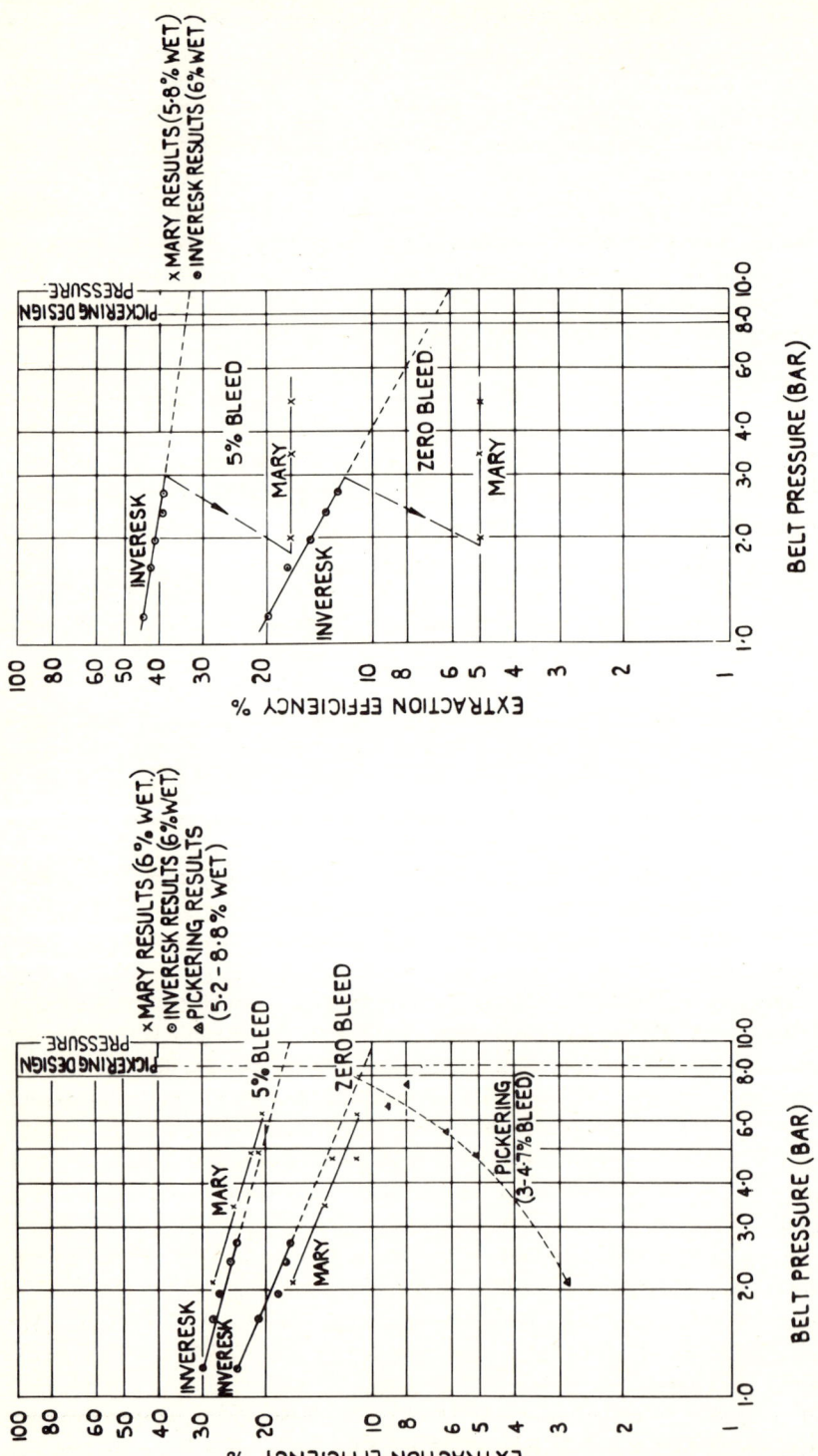

FIG. 5.2.10 Influence of pressure and bleed on efficiencies of louvred extraction belt.

FIG. 5.2.9 Influence of pressure and bleed on efficiencies of open extraction belt.

expectation, the greater proportion of the water was found to be close to the cylinder wall.

The change in performance of the louvred belt in the Mary turbine is currently thought to be associated with the increase in louvre-surface water loading over Inveresk, a factor that tends to increase with steam pressure at a given wetness. This probably causes a deeper water film to form between the slots, which can be drawn more readily past the slots under the near-axial drag of the steam, as well as permitting greater surface reentrainment back into the steam flow.

The HP turbine of the first Pickering machine was fitted with an identical belt to the Mary turbine belt after the 11th stage, where the full load pressure level was 9 bars and the design equilibrium wetness 12%. Ontario Hydro engineers subsequently estimated the extraction efficiencies by heat balance over a range of loads, and these have been superimposed over the Inveresk and Mary values (Fig. 5.2.10). It will be noted that the extraction efficiency rises with pressure (and load) until it is estimated to reach 11% at full pressure with a bleed of 4.7%. The full-scale result consequently falls above the model zero-bleed condition, but is some 7 points less than had been predicted with bleed.

On the strength of the Inveresk results, louvres were fitted into pre-machined circumferential grooves on the sides of the extraction belt in the second Pickering machine, their efficiency again being estimated by heat balance at various loads. From the one set of readings taken, it was estimated that the extraction performance was similar to the open belt of the first machine, a result that was later confirmed from the Mary turbine (Fig. 5.2.10). Although both Pickering machines comfortably met their guaranteed outputs and heat rates, it was considered that the reduction in HP turbine water-extraction rates could accelerate erosion damage to the moving blades downstream from the belt. This fear proved groundless, however, since the blade inspection on the first machine after 7900 hr showed the HP as well as the LP blading to be in pristine condition. Even in the thin-tipped area at the moving-blade tips, where one might have expected greater evidence of erosion than elsewhere, pitting was totally absent; in fact, the only visual evidence of water was a degree of worming erosion on a mild-steel cylinder-blade packing section that had inadvertently been used during construction.

After allowing for the coarse water extracted from the belt and the additional work extracted from the remaining HP stages, the moisture content of the steam at entry to the external separators approaches 12%, which has to be removed before reheating the steam for the LP turbines. Straight-through cyclone-type separators were fitted horizontally in each of the four HP turbine exhaust ducts at Pickering to remove this moisture before the steam was reheated. These were fitted with stationary swirler vanes followed by a louvred cylindrical liner, the system from which the louvred cylinder extraction belts had been derived (Fig. 5.2.11). Using such liners to protect

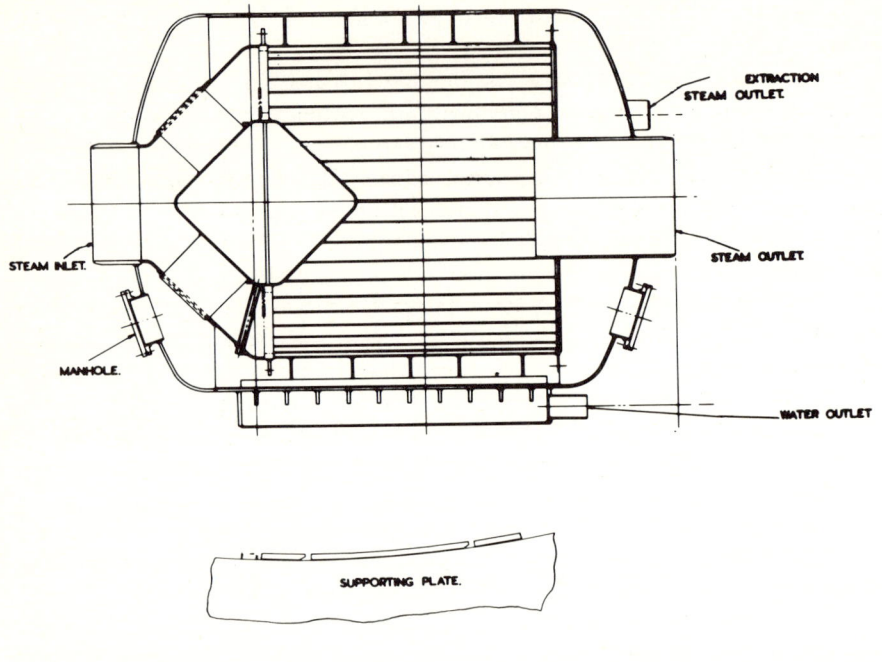

FIG. 5.2.11 Cyclone separator.

the detrained water from the reatomizing influences of the main steam flow, it had been demonstrated on a fifth scale model supplied with naturally wetted steam from the Inveresk turbine that efficiencies of over 95% could be realized, with a tenfold increase in flow rating over a conventional cyclone separator in droplet-size ranges above 30 μm SMD.

The proportions of the full-scale separators were designed on the expectation that the majority of the water would be in the form of second-generation droplets, and were therefore based on the stripping sizes from the last blade row of the HP turbine [5-15]. This approach was subsequently justified by the efficiency of 97% recorded by the Ontario Hydro engineers.

Both woven-wire mesh and impaction-type corrugated-plate separators have been used by other manufacturers, but their requirement for low specific momentum of the steam at inlet makes them bulky, so that they are usually combined with the reheater in the same vessel. Test results would indicate that such units can be effective if their inlet-velocity distribution is satisfactory.

5.3 EROSION PROTECTION

The protection of turbines from water erosion falls essentially into two parts, the first being concerned with the protection of cylinders and liners from worming erosion [5-19] by free water, and the second, blade protection from collision with droplets in the vapor.

The first condition is usually confined to HP wet-steam turbines where, if double-casing construction is employed to reduce cylinder stresses, a high-pressure differential can arise across the supporting faces between the inner and outer cylinders, or across horizontal joints. Water reaching these interfaces can be close to the prevailing saturation temperature so that it flashes into steam as the pressure is reduced in leaking through any small clearances. The resulting high velocity of the unevaporated water can cause damage to the metal interfaces. Turbines supplied from both BWRs and PWRs are prone to this worming or "wire drawing" erosion, as it is sometimes called, but an additional hazard of the BWR and SGHWR systems is that the irradiated steam that passes into the turbines has an abnormally high oxygen content, so that the erosion process is supplemented by significant corrosive attack.

Autoclave tests were conducted on many materials prior to construction of the Dresden BWR plant [5-19], so that the turbine cylinder interfaces could be protected from these corrosive and erosive forms of attack. Monel and Stellite were found to be the most satisfactory, probably because of their high corrosion resistance, but steels with low nickel and chrome contents were noted to be generally more resistant to attack than steels with molybdenum and high carbon content.

At Pickering, the HP turbine cylinders were made of $2\frac{1}{4}\%$ chrome steel, the joints of the inner cylinder and the seatings between the inner and outer casing being protected by 12% chrome weld deposits, so as to retain the same thermal expansion properties to the ferritic cylinder steel and avoid cracking at weld interfaces by thermal cycling. The rotor was also forged from $2\frac{1}{4}\%$ chrome steel, the hollow drum construction enabling the stub ends, carrying the journals, to have a lower chrome content to avoid "wire wool"-type bearing failures. Inspection after 7900 hr revealed no defects from worming erosion, and a further 17,000 hr of high-load running have since been satisfactorily completed.

An early but accurate account of the blade-erosion process was made by F. W. Gardner [5-20] in 1932, this being followed by a more detailed physical account of events leading to erosion by G. C. Gardner [5-21] in 1962.

Blade erosion is currently visualized as being preceded by the collection of relatively large water droplets on the concave pressure surfaces of blades, where they form a film. In the moving blades this film tends to be centrifuged to the cylinder wall, but in the stationary nozzles water is drawn from this film toward the trailing edges by the drag of the steam. Here the film grows in thickness and may even be sucked around the trailing edge onto the suction surface in cases of separated flow [5-22], before being torn away by the steam drag forces. Much larger drops than those deposited are thus formed, and because of their mass they take longer to accelerate back to the free-stream vapor velocities; consequently they are struck by the backs of the following moving blades which, in the case of very large drops, would correspond to an impact velocity approaching the peripheral speed of rotation (Fig. 5.3.1). That such drops are responsible for erosion is confirmed by damage to the following moving blades being largely confined to the suction surface near the leading edges. Experiments on the Bess turbine, as well as at site, would indicate however that these large drops account for only a small proportion of the total droplet population; Thorpe and Wood [5-23] suggest that they account for 1-3% of the total water content, as compared with Bess model measurements of $\frac{1}{3}$%.

The initial blade droplet collection process is somewhat obscure, since the droplet sizes are difficult to assess with accuracy in a turbine. Gardner [5-21], however, has suggested eddy diffusion through the blade surface boundary layer as a possible process, this being followed by inertial impact when the droplet masses are sufficiently large for their trajectories to differ from that of the steam when being turned through blading.

Freudenreich [5-24] estimated the maximum stable droplet size of atomized water by equating the pressure created by steam drag on spherical droplets with the internal pressure maintained by surface tension. The ratio of these dynamic-to-surface-tension forces is now known as Weber number (We = $\rho C^2 d_c / \sigma$), which has been used as a basis for droplet-size correlation in the stripping process of water films from blading by the steam. The distance between the nozzle and the following moving-blade row, however, is usually insufficient to allow complete break-up to occur, since the water is initially concentrated in the momentum-defect zone of the blade trailing-edge wakes. Consequently, experimental determination of these droplet sizes has been considered necessary [5-15].

Collision between the stripped water and the following moving blades constitutes the final phase of the erosion process, the stresses in the surface of

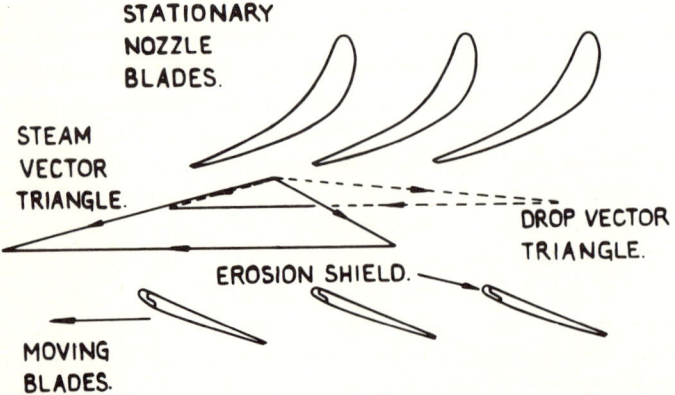

FIG. 5.3.1 Angle of impingement of droplets onto tips of moving-blade row.

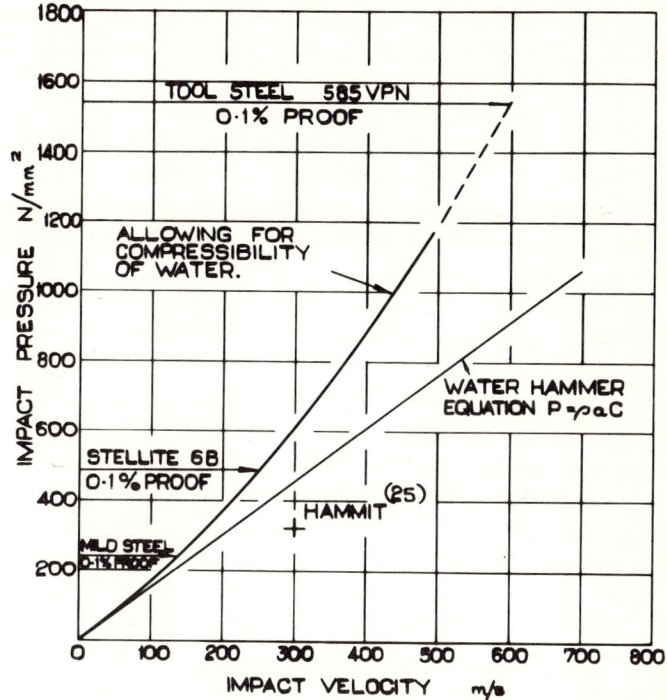

FIG. 5.3.2 Variation of impact pressure with velocity.

the material normally being identified with the pressures produced by the water-hammer relationship ($P = \rho a u$). Although this is a crude approximation, because of factors such as the shape and size of drop [5-25], the elasticity of the material under impact [5-26], the condition of the metal surface, and the hydrodynamic and thermodynamic behavior of water under high rates of compression, it still serves to illustrate the order of stress levels (Fig. 5.3.2). At 600 m/s, the tip speed of many high-speed LP turbines, impact pressures rise to 1000 N/mm², which with the exception of a few special steels is considerably in excess of the yield stress of most materials.

Although designers can limit erosion damage either by employing large axial spacing between the LP nozzle and moving-blade rows, to allow greater time for droplet break-up and acceleration, or by fitting shields of suitable material along the moving-blade leading edges, it is inevitable that some degree of damage will be suffered in the 20-yr lifespan of a turbine.

Experiments to establish the relative erosion resistance of materials to droplet impact [5-27] indicate that damage does not occur until the normal component of the impact velocity u has reached a critical value u_c, above which the rate of weight loss per unit mass of impacting water increases rapidly, following a power relationship of the form $dm/dw = K(u \sin \theta - u_c)^z$.

The use of an experimental value of 3 for the exponent z resulted in the life prediction of three months for the erosion shields of the last-row LP blading of the CEGB's 500-MW sets (see p. 269).

A contrarotating erosion machine built at Heaton Works [5-28] was capable of checking the behavior of erosion-shield specimens up to impact speeds of 610 m/s. The contrarotating principle was employed to reduce the disk stresses, four specimens being equally spaced around the periphery of one disk, while a second carried two pressure-atomizing jets rotating in the opposite direction. Both were contained in a vacuum chamber fitted with antiswirl and water-deentrainment devices. The impact angle onto the specimens could be varied by locking them at different angles to the radial direction in their holders; moreover, some measure of droplet-size control was also possible, by changing either the jet size or water pressure by adjusting the relative shaft speeds.

Specimens were removed and weighed at prescribed intervals, keeping the impact speed and water supply constant, from which it was established that materials did not erode continuously at the same rate when subjected to impact velocities above the critical value. In the initial or incubation zone, no weight loss could be measured, this being followed by a secondary zone where the rate of weight loss rose to a maximum, and finally by a tertiary zone where the rate fell to a substantially steady-state value that could be less than half the maximum rate (Fig. 5.3.3). At Heaton Works, materials have subsequently been compared on their performance in this tertiary zone, since it is there that turbines are considered to spend most of their lives. Others, however, have considered maximum rate to be a better criterion [5-29].

Discrepancies between the erosion rates of similar materials from tests on different rigs ultimately led to comparisons being made on three standardized materials on the four rotating impact rigs being used in the United Kingdom [5-30]. These showed that droplet size played almost as decisive a role as impact speed in determining the rate of erosion damage, as well as influencing the duration of the respective zones.

Duration of the incubation period was found to be extended by increasing the size of the impacting drops, this being identified with a fatigue process, since for a given mass of impacting water the number of stress cycles was diminished. Micrographic examination of the specimen surfaces also revealed cracks developing from the specimen surface, their depths being coupled to the size of impacting drops. These cracks subsequently appeared to extend under further impact and then intersect to release material from the surface in the secondary zone, causing rapid weight loss, the rate of loss again being related to crack depth and therefore the drop size. Explanations for the reduction in the rate of weight loss to a near steady-state value in the tertiary zone, however, have not been satisfactory; suggestions that the retention of protective water films in the craters, or attenuation of the stress levels by oblique impact on the slopes of these craters [5-29], are not entirely convincing.

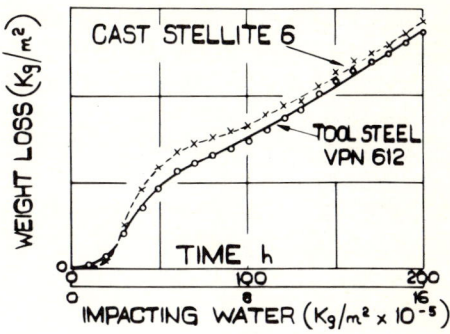

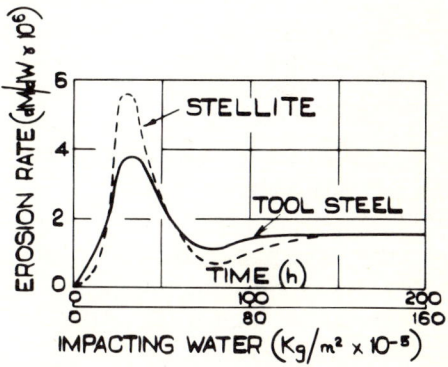

FIG. 5.3.3 Comparison of weight loss and erosion rates of tool steel and Stellite 6 with time.

Whereas the erosion process is now known to differ considerably in the four test rigs because of differences in the spray, the comparative order of erosion resistance of the three test materials was found to be similar. Many materials have since been compared in these rigs, and some, such as wrought Stellite 12 and 10W-10Co-4Cr tool steel, would appear to offer greater erosion protection than wrought Stellite 6 or the hardened 18W-6Cr tool steels in common use.

Frequently hardness has been considered the prime consideration in determining erosion resistance, but the similarity in performance of Stellite and tool steel with respective hardnesses of 500 and 700 VPN tends to contradict this. Such values, however, tend to reflect the hardnesses of their base matrices and not the hardness of their carbide content, which may exceed 1000 VPN. It is these carbides, in combination with a fatigue-resistant matrix, that are thought to control erosion resistance. Since the carbide content of tool steel is directly linked with hardness, this explains why this property was initially considered to be so important in judging materials.

Some manufacturers prefer to protect the leading edges of their LP moving blades by local hardening because of difficulties that can arise in providing a bond of high mechanical integrity between the shield and the blade. Parsons still prefers to employ hardened tool steel shields, although they tend to be less erosion resistant than the best of the Stellite at practical hardness levels. The reason for this preference lies in the similarity of the linear expansion coefficients of tool steels with the blade stock material, so that thermal stresses produced during the brazing attachment process are minimized. Stellites, which because of their cobalt base have a 12% greater linear expansion than steels, consequently require greater care during brazing, if shield cracking or detachment is to be avoided.

As LP blade-tip speeds rise with machine size, the higher expansion rates tend to reduce primary droplet sizes, so that the deposition rate of potentially damaging water in nozzle blade surfaces tends to be reduced. In addition to the reduction in free water, both the size and duration of impact of stripped droplets tend to be reduced, so that with reasonable axial spacing between the last-stage nozzles and moving blades to permit acceleration and reatomization, erosion problems are not expected to increase significantly with machine size.

NOMENCLATURE

a	acoustic velocity in water
C	blade efflux velocity
C_0	velocity corresponding to row isentropic static heat drop
d	droplet diameter
d_c	maximum stable droplet diameter
d_{32}	Sauter mean diameter (SMD = $\Sigma d^3 / \Sigma d^2$)
K	constant
m	mass of specimen
o	blade opening
o/p	blade-opening coefficient
p	blade pitch
P	static pressure
$\dot{P}$	expansion-rate parameter $(1/P)(dP/dt)$
t	time
u	water-drop absolute velocity
u_c	critical impact velocity
u/C	ratio of moisture-to-steam velocity into next blade row
U	blade speed
U/C	blade-to-steam velocity ratio
U/C_0	blade-to-steam isentropic velocity ratio
w	mass of impacting water
We	Weber number
Y_m	arithmetic mean wetness fraction
z	impact velocity exponent
α	wetness-loss coefficient
Δhd	actual heat drop in dry zone
Δhw	actual heat drop in wet zone

η_d dry isentropic efficiency
η_0 isentropic efficiency over both wet and dry expansion
η_w wet isentropic efficiency
ρ steam density
σ surface tension
θ angle of impact to surface

REFERENCES

5-1 Baumann, K.: Some Recent Developments in Large Steam Turbine Practice, *J. Inst. Elec. Eng.*, vol. 59, p. 565, 1921.
5-2 Stodola, A.: "Steam and Gas Turbines," p. 313, McGraw-Hill, New York, 1927.
5-3 Soderberg, C. R.: Importance of Moisture in Steam Mixtures, *Elec. J.*, vol. 31, p. 285, July, 1934.
5-4 Traupel, W.: "Thermische Turbomaschinen," vol. 1, p. 297, Springer, Berlin, 1958.
5-5 Miller, E. H. and Schofield, P.: The Performance of Large Steam Turbines with Water Reactors, *ASME Winter Ann. Mtng.*, 1972.
5-6 Gyarmathy, G., Lesch, F., and Siegenthaler, A.: Spontaneous Condensation of Steam at High Pressures (first experimental results), *Inst. Mech. Eng., T & FM Conf. Publ.* 3, p. 186, 1973.
5-7 Moore, M. J., Walters, P., Crane, R. I., and Davidson, B. J.: Prediction of Fog Drop Size in Wet-Steam Turbines, *Inst. Mech. Eng., T & FM Conf. Publ.* 3, p. 107, 1973.
5-8 Smith, A.: The Influence of Moisture on the Efficiency of a One-Third-Scale Model Low-Pressure Turbine, *Proc. Inst. Mech. Eng.*, vol. 180, pt. 3J, 1965-1966.
5-9 Parsons, R. H.: "The Development of the Parsons Steam Turbine," p. 228, Constable, London, 1936.
5-10 Harris, F. R.: Nuclear Wet-Steam Turbines, *Proc. Inst. Mech. Eng.*, vol. 183, pt. 30, p. 22, 1968-1969.
5-11 Deich, M. E., Filippov, G. A., Pryakhin, V. V., and Povarov, A.: Experimental Investigation of Moisture Separation in Turbine Stages of Average Divergence. *Therm. Eng.*, vol. 1, p. 65, 1968.
5-12 Christie, D. G., and Hayward, G. W.: Observation of Events Leading to the Formation of Water Drops which Cause Turbine-Blade Erosion, *Phil. Trans. Roy. Soc. (London)*, ser. A, no. 1110, vol. 260, p. 183, 1966.
5-13 Ryley, D. J., and Parker, G. J.: The Removal of Water from Low-Pressure Steam-Turbine Blades by Trailing-Edge Suction Slots, *Inst. Mech. Eng., Wet Steam 2*, The Institution, p. 9, 1970.
5-14 Kirillov, I. I., Nosovitskii, A. I., Shpenzer, G. G., and Naumchik, B. V.: Increasing the Separation Efficiency in the End-Stage Blade Passages of Large Steam Turbines, *Izv. VUZ-Energetika*, no. 11, pp. 122-126, Nov., 1969.
5-15 McAllister, D. H., and Moore, C. T.: Water Drainage from a Cascade of Hollow-Slotted Low-Pressure Turbine Blades, *Inst. Mech. Eng., Wet Steam 2*, The Institution, p. 69, 1970.
5-16 Todd, K. W., and Gregory, B.: An Experiment on Erosion Control, Using a 60 MW Steam Turbine, *Inst. Mech. Eng., Wet Steam 2*, The Institution, p. 51, 1970.
5-17 Parsons, N. C.: The Development of Large Wet-Steam Turbines, *Trans. N. E. Cst. Eng. Inst. Shipbuilders*, vol. 89, pp. 31-42, Dec., 1972.
5-18 Gyarmathy. G.: Basis for a Theory for Wet Steam, *Inst. Therm. Turbomach., Bull.* no. 6, E.T.H., Zurich, 1962.

5-19 Cataldi, C., Cheng, C. F., and Musick, V. S.: Investigations of Erosion and Corrosion of Turbine Materials in Wet Oxygenated Steam, *Trans. ASME,* vol. 80, p. 1465, 1958.

5-20 Gardner, F. W.: The Erosion of Steam-Turbine Blades, *The Engineer,* vol. 153, pp. 146-147; 174-176; 202-205, 1932.

5-21 Gardner, G. C.: Events Leading to Erosion in the Steam Turbine, *Proc. Inst. Mech. Eng.,* vol. 178, no. 23, p. 593, 1963-1964.

5-22 Christie, D. G., Hayward, G. W., and MacDonald, A. N.: The Formation of Water Drops which Cause Turbine Blade Erosion, *Proc. Inst. Mech. Eng.,* vol. 180, pt. 30, p. 13, 1965-1966.

5-23 Thorpe, A. D., and Wood, M. R.: Coarse-Water Distribution within the Low-Pressure Stages of a 350 MW Steam Turbine, *Inst. Mech. Eng., Wet Steam 2,* The Institution, p. 19, 1970.

5-24 Freudenreich, J. V.: Einfluss der Dampfnasse auf Dampfturbinen, *Zeits VDI,* pp. 664-667, May 14, 1927; *Brown Boveri Review,* vol. 14, pp. 119-124, May 27, 1927.

5-25 Hammitt, F. G., and Huang, Y. C.: Liquid Droplet Impingement Studies at the University of Michigan, *Inst. Mech. Eng. T & FM Conf. Publ.* 3, p. 237, 1973.

5-26 Hancox, N. L., and Brunton, J. H.: The Erosion of Solids by Repeated Impact of Liquid Drops, *Phil. Trans. Roy. Soc. (London),* ser. A, no. 1110, vol. 260, p. 121, 1966.

5-27 Baker, D. W. C., and Eaton, J. L.: Steel to Combat Turbine-Blade Erosion, *Engineering, London,* vol. 195, p. 302, Feb. 22, 1963.

5-28 Smith, A.: Physical Aspects of Blade Erosion by Wet Steam, *Phil. Trans. Roy. Soc. (London),* ser. A, no. 1110, vol. 260, p. 209, 1966.

5-29 Smith, A., Kent, R. P., and Armstrong, R. L.: Erosion of Steam-Turbine Shield Materials, *ASTM, STP,* no. 408, p. 125, 1967.

5-30 Elliott, D. E., Marriott, J. B., and Smith, A.: Comparison of Erosion Resistance of Standard Steam-Turbine Blade and Shield Materials on Four Test Rigs, *ASTM, STP,* no. 474, p. 127, 1970.

CHAPTER 6

Operating Experience of Wet-Steam Turbines

W. Engelke

6.1 EROSION–CORROSION

The phenomenon of erosion-corrosion is a vitally important subject for turbines in light-water reactor plants. Although the effects of erosion-corrosion bear a resemblance to those of end-blade erosion, they are due to a completely different phenomenon. Erosion-corrosion takes place only in the temperature range 40–260°C with saturated steam or water in turbines, valves, moisture separator-reheaters, and piping systems.

6.1.1 Causes of Erosion-Corrosion

Erosion-corrosion is caused by Fe^{2+} ions being dissolved by pure water and forming a solution of iron hydroxide. Figure 6.1.1 shows erosion-corrosion of an unprotected horizontal joint surface of a very early turbine for a nuclear research center after 23,000 operating hr. The pressure differential at this location is 30-2 bar and the steam moisture 2.5%. Figure 6.1.2 is a photograph of a badly erosion-corroded area of the turbine blade carrier after the same length of service, where the steam is at a pressure of 4.5 bar and has 10% moisture.

Erosion-corrosion is prevented in newer nuclear turbines either by using high-alloy chromium steel for endangered parts or by armoring exposed areas with similar material. Erosion-corrosion of 2.5% chromium steel drops to about one-quarter of that to be expected with nonalloy steel, whereas 12–13% chromium steel suffers none at all. Figure 6.1.3 shows the 12% chromium inner casing of a wet-steam turbine after 8000 operating hr, which shows no influence of erosion-corrosion. This is due to the different chemical reaction of alloy steel. Other material properties, such as hardness, do not combat the erosion-corrosion attack.

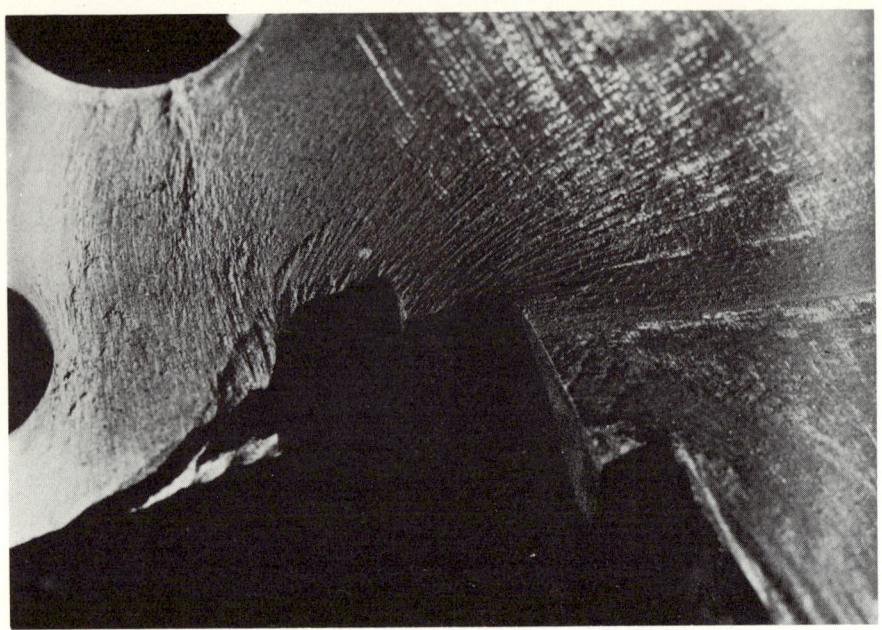

FIG. 6.1.1 Erosion-corrosion of an uncoated horizontal turbine joint after a service time of 23,000 hr (steam conditions: $p_1 = 30$ bar, $y_1 = 2.5\%$, $p_2 = 2$ bar).

FIG. 6.1.2 Erosion-corrosion of an uncoated blade carrier above the moving blades after a service time of 23,000 hr (steam conditions: $p = 4.5$ bar, $y = 10\%$).

FIG. 6.1.3 A 12% chromium inner casing of a wet-steam turbine after 8,000 operating hr ($p = 9$ bar, $y = 12\%$).

Early research indicated that erosion-corrosion can be reduced to negligible values if the pH value of the water film, to which the iron surface is directly exposed, is maintained higher than 9.5. With steam turbines and their accessories, however, where the attacking medium is not just water but a combination of mostly steam and relatively little water, the pH value of the feedwater was found not to be a reliable criterion. A logical explanation for this seems to be that the water that separates out of the steam in the course of expansion below the saturation line has a much smaller alkalinity than that of the hydrazine-dosed feedwater.

6.1.2 Evaluation of Erosion-Corrosion[*]

In order to evaluate the erosion-corrosion attack on nonalloy steel, four factors had to be taken into account: temperature $f(\vartheta)$, velocity $f(c)$, moisture $f(y)$, and flow configuration $f(K_c)$. The erosion-corrosion effect can be formulated as

$$s = f(\vartheta)c\sqrt{y}\,K_c - \frac{1 \text{ mm}}{10{,}000 \text{ h}}$$

[*]Based on [6-1].

where s is the maximum local abrasion depth in the material in mm/10,000 hr of operation.

In Fig. 6.1.4, the influence of the steam or water temperature on erosion-corrosion is plotted. The dimensionless factor $f(\vartheta)$ exists to an appreciable extent only within the temperature range 40–260°C. The severest damage occurs at about 180°C. Erosion-corrosion falls off at lower temperatures because of the slower rate of chemical reaction, and at higher temperatures because of the formation of a protective layer of magnetite on the metal surface.

The second factor is the steam velocity. Erosion-corrosion has been observed to be proportional to the velocity with geometrically and hydraulically similar flows. It is a well-known phenomenon that the speed of chemical reaction increases directly with the velocity of the medium. As is shown in Fig. 6.1.5, pt. 1, an increase of velocity leads to a bombardment of the pipe wall with droplets of low ion concentration from the primary steam flow. The differences due to dissimilar flows in various turbine parts are taken into account by the configuration factor K_c. Consequently, the steam velocity c in m/s is used directly as the value for the influence of the velocity on the erosion-corrosion effect.

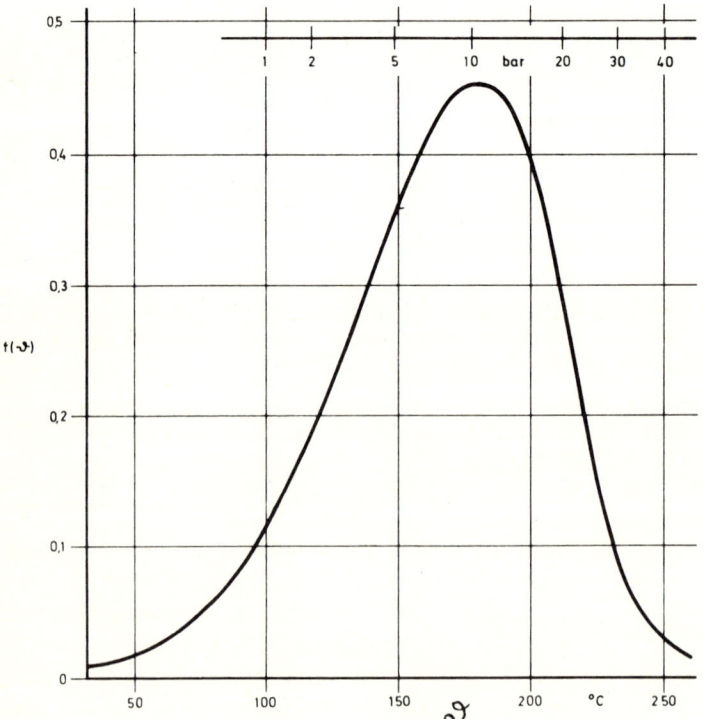

FIG. 6.1.4 Influence of temperature on erosion-corrosion.

1. Influence of velocity of saturated steam — f(c)

Waterfilm with high ion-concentration at the walls

At low velocity less water droplets travel with the steam flow and the water film remains on the wall

Within the same time at high velocity more on the wall and water with low ion concentration makes contact with the wall

2. Influence of moisture of saturated steam — f(x)

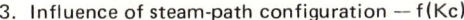

Small amount of moisture leads to less water with low ion concentration making contact with the wall, despite the thin water film

High moisture means more droplets of low ion concentration penetrate the water film on the wall and make contact with the metal

3. Influence of steam-path configuration — f(Kc)

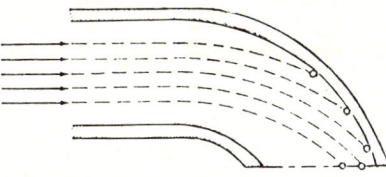 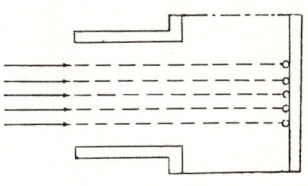

Elbow pipe with a large radius leads to only a small number of droplets hitting the wall. Droplets dip smoothly into the water film

T-shaped pipe completely disturbs the water film on the wall. Droplets hit the wall at high velocity, penetrate the water film and produce partial flows of water of low ion concentration along the wall

FIG. 6.1.5 Influence of steam velocity, moisture, and configuration on erosion-corrosion.

The influence of moisture on erosion-corrosion, just like that of temperature and configuration, was ascertained empirically by investigation of the first turbines for nuclear research centers. It was found that even a low amount of moisture leads to erosion-corrosion, which can be explained by the fact that a very thin water film is not sufficient to prevent the water droplets in the steam flow from coming into contact with the surface of the metal. The increase of erosion-corrosion with greater moisture is evident (Fig. 6.1.5, pt. 2) but is not directly proportional. From an examination of wet-steam turbines, as well as conventional nonreheat machines, the influence of moisture was found to approximate best to the square root of the moisture content as $\sqrt{y}$.

Empirical values of K_c for some pipe arrangements and turbine parts are tabulated in Fig. 6.1.6. The worst flow conditions in pipes occur in T-shaped configurations, with the water droplets of low ion concentration from the primary flow impinging directly on a wall and producing partial flows along the wall in all directions (Fig. 6.1.5, pt. 3). The least harmful flow, or the lowest K_c value of only 0.04, occurs in a straight pipe where only marginal vortices can cause an exchange of the water in the film on the wall surface.

It was also found that due to the normal passivation of the cycle, erosion-corrosion takes place only if a certain threshold value is exceeded. This is covered in our evaluation by the constant term 1 mm/10,000 hr, which means that if the multiplication of the four factors results in less than 1 mm/10,000 hr, there is no danger of an erosion-corrosion attack.

6.1.3 Agreement of Predictions with Measurements

In Fig. 6.1.7, the values of s that were predicted by these calculations are compared with values s that were measured during inspections of turbines or associated parts. The measured maximum local abrasions lie, with few exceptions, within an expected tolerance band delineated by hatched lines. In two cases a wall was actually eaten through by erosion-corrosion, but no leakage external to the turbine occurred, so that the punctures were discovered only during routine maintenance outages. The approximately realistic values of s in these two cases are indicated by arrows.

When one considers the complexity of the erosion-corrosion phenomenon, the quantitative-evaluation method has yielded remarkably accurate predictions. Most of the data that were used to establish the various factors for such a quantitative approach were collected from measurements made on the first turbines to operate with research light-water reactors. These turbines were largely unprotected against erosion-corrosion and thus clearly revealed the difference in attack intensity at various locations.

Reliable quantitative predictions of erosion-corrosion in all parts of a turbine and, even more important today, in all its accessories, are a most valuable guide to the turbine designer in making material selections and providing suitable armoring at critical locations to ensure the integrity of all parts in erosion-corrosion environments.

As a result of this analysis, today's wet-steam turbines and piping can be designed with a high grade of reliability against erosion-corrosion by selecting

	FLOW PATTERN		REFERENCE VELOCITY	K_c
PRIMARY FLOW STAGNATION POINTS	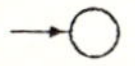	AT PIPES	VELOCITY OF INITIAL FLOW (UPSTREAM OF STAGNATION OBSTACLE)	1
		AT BLADES		1
	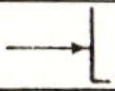	AT PLATES		1
		IN PIPE JUNCTIONS		1
				 0.8
SECONDARY FLOW STAGNATION POINTS	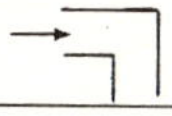	$R_{MEAN}/D = 0.5$	FLOW VELOCITY	0.7
		$R_{MEAN}/D = 1.5$ IN ELBOW PIPES		0.4
	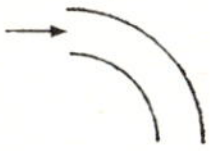	$R_{MEAN}/D = 2.5$		0.3
	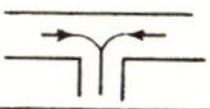	BEHIND PIPE JOINTS		0.2
STAGNATION POINTS DUE TO VORTEX FORMATION	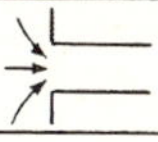	BEHIND SHARP EDGED ADMISSION PIPES	FLOW VELOCITY	0.2
	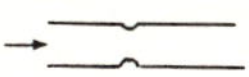	AT AND BEHIND BARRIERS		0.2
NO STAGNATION POINTS		IN STRAIGHT PIPES	FLOW VELOCITY	0.04
		IN UNTIGHT HORIZONTAL TURBINE JOINTS	VELOCITY CALCULATED FROM PRESSURE DROP	0.08
COMPLICATED FLOW THROUGH TURBINE PART	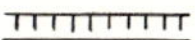	IN TURBINE GLAND SEAL	VELOCITY CALCULATED FROM PRESSURE DROP	0.08
	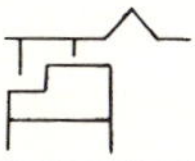	AT AND ABOVE TURBINE BLADES, AND AT DRAINAGE COLLECTING RINGS	AVERAGE CIRCUMFERENTIAL BLADE VELOCITY	0.3

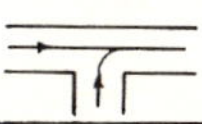

FIG. 6.1.6 Influence of steam-path configuration on erosion-corrosion (K_c in mm·s/m·10,000 hr).

the suitable material. For example, a 12% chromium steel is used for inner casing, blades, and gland rings in the high-pressure range. Critical areas of the piping system exposed to wet steam, such as elbows and pipe junctions, are constructed from 2.5% chromium alloy.

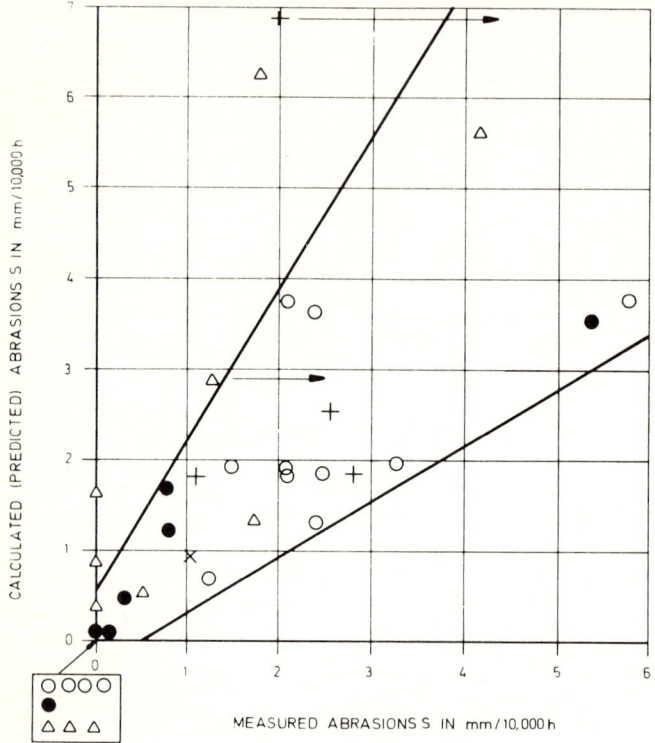

○ EROSION-CORROSION AT UNTIGHT HORIZONTAL JOINTS
● EROSION-CORROSION AT AND ABOVE BLADES AND AT DRAINAGE COLLECTION RINGS
△ EROSION-CORROSION IN ELBOW, JOINT AND JUNCTION PIPES
+ EROSION-CORROSION AT STAGNATION AREAS LIKE ACROSS STEAM PATH PIPES

FIG. 6.1.7 Comparison of predicted and measured values of erosion-corrosion.

6.2 EROSION OF LP BLADES

The development of large LP turbine end blades and the desire to use low-temperature cooling water instigated early investigations as to how to classify and minimize the effects of water erosion on these blades.

6.2.1 Causes of Erosion

All the experience and investigations revealed that water droplets within a definite size range are responsible for erosion. It was discovered that droplets with a diameter smaller than 50 μm are harmless. Further, droplets with a diameter larger than 400 μm do not occur in this LP turbine area.

Consequently, the water droplets that had to be observed were those of 50–400 μm dia.

This water-droplet size was found to form only downstream of the stationary blade trailing edges. The water being carried with the steam through the stationary-blade area is in the form of a fine spray of minute droplets. Knowledge of these effects allowed the factors influencing erosion to be narrowed down and the danger of erosion to be empirically calculated.

6.2.2 Erosion Coefficient

The severity of erosion as a function of operating conditions can be ascertained for any particular turbine [6-2] from

$$E = \frac{y_0^2 u^3 K}{p_0}$$

where E = erosion coefficient in $m^4/s \cdot kg$
 y_0 = moisture content (−) ahead of the last stage
 u = tip velocity of the end blades in m/s
 p_0 = pressure ahead of the last stage in N/m^2
 K = influence of blade and casing design

Knowledge about the value of the erosion coefficient E at rated flow conditions, the mode of operation, and the degree of partial-load service helps the designer decide on the different provisions necessary to prevent the turbine blade from excessive erosion.

Figure 6.2.1 shows the erosion coefficient as a function of load for fossil and nuclear turbines, and for different modes of operation. A higher value of E for wet-steam turbines is permissible because of their excellent partial-load erosion characteristics, which are mainly due to the fact that the expansion lines of wet-steam turbines terminate in drier steam areas of the Mollier chart than the expansion lines of fossil turbines as the load is decreased. The curves in Fig. 6.2.1 are based on the assumption of constant circulating-water flow through the condenser, with a resulting decrease in condenser pressure as the load is reduced.

The effects of the individual parameters on the occurrence of end-blade erosion follow.

Moisture content The water-droplet size, which is responsible for erosion, appears only downstream of the stationary-blade trailing edge, from which the water film is sprayed off. This water film occurs on either side of the stationary blade for different reasons. The film collects on the concave side from water droplets that are centrifuged as a result of a change in direction of the steam streamlines. The film on the convex side is formed from water droplets that are centrifuged from the next-to-the-last row of moving blades. This means that the erosion of the end blades depends on the moisture

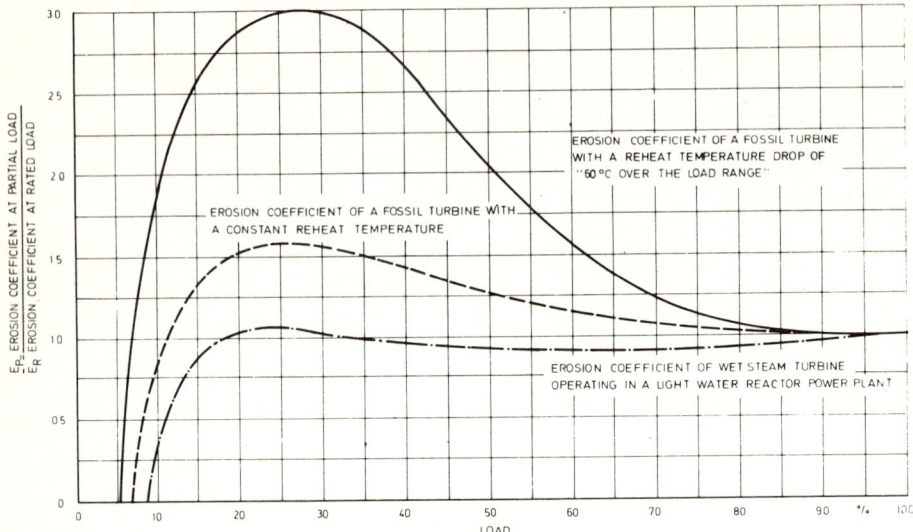

FIG. 6.2.1 Erosion coefficient as a function of load and type of steam cycle.

content of the steam entering the last stationary blades, because this content determines how much water will build up on both sides of the stationary blades (Fig. 6.2.2).

Tip velocity of the end blades With increasing tip speed the steam velocity also increases, but the velocity of the water droplets that leave the

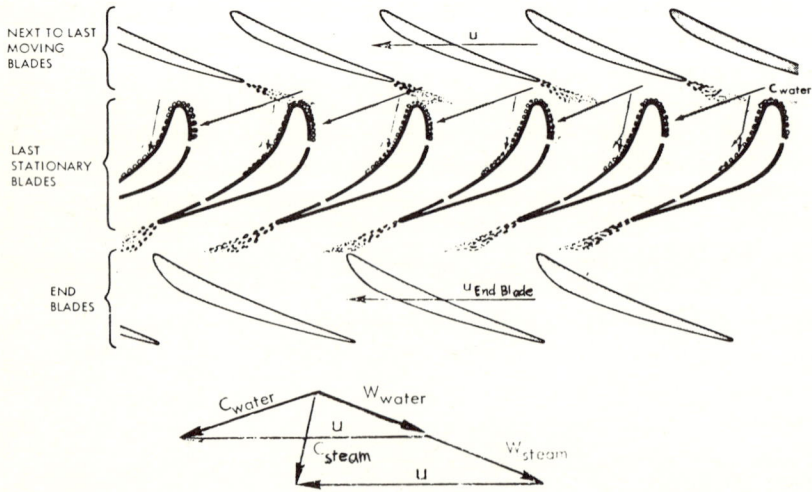

FIG. 6.2.2 End-blade erosion as a function of moisture in front of the last stationary blades.

stationary-blade trailing edges is always about zero. If the acceleration of the water droplets being sprayed off the stationary blades remains the same for all end-blade lengths, the erosion increases as a function of the tip velocity. Only if there is no relative velocity between the steam and the water droplets does the water flow with the steam through the end-blade area without causing erosion.

Steam density The steam density between the stationary- and moving-blade rows of the last stage affects the acceleration and atomization of the water droplets that are sprayed from the trailing edges of the stationary blades. High steam density leads to high acceleration and small water droplets. The steam density, and thus this influence, are directly related to the steam pressure between the stationary- and end-blade rows (Fig. 6.2.3). The erosion coefficient is calculated on the basis of the pressure p_0 upstream of the

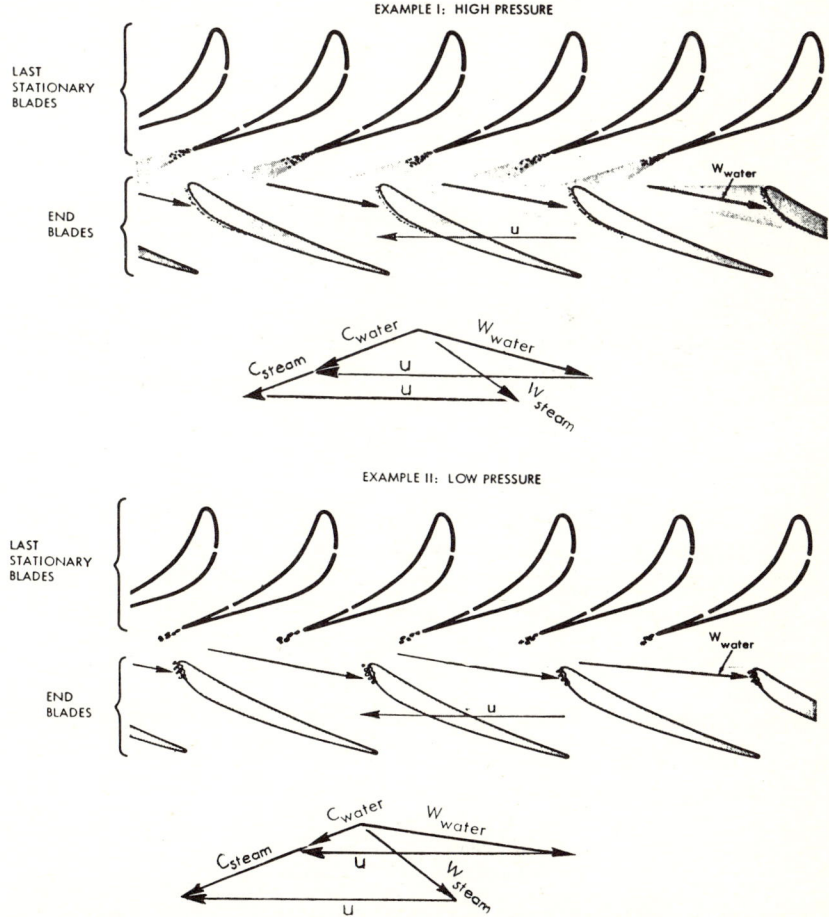

FIG. 6.2.3 Influence on erosion of density between stationary and rotating blade.

stationary-blade row; though not quite correct, this is sufficiently accurate for an empirical determination of the severity of erosion.

Influence of blade and casing design Other factors that influence the degree of erosion are dependent not upon operating conditions but upon suitable design of the turbine's last stages. Some of these factors are:

1. Large clearance between the stationary- and moving-blade rows, so that the steam can accelerate the water droplets. Water droplets that impinge on the blades at low relative speed do no harm.
2. Sharp trailing edges of the stationary blades, so that the water films are thin when leaving the blade. Of course a compromise with the mechanical behavior of the blade is necessary.

The influence of the two main design parameters is shown in Fig. 6.2.4, based on experience with several LP stages.

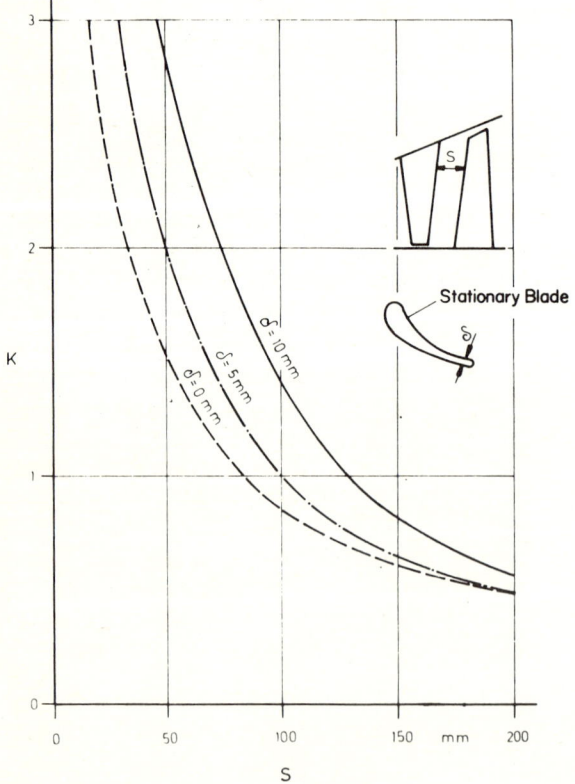

FIG. 6.2.4 Influence of design parameters on erosion coefficient.

TABLE 6.2.1 Erosion in two-flow and four-flow LP turbines

	Four-flow	Two-flow	% Change
Annulus area	4×5 m²	2×10 m²	0
End-blade height	676 mm	1080 mm	+60
Tip speed	443 m/s	583 m/s	+32
Erosion coefficient	30.5 m⁴/(s·kg)	32.5 m⁴/(s·kg)	+7

6.2.3 Examples of Erosion with Different Machines

Using the erosion coefficient, it is possible to forecast what degree of erosion will occur in different machines at different operating conditions. A comparison between a two-flow and a four-flow turbine for the same application gives the results in Table 6.2.1.

The end-blade erosion of these two turbines will be about the same, despite the fact that the blade tip speed for the two-flow machine is 32% higher. The reason for this surprising result is that increased tip speed means increased tangential velocity u, which leads to higher steam velocity c for stage efficiency reasons, and consequently to a higher enthalpy stage drop Δh. With a given backpressure, this higher enthalpy drop results in higher steam pressure and density upstream of the last stage, which reduces erosion of the moving end blades.

6.2.4 Experience with Erosion on 750-mm LP Blades

The accuracy of predictions such as those in Sec. 6.2.3 has been checked by measuring the water erosion that has actually taken place on end blades in service, and correlating these measurements with the appropriate calculated erosion coefficients. Figures 6.2.5 and 6.2.6 show an example of the excellent agreement between the actual erosion attack and the calculated erosion coefficient in the case of the same 750-mm end blades in service for about equal lengths of time under different operating conditions.

The severely eroded blade (Fig. 6.2.5) is from a fossil-fired reheat turbine that was installed in Finland and operated at extremely low backpressure. The slightly eroded blade of the same design (Fig. 6.2.6) is from a similar turbine in Germany that is run in conjunction with a cooling tower.

6.2.5 Design Principles

Using the evaluation of the erosion coefficient as explained above, the provisions in Table 6.2.2 are normally made.

6.2.5.1 Drainage through Slits in Surface of Hollow Stationary Blades

The reduction in erosion due to using hollow last-stage stationary blades with slits that suck the water film off the blade surfaces to the condenser is shown

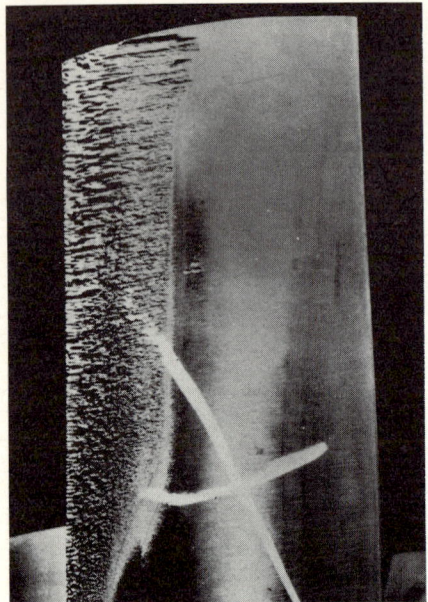
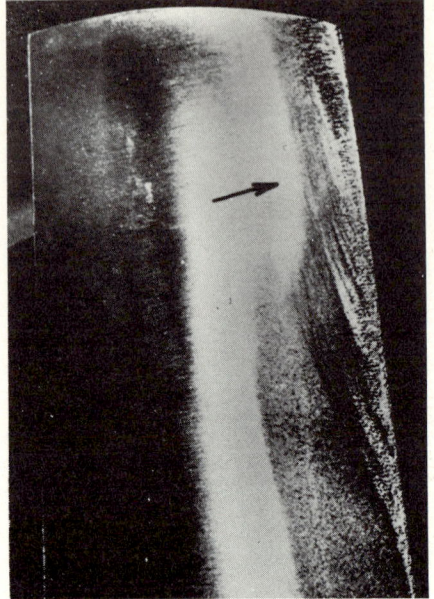

Blade motion ←   Blade motion →

FIG. 6.2.5 750-mm end blade tip (back-pressure 20 mbar, service time 23,500 hr, erosion coefficient 80 m⁴/s·kg).

FIG. 6.2.6 750-mm end blade tip (back-pressure 86 mbar, service time 21,500 hr, erosion coefficient 35 m⁴/s·kg).

in Figs. 6.2.7 and 6.2.8. Both these 750-mm end blades are from different rows of the same nuclear-fueled six-flow turbine. For experimental comparison purposes, only one of the three two-flow LP turbines was provided with suction slits in the hollow last-stage stationary blades; the other two turbines are without this feature. The beneficial effect of moisture removal through hollow stationary blades can be observed by comparing the degree of erosion on a blade from Flow No. 2 (Fig. 6.2.7) that is equipped with hollow last-stage stationary blades, and on a blade from Flow No. 4 (Fig. 6.2.8) without such a moisture-removal system.

The appropriate position of the slits on the pressure and suction side of the blade is shown in Fig. 6.2.9. To prevent a short circuit of steam flow through the stationary blade, it is necessary to cut the slits at points of the same pressure level.

TABLE 6.2.2 Turbine blade design provisions

Erosion coefficient	Provision
$E < 8$ m⁴/(s·kg)	None
$E > 8$ m⁴/(s·kg)	Stationary blades with suction slits or heated stationary blades Hardening of leading edges of rotating blades

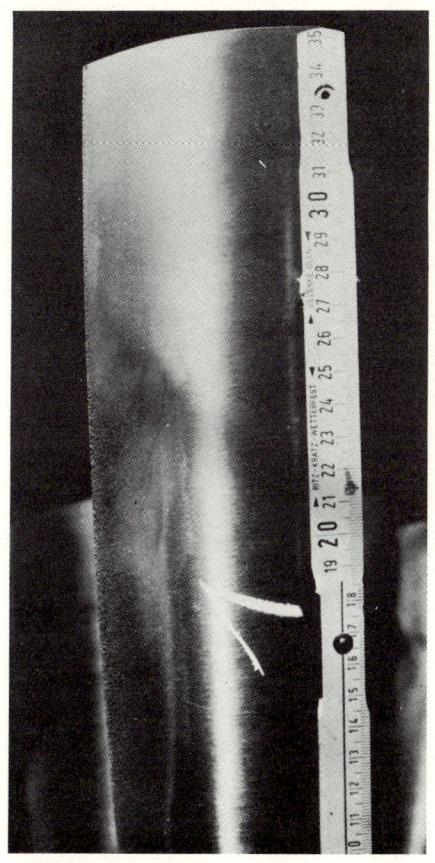 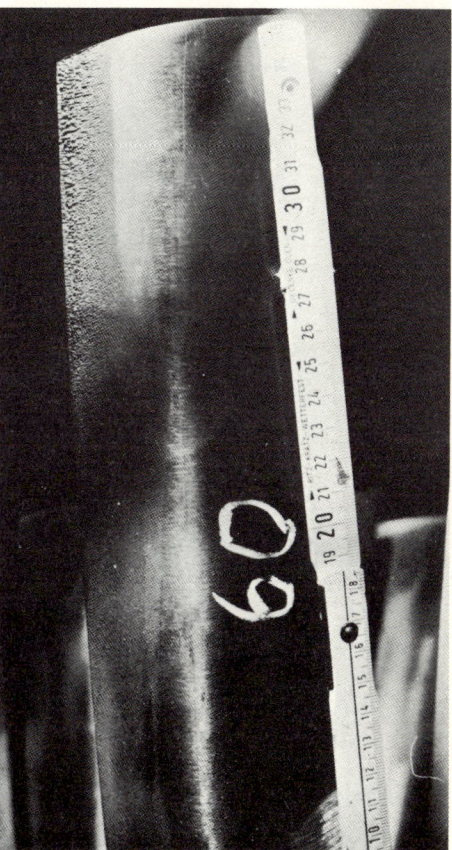

FIG. 6.2.7 750-mm end blade tip. End blade from Flow 2 with moisture removal through hollow last-stage stationary blades after 12,500 service hr.

FIG. 6.2.8 750-mm end blade tip. End blade from Flow 4 without moisture removal through stationary blades after 12,500 service hr.

For good efficiency, a favorable shape of the slits was developed. A simple rectangular saw cut, as shown in Fig. 6.2.10, gave poor efficiency. A much higher efficiency was reached with a shape as shown in Fig. 6.2.11. This shape can easily be machined with a spark-erosion machine.

6.2.5.2 Steam-Heated Stationary Blades

Since 1972, operating experience has been gained in heating the last-stage hollow stationary blades to ensure that the moisture droplets impinging on the stationary-blade walls are immediately evaporated, before larger drops or films can form and water streaks occur. By preventing moisture films from building up on the stationary-blade surfaces, by either suction removal or evaporation, the amount of water being shed from the stationary blades toward the moving

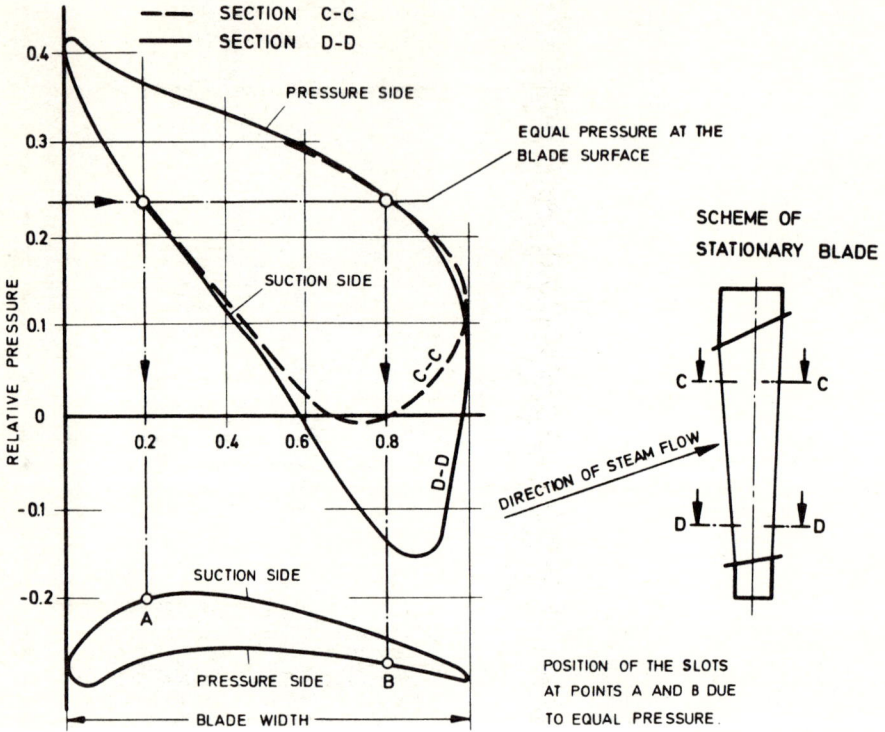

FIG. 6.2.9 Location of slits in stationary blades for water suction.

IN THE AREA OF FLOW SEPARATION WATER ACCUMULATES AND PASSES OVER THE SLOT. AN UNFAVORABLE SLOT SHAPE CAUSES HIGH VELOCITY, TURBULENCE AND LESS MOISTURE REMOVAL.

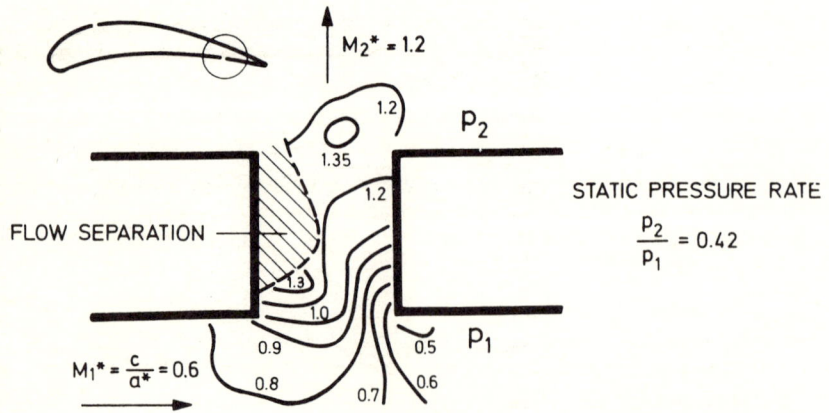

CURVED LINES ARE LINES OF EQUAL STEAM VELOCITY (MACH NUMBER)

FIG. 6.2.10 Improper shape for suction slits.

MAXIMUM MOISTURE REMOVAL IS REACHED AT LOW PRESSURE DIFFERENCE ACROSS THE SLOT

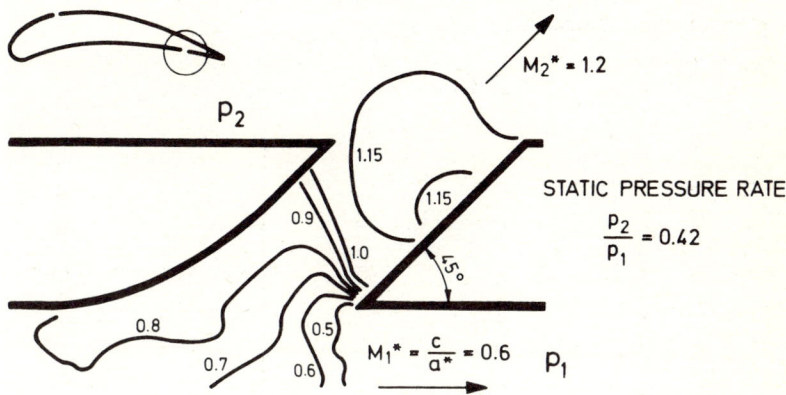

FIG. 6.2.11 Proper shape for suction slits.

blades is minimized, thus sharply reducing erosion of the last-stage moving blades. The development of these steam-heated stationary blades was started in the laboratory.

Figure 6.2.12 is a photograph of the test rig featuring three blades. Steam of variable pressure and temperature was allowed to flow through the blades with water droplets being formed by a nozzle upstream of the blades. A separate, also variable source of steam supplied heating steam to the hollow central blade. The moisture on the surfaces of the heated blade was observed through inspection ports under different simulated operating conditions. The modifications to the hollow-blade inner and outer contours (ratio of internal to external surface), as well as the relationship of the steam conditions (heating steam to working steam) necessary to ensure almost totally dry blade outer surfaces with no shedding of moisture from the trailing edge, were determined in a lengthy series of tests.

The first application of the steam-heated stationary blades was a 150 MW single-reheat unit that was designed to operate with an average cooling-water temperature of 8°C, resulting in a backpressure ranging from 27 mbar at full load to 18 mbar at half load. With this unusually high erosion coefficient, it was decided to fit the two-flow LP turbine with special provisions to protect the 8 m^2 type last-stage moving blades from excessive erosion.

It was further decided to employ the already-proven moisture-removal technique on the last-stage stationary blades, with suction slits at the generator end of the turbine, and the new stationary blades at the other end of the LP turbine. In this way, a true comparison could be gained of the effectiveness of both methods in combating erosion under the same operating conditions.

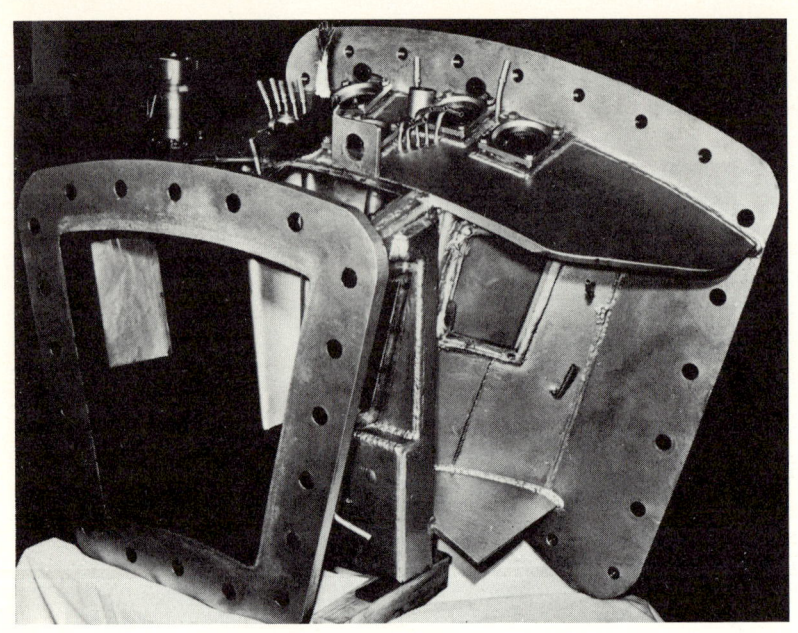

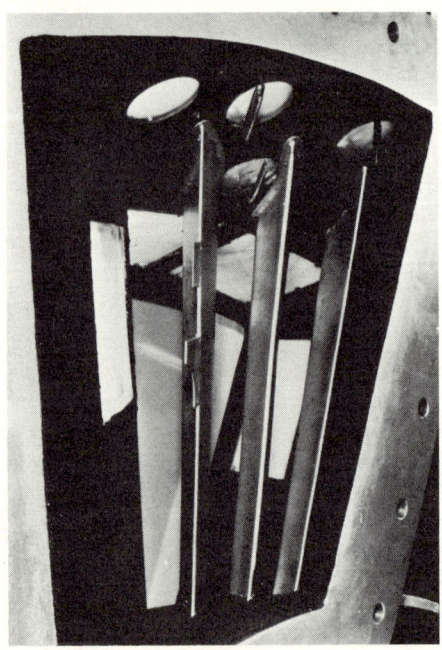

FIG. 6.2.12 Steam inlet and exhaust ends of test rig with three last-stage hollow stationary blades.

Figure 6.2.13 shows one of the last-stage stationary-blade row halves designed for internal steam heating of the hollow blades being tested in the factory. During hydrostatic tests to establish the tightness of all welded joints, distortion of the assembly was observed by dial indicators; the material stressing was measured by strain gauges located at critical and representative points on the blade rings and on the blades themselves. Another series of tests involved supplying the interior of the assembly with water at different pressures and steam at various temperatures, to ascertain the adequacy of the construction under all operating conditions, and to study the relative movements of component parts of the assembly.

The fully tested last-stage stationary-blade row was installed in the LP turbine and connected in the field to a steam extraction line. The turbine extraction point selected for blade-heating purposes is located at the IP turbine exhaust to the LP steam cross-over system. Since the whole system is under practically the same pressure and steam is drawn into the ring manifold at the rate condensate is formed, no special heating-steam admission control is necessary. The condensate from the heated-blade row is returned to the heater associated with the same turbine extraction.

FIG. 6.2.13 Testing of upper half of last stage row of internally steam-heated stationary blades for a 150-MW turbine.

It is estimated that the modifications to the blade profile and the use of extraction steam for blade-heating purposes result in a heat-rate deterioration of about 0.15% and a capability reduction of approximately 230 kW at full load.

During a planned outage of the unit in November 1973, the blades of the LP turbine were inspected. By then the turbine had accumulated 2240 service hr largely at partial load: 16% of the operational time at about 60-80 MW, 48% of the operational time at about 100 MW, and 36% at about 150 MW. The unit had been subjected to a total of 40 startups, three from cold, the remaining 37 from hot or warm conditions. The steam heating of the stationary blades was not placed in service during the first 100 hr of unit operation.

A distinct difference could be discerned between the erosion of the last-stage moving blades at the generator end of the turbine, where moisture was removed through suction slits in the hollow stationary blades, and of the same blades in the other flow, which were protected by the internally steam-heated stationary blades.

Figure 6.2.14 shows photographs of the most and least severely eroded blades in the flow with moisture removed by suction slits. The very much less severe attack by erosion on the blades in the other flow with moisture

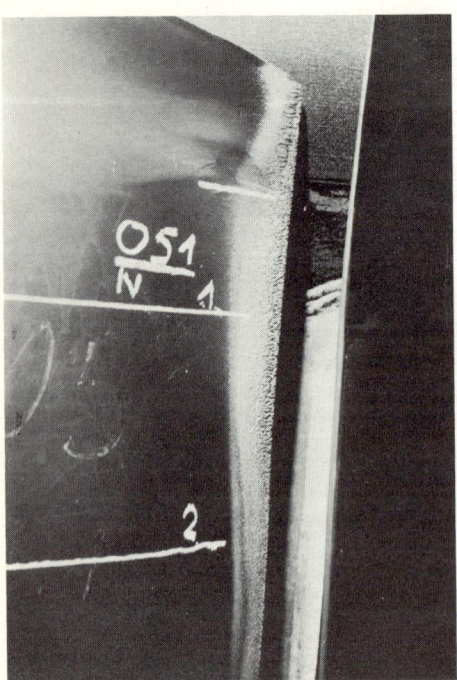 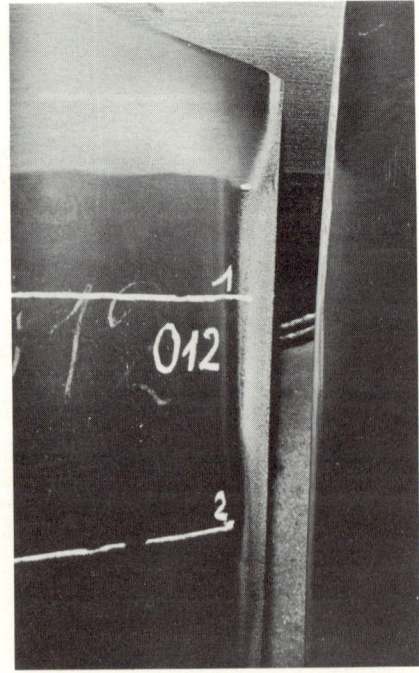

FIG. 6.2.14 Most and least eroded last-stage moving blades associated with hollow stationary blades with moisture-removal suction slits.

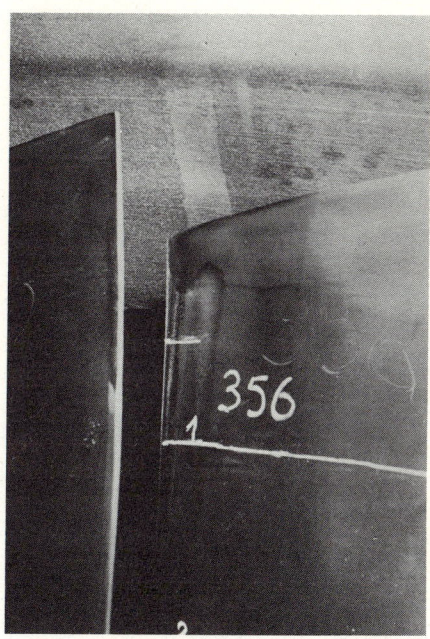

FIG. 6.2.15 Most eroded last-stage moving blade associated with heated stationary blade.

FIG. 6.2.16 Least eroded last-stage moving blade associated with heated stationary blade.

removed by steam heating is apparent from Figs. 6.2.15 and 6.2.16, which show the blades with the most and least erosion in the row. The scale of the photographs can be judged from the fact that the distance between the blade tip and the horizontal chalked line marked 1, and between the lines marked 1 and 2, is in each case about 100 mm.

Despite the short period of operation, it has already become evident that internal steam heating of last-stage stationary blades provides an even more effective means of protecting the end blades against erosion than removing the moisture through suction slits in hollow stationary blades. For this reason, it has been decided to evaporate the moisture on the stationary-blade surfaces by steam heating the last-stage row in each of the two LP flows for the second 150 MW unit, to be installed in the same power plant. Even so, with more successful operational experience and design improvements in the future, internal steam heating of last-stage stationary blades will most likely be resorted to only in turbines for which abnormally high erosion coefficients have been calculated. Such cases of extremely erosive back-end environment generally occur only in turbines being operated at backpressures less than 35 mbar.

6.2.5.3 Protecting of Leading Edges of Rotating Blades

In addition to the provisions already described, erosion wear can be reduced by protecting the leading edges of the rotating blades. Basically there are two

FIG. 6.2.17 Automatically operated flame-hardening facility.

possibilities for protection: Stellite strips, and flame hardening. Both procedures result in nearly the same Vickers hardness of 4.5 kN/mm^2. However, Stellite is a very brittle material, and cracks may occur due to temperature changes or steam-load changes during operation; these cracks in the strips can initiate cracks in the blade material.

Flame hardening does not present these problems, the hardened blade being very ductile and having no stress concentrations due to grooves for brazed or welded pieces. The hardening process is done on an automatic facility as shown in Fig. 6.2.17.

6.3 OVERSPEED AFTER MAJOR LOAD REJECTION

The overspeed problem after sudden load rejection is not a specific problem of wet-steam turbines, but load rejection tests for wet-steam turbines showed a different result from what was expected on the basis of experience with conventional steam turbines. The difference was due to the stored energy in the enclosed water.

6.3.1 Calculation Method

The overspeed calculation for the first wet-steam turbines was based on a static energy balance. The accuracy of this method in comparison with test results was poor, so it was necessary to introduce a new calculation method, based on the dynamic behavior of the steam flow in the turbine [6-3]. The advantage of this method is that it enables the overspeed to be calculated by taking into account these influences:

1. Load rejections to part load
2. Change of efficiency under transient conditions
3. Evaporation of water films
4. Operation of valves in the cycle

Unknown influences in the calculation were the enclosed water mass in the cycle, and the possibility of energy transportation from the casing and pipe walls into the cycle.

Measuring the water film thickness on pipe walls with ultrasonic probes has so far given incorrect results, due to the very turbulent surface of the water film and size of the probes. But intensive laboratory work on this subject gives hope that better results may be achieved. From correlation of calculation and test it was found that an average film thickness of 3 mm on all surfaces in the wet-steam region gives the best results and includes a small safety margin.

6.3.2 Test Results

Figure 6.3.1 shows a good agreement between the overspeed calculation and test for a 662-MW wet-steam turbine with a rated speed of 25 s^{-1}. Figure 6.3.2 shows the same for a 477-MW wet-steam turbine with a rated speed of 50 s^{-1}.

6.3.3 Measurements to Prevent Excessive Overspeed

There are four methods to reduce the overspeed after load rejection:

1. *Overspeed trip device*: Normally the set point of the overspeed trip device is at 10% overspeed. In the case of a load rejection, the stop valves are closed if the overspeed exceeds 5% and if the control valves are not closed. This reduces the maximum overspeed by approximately 5%.

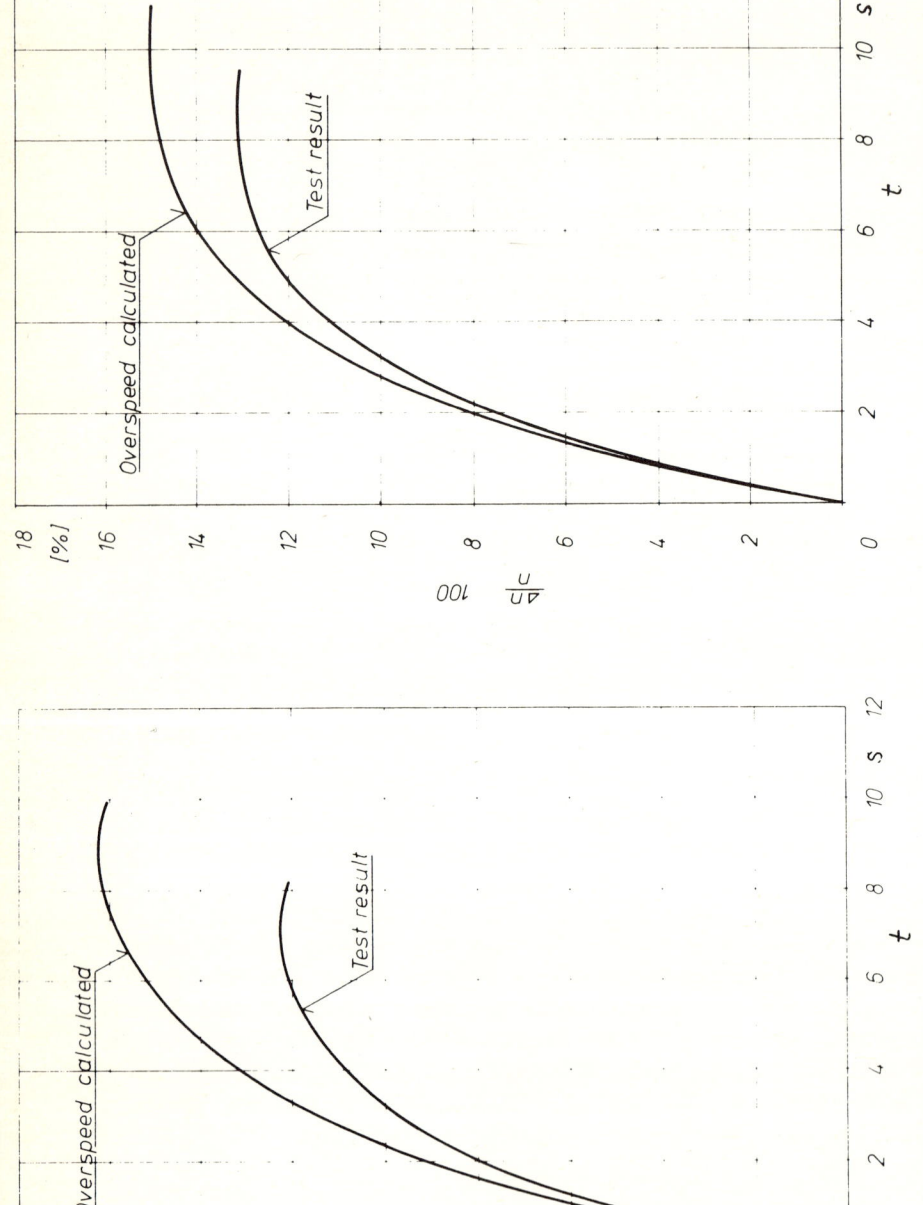

FIG. 6.3.1 Comparison of calculation and test results for the overspeed after a full-load rejection by a 662-MW wet-steam turbine.

FIG. 6.3.2 Comparison of calculation and test results for the overspeed after a full-load rejection by a 477-MW wet-steam turbine.

2 *Bypass valves*: In case of a load rejection, bypass valves from the cold reheat blow down to the condenser and thus the energy does not accelerate the LP turbine.
3 *Controlled drainage*: The water level in heaters, in the moisture separator, etc., is controlled by special probes, and the load on the turbine is reduced if the stored water level is above the set point.
4 *Intercept valves*: If these three measures cannot prevent excessive overspeed, it is necessary to provide intercept valves ahead of the LP turbine. This is the most expensive method.

From these methods the most economical solutions can be selected. Depending on the calculated overspeed—basically a function of condenser pressure, system of moisture separator, and size of reheater—a compromise between the selection of overspeed trip device and the provision of two intercept valves in tandem is chosen. Additional laboratory work is necessary to predict the thickness of the water film on wet surfaces.

REFERENCES

6-1 Keller, H.: Erosionskorrosion an Heissdampfturbinen VGB Kraftwerkstechnik 1974, Heft 5.
6-2 Gloger, M.: Probleme der Wasserabscheidung in Nassdampfturbinen BWK Bd. 22 (1970) Nr. 9 Seite 417/20.
6-3 Traupel, W.: Thermische Turbomaschinen Dampfturbinen, Gasturbinen, Turboverdichter Bd. 1.

CHAPTER 7

External Water Separators

7.1 PERFORMANCE OF KNITTED WIRE MESH AND CORRUGATED PLATE SEPARATORS

G. C. Gardner

7.1.1 Introduction to Basic Concepts

The role and importance of moisture separators in wet-steam turbine technology will be described in Sec. 7.2 by Coit et al. This section discusses the performance of the two most widely used separators: knitted wire mesh, which is sometimes called a demister, and corrugated plates, which are sometimes called chevron plates, louvre plates, or venetian-blind separators.

In an ideal world, we would know the moisture concentration and drop-size distribution of the moisture entering the separating device. Then, if the separation efficiency with respect to drop size (fractional separation efficiency) were known, the separation performance could be calculated. In practice, we have only a shrewd idea of the size distribution, and an accurate calculation of separation performance is not possible. In consequence, we will find that the study of the fractional separation efficiency has been taken only to the point where approximate estimations of its value can be made. The design approach is simply to state from our shrewd idea of the drop-size distribution that we must eliminate with high efficiency drops of more than a certain size, and then design equipment with adequate margin for error to meet this demand.

Occasionally, the fortunate situation arises where we can eliminate drops of less than a certain size from consideration. A simple example occurs in boiler drums where at, say, 60 bar, the steam has a residence time of about 0.5 s before entering the final separators. It may then be calculated by the methods given by Gardner [7-1] that drops of size less than 1 μm will have evaporated completely due to the depression of the boiling point by surface-tension effects.

The two types of separator under discussion are called inertial separators, since the inertia of the drops prevents them from following the steam streamlines, and they are flung against the collecting elements. Obviously the collection efficiency improves with increasing drop size and steam velocity, and with decreasing size of the collection element. For this reason knitted wire mesh, where the element is a wire of about 0.1 mm dia., is found to have better performance than corrugated plates, where the element size is effectively the half wavelength of the corrugations, not normally less than 10 mm. However, this statement must be qualified by an examination of the so-called breakthrough or loading characteristics.

If the steam velocity is increased for any given separator, a point is reached at which deposited water is reentrained. The reentrainment is so strong that it is pointless to exceed the critical or breakthrough velocity. This most important factor in design can be determined accurately for a given device because it is dependent only upon the water loading and steam velocity, and not dependent upon the drop size. If the steam flow is vertically upward, as is normally the case for knitted wire mesh, the critical velocity is independent of the depth of the mesh. If the steam flow is horizontal, as is usually the case for corrugated plates, the breakthrough velocity is dependent upon the manner in which the plates are drained and also upon the height of the plates.

The breakthrough velocity of corrugated plates is higher than that of knitted wire mesh so that there is some compensation for the lower efficiency at a given velocity. Of more importance is the consideration that the higher pemissible velocity allows smaller equipment to be designed, which is a great advantage in arranging the separating device within the vessel that also normally includes the tube bank for reheating the steam. If corrugated plates do not have adequate separating performance, there is always the option of preceding them with a layer of knitted wire mesh, whose breakthrough velocity may be exceeded but from which only large drops that can be separated by the plates will be entrained.

7.1.2 Fractional Separation Efficiency

7.1.2.1 Knitted Wire Mesh

Correlation of the efficiency of knitted wire mesh starts from consideration of the fractional separation of a single wire normal to the steam flow. Potential flow of the steam is usually assumed and the efficiency, as calculated by Brun [7-2, 7-3], is shown in Fig. 7.1.1. Here η_w is the fraction of drops collected from the steam approaching the wire within its projected area, and the Stokes number is

$$\text{Stk} = \frac{2l}{D} \qquad (7.1.1)$$

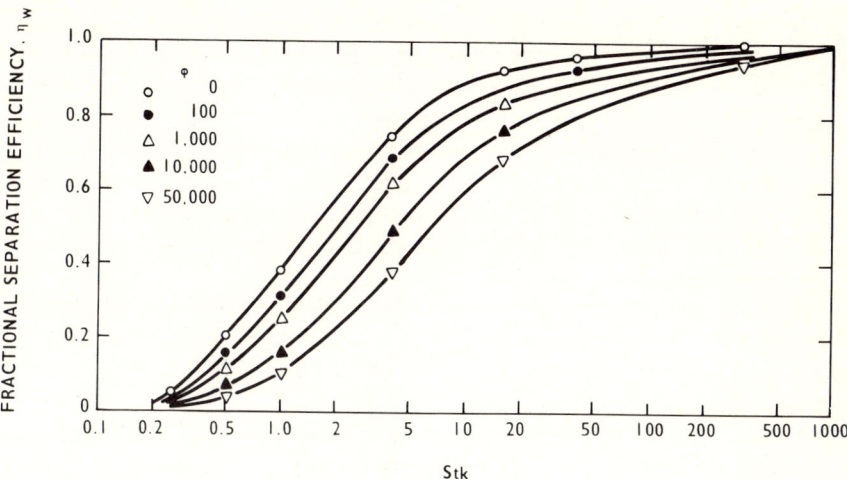

FIG. 7.1.1 Fractional separation efficiency for potential flow round a wire [7-2, 7-3].

where l is the stopping distance (which can be taken from Fig. 7.1.2), and D is the wire diameter. When a Stokes law resistance of the drop to motion can be assumed

$$\text{Stk} = \frac{\rho_f d^2 U}{9 \mu_g D} \tag{7.1.2}$$

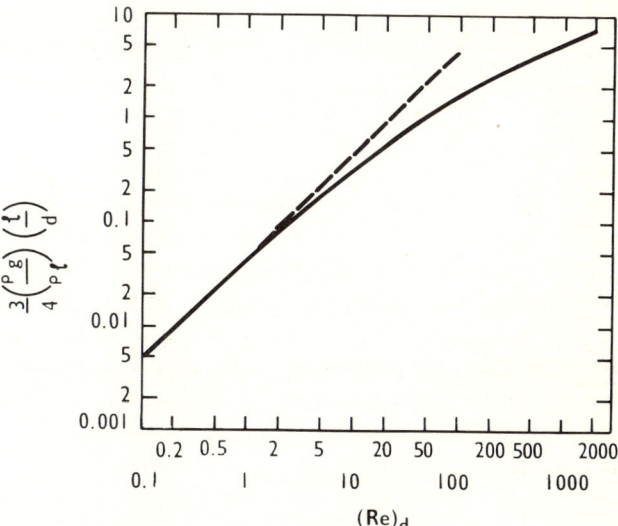

FIG. 7.1.2 Stopping distance of drop [7-4].

where ρ_f is the drop density, d is the drop diameter, U is the approach velocity of the steam, and μ_g is the steam viscosity. Figure 7.1.2 gives the correction necessary to Eq. (7.1.2) when Stokes law drag cannot be assumed, and the correction is of course a function of

$$\mathrm{Re}_d = \frac{dU}{v_g} \qquad (7.1.3)$$

where v_g is the kinematic viscosity of the steam.

A family of curves of fractional separation efficiency occur in Fig. 7.1.1 because the efficiency remains dependent upon the drag law, even in the chosen coordinates. Therefore the curves are differentiated by the parameter

$$\phi = \frac{9\rho_g DU}{\rho_l v_g} \qquad (7.1.4)$$

Figure 7.1.1 has been described in detail to indicate the many factors influencing the separation efficiency. Others are the Reynolds number of the steam flow, the size of the drop relative to the wire diameter, the size of the drop relative to the mean free path, and the diffusion of the drops. Detailed discussions of these factors can be found in Fuchs [7-4] and Strauss [7-5], but here it need only be stated that the curve in Fig. 7.1.1 for $\phi = 0$, when the Stokes number is defined by Eq. (7.1.2), is normally assumed to be valid, since we are usually dealing with drops that obey the Stokes drag law, and they are small compared to the wire diameter and large compared to the mean free path. However, the potential steam flow pattern may sometimes not be completely valid. It is also noted that theoretically the value of η_w will be zero when Stk = 0.125, but then either diffusion might play an important role or the imperfections of flow in a knitted wire mesh might allow an appreciable separation efficiency.

Experimental results of Wong and Johnstone [7-6] and May and Clifford [7-7] for single wires are compared in Fig. 7.1.3 with the theoretical values from Fig. 7.1.1 for $\phi = 0$. The comparison gives confidence in the theoretical line, especially since the most important part is that for small values of η_w, as will become clear later.

We can now formulate an expression for the efficiency η_k of knitted wire mesh in the same manner as Stairmand [7-8] or Carpenter and Othmer [7-9]. Imagine the wire to be arranged to form a cubic lattice of side L. Two-thirds of the wire is thus assumed to be transverse to the steam flow and offers a surface for drop collection. It is readily calculated that

$$L = \left[\frac{3\pi}{4(1-\epsilon)}\right]^{1/2} D \qquad (7.1.5)$$

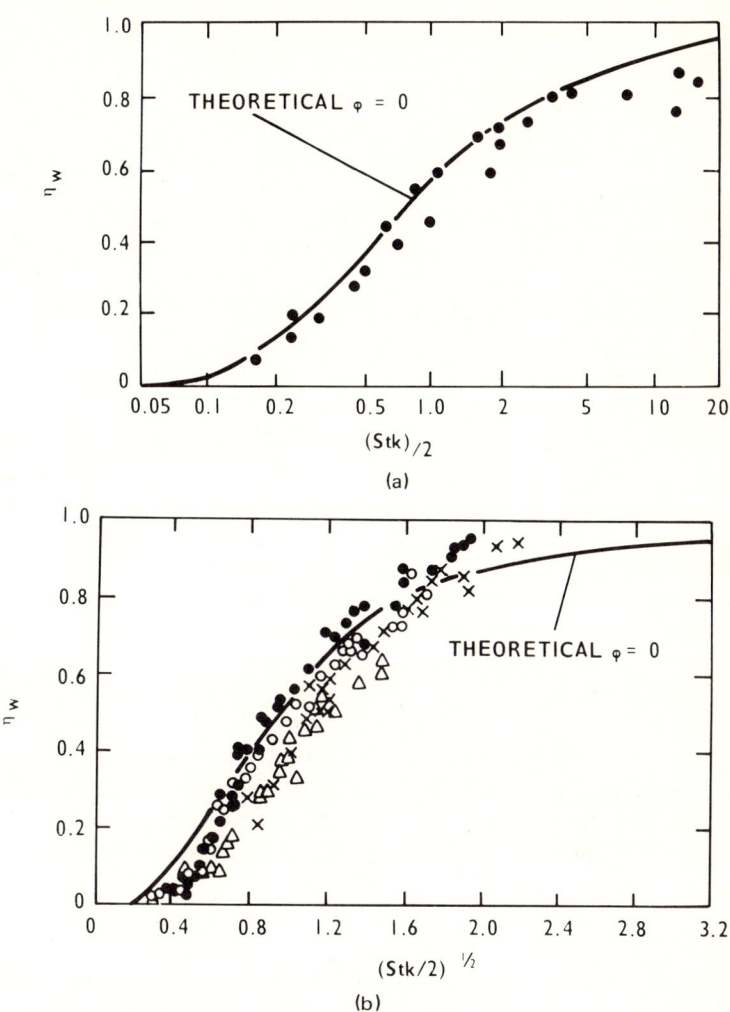

FIG. 7.1.3 Comparison of experimental results of single-wire separation efficiency with theory: (a) May and Clifford [7-7] and (b) Wong and Johnstone [7-6].

where ϵ is the voidage of the mesh. The efficiency of one layer of wires transverse to the steam flow is $2DL\eta_w/L^2$, and the number of layers is the total depth of the knitted wire mesh H_k divided by L. We assume that the flow is completely mixed between layers, and thus a binomial expression for the overall efficiency η_k is obtained. Since there are a large number of layers, this can be approximated by

$$\eta_k = 1 - \exp\left[-\frac{8}{3}\frac{(1-\epsilon)\eta_w H_k}{\pi D}\right] \qquad (7.1.6)$$

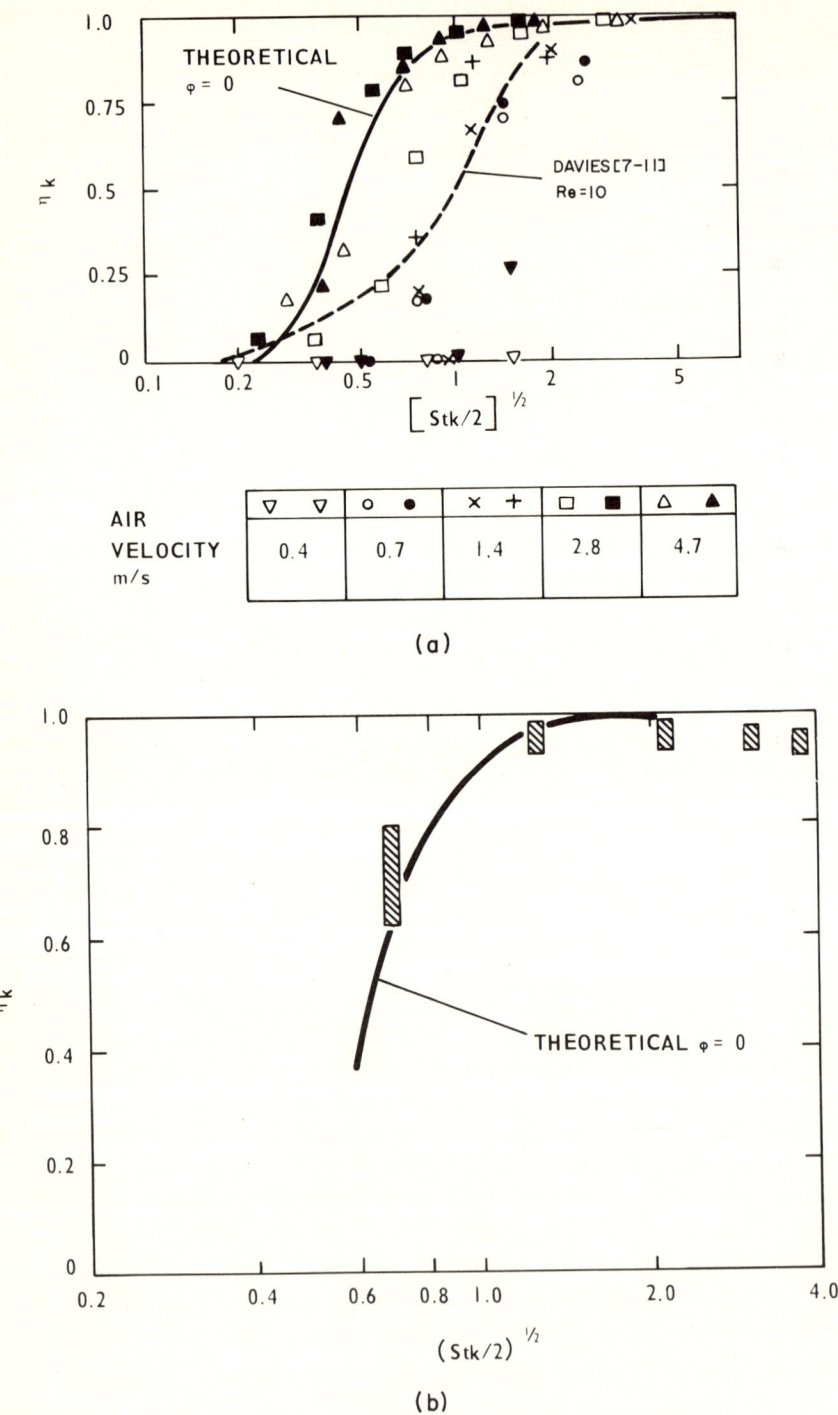

FIG. 7.1.4 Knitted wire mesh separation efficiency: (*a*) Bürkholz [7-10] and (*b*) Katz [7-12].

Equation (7.1.6) was tested by Bürkholz [7-10] against experiments using sulfuric acid drops in air for a knitted wire mesh with 0.27 mm wires. There were 30 effective layers, each with a blockage factor of 0.17, and it can therefore be deduced that $\epsilon = 0.983$. The result of the comparison is given in Fig. 7.1.4a. It is seen that the equation is good for a velocity of greater than 2.8 m/s but the efficiency is overestimated for lower velocities; this difference is probably attributable to the potential flow assumption breaking down. A second theoretical curve due to Davies and Peetz [7-11] is given in Fig. 7.1.4a for a Reynolds number of 10, which is equivalent to an air velocity of 0.6 m/s. However, it is noted that the curve from Fig. 7.1.1 is of sufficient accuracy for velocities of practical interest.

Earlier data of Katz [7-12] are shown in Fig. 7.1.4b. Oil drops in air were used. The knitted wire mesh was composed of wire with $D = 0.15$ mm; the mesh had a voidage of 0.986 and was 150 mm thick. The comparison with the theoretical curve is good.

Finally, some results of Bürkholz [7-13] are presented in Fig. 7.1.5 for collection on a staggered array of parallel wires. The dimensions of the array are given in the figure. Comparison is made with Eq. (7.1.11), which was derived for corrugated plates by the writer, as described in the next section. The flow is assumed to pass through the array as a set of sinusoidal streams, and mixing of the flow can be assumed to occur after every row of wires ($a = 20$) or every other row ($a = 10$). The theory satisfactorily describes the experimental results.

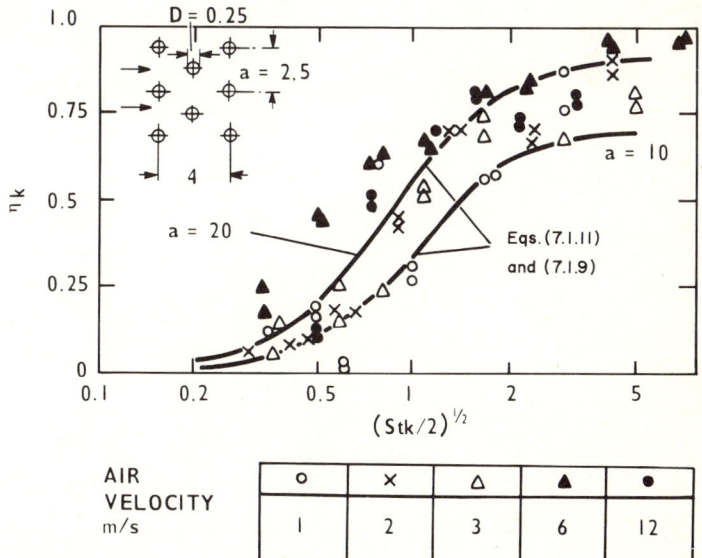

FIG. 7.1.5 Array of parallel wires: results of Bürkholz [7-10] (dimensions in mm, 20 rows of wires).

7.1.2.2 Corrugated Plates

There is no theory completely describing the separating performance of corrugated plates, but a useful way of examining the important parameters was developed by the writer.

Most plates can be approximately represented by the sine-wave form shown in Fig. 7.1.6. They have a wave amplitude y_0, a wavelength P, and a gap between plates h. The forward velocity within the plates is constant at U, and the transverse velocity fluctuates. Drops are assumed to have a Stokes law resistance to relative motion with the steam; their trajectories are readily calculated and found to have a sine-wave form, with the same wavelength as that of the plates but with a phase lag. The fractional separation efficiency is found to be

$$\eta_p = \frac{2}{H}(1 + \text{Stk}_p^{-2})^{-1/2} \tag{7.1.7}$$

$$H = \frac{h}{y_0} \tag{7.1.8}$$

$$\text{Stk}_p = \frac{\pi}{9} \frac{\rho_f d^2 U}{\mu_g P} \tag{7.1.9}$$

The characteristic dimension corresponding to the wire diameter in knitted wire mesh is seen to be the wavelength, while the gap between plates and the wave amplitude appear in a separate parameter H, which may be termed the blockage parameter. When $H < 2$ you cannot see through the stack.

Equation (7.1.7) implies that efficiency is independent of the number of wave cycles m since the flow is assumed laminar. Obviously allowance must be made for remixing of the flow, and therefore the overall efficiency is written as

$$\eta_{po} = 1 - (1 - \eta_p)^{am} \tag{7.1.10}$$

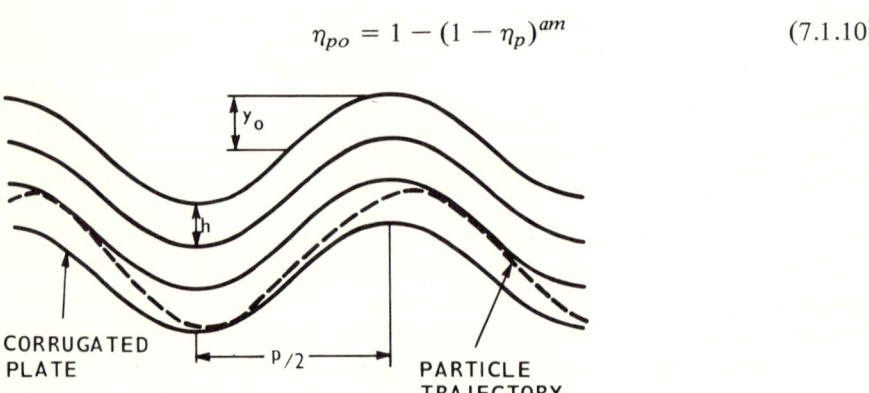

FIG. 7.1.6 Plate form assumed for theory of Eq. (7.1.7).

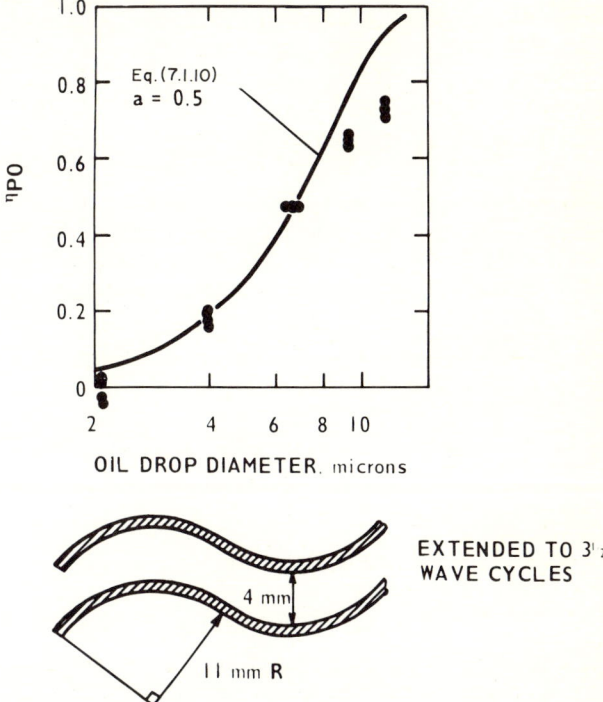

FIG. 7.1.7 Corrugated plates: results of Katz [7-12].

where a is a factor to be determined by experiment. It is not usually justified to approximate Eq. (7.1.10) with an exponential expression, as in the theory of knitted wire mesh, because the product am is not normally large.

Figure 7.1.7 shows results obtained by Katz [7-12] for oil drops. They indicate that for the simple plate form illustrated in the figure $a = 0.5$, which is much smaller than would be expected from the results for knitted wire mesh where the flow is assumed to remix after every wire. Two factors may account for this. The first is that the blockage factor for wires in one layer is small, and the probability that a wire would be trying to collect drops from the stream already treated by wires in the preceding layer is small. The other factor is that the flow in corrugated plates is often laminar, even though the passage Reynolds number appears high. It is well known that curved passages tend to delay the onset of turbulence, and this has been observed experimentally by the writer in work with corrugated plates.

Figures 7.1.8 and 7.1.9 show results obtained by the writer for simple plate forms but with widely different values of H. The data are fitted with $a = 0.53$ and $a = 0.63$, and therefore it appears that the use of $a = 0.5$ in estimations with this plate form provides a conservative approach. It must however be pointed out that there were unexplained variations of efficiency with velocity,

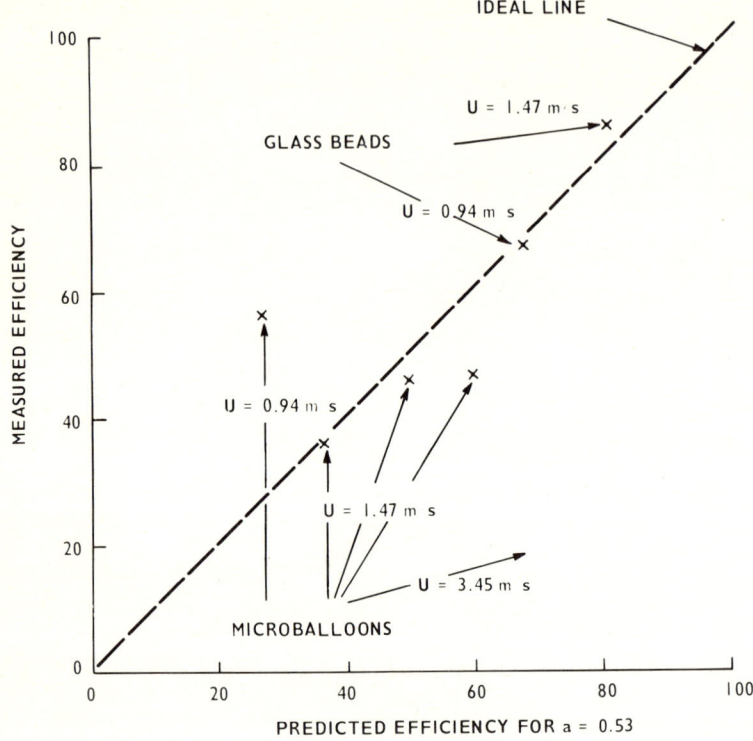

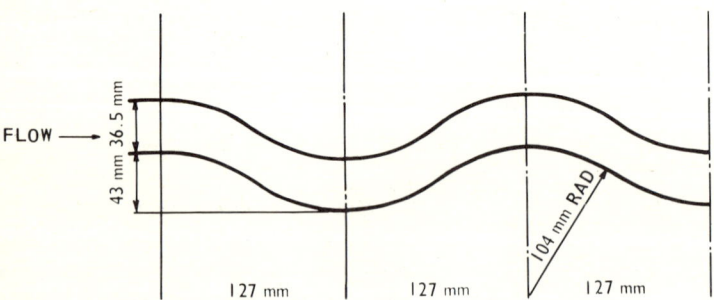

FIG. 7.1.8 Corrugated plates: results of Gardner and Neller.

and for this reason the data points in Figs. 7.1.8 and 7.1.9 are differentiated by the velocity employed. Incidentally, the microballoons referred to in the figures were very light hollow plastic spheres of about 50 μm dia.

Results for a type of corrugated plate commonly used in wet-steam turbine technology are shown in Fig. 7.1.10; they are due to Bürkholz and Muschelknautz [7-13]. The theoretical lines in the figure were calculated by making allowance for the drainage scoops with respect to the values of both H

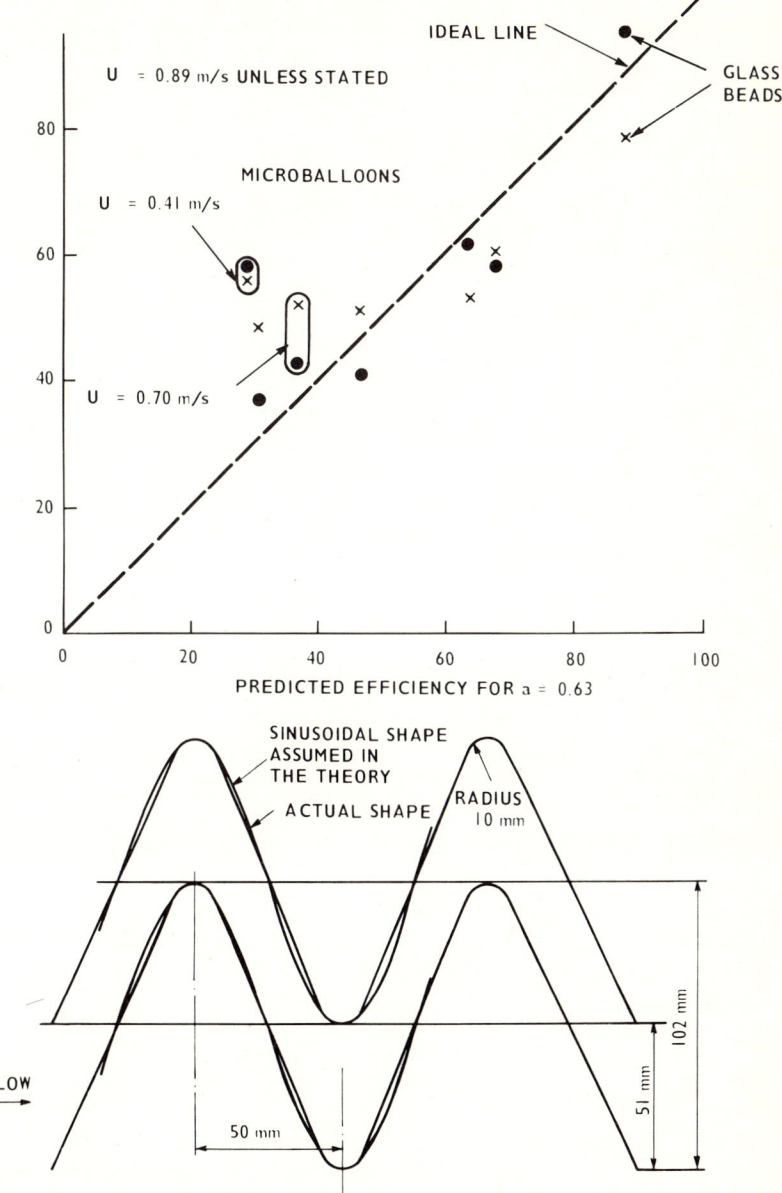

FIG. 7.1.9 Corrugated plates: results of Gardner and Neller.

and U. It is reasonable to expect that the remixing factor a should be larger than for the simpler plate forms, because the scoops tend to promote turbulence. However, a choice is difficult because there is again an unexplained influence of the velocity U. The choice of $a = 1$ appears to be conservative.

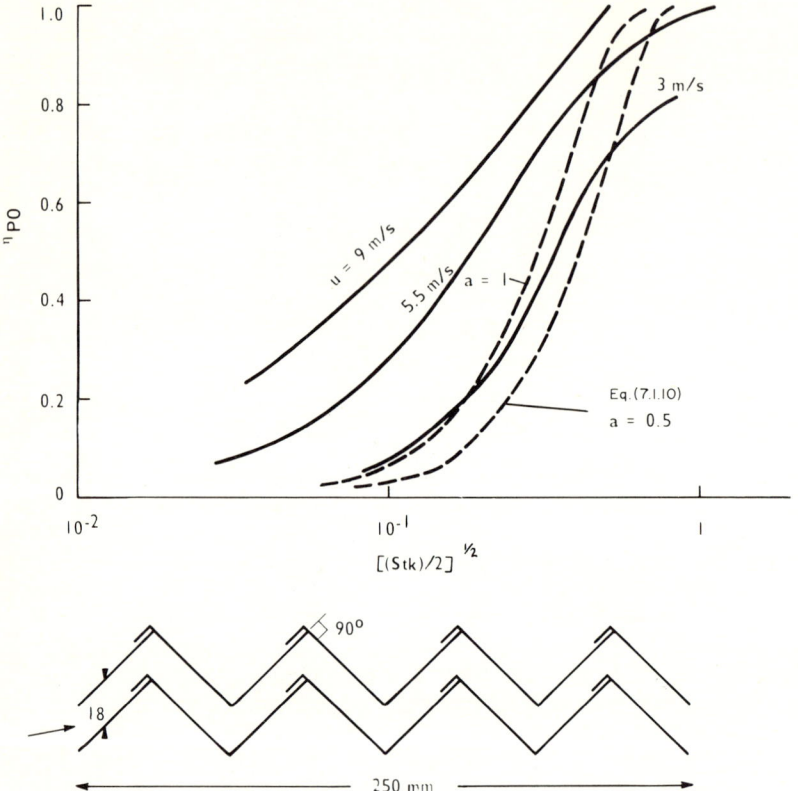

FIG. 7.1.10 Corrugated plates: results of Bürkholz and Muschelknautz [7-13].

Finally, it is to be noted that Eqs. (7.1.7) and (7.1.10) can be applied to a tube or wire bank in cross flow. The assumed sinusoidal flow stream is shown in Fig. 7.1.11. The result is

$$\eta_T = \frac{D}{G}(1 + \text{Stk}_p^{-2})^{-1/2} \qquad (7.1.11)$$

where G is the gap between tubes or wires and Stk_p is as already defined in Eq. (7.1.9). The overall efficiency of a tube bank can be calculated by replacing η_p with η_T in Eq. (7.1.10). Comparison of this analysis with results of Bürkholz [7-10] has already been made in the last subsection. It is noted that Eq. (7.1.11) makes allowance for the tortuosity of the flow in the bank, and in that sense is preferable to an analysis that considers each wire to be isolated in an infinite volume. Indeed, the equation can be applied to knitted wire mesh and compares well with experiments for practicable voidages.

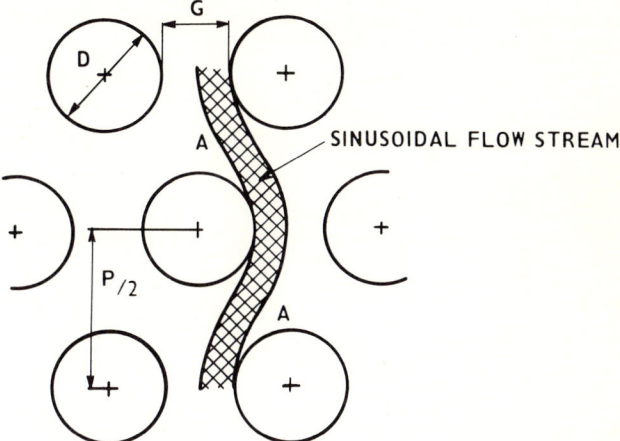

FIG. 7.1.11 Sinusoidal flow stream assumed in wire or tube bank.

7.1.3 Breakthrough Characteristics

A distinction is made between separators in which the steam flows vertically upwards and those in which the flow is horizontal or inclined to the horizontal. In the vertical case, water is distributed uniformly normal to the flow, but larger concentrations collect near the bottom of the separator prior to drainage. When the steam velocity is raised, a critical value is attained when the water distributes more uniformly through the whole depth, the pressure drop increases dramatically, and reentrainment starts. However, the depth of the separator plays no part in determining the critical velocity.

Water still drains to the bottom of the separator when the flow is horizontal, and more resistance will be offered to the flow passing through the lower parts of the device. Therefore there is a tendency for the steam flow to become nonuniform, and this tendency is usually greater for knitted wire mesh than for corrugated plates. Mesh is therefore normally avoided in this configuration. It is also clear that, if the critical reentrainment velocity is dependent upon the amount of water present locally, its value will be dependent upon the height of the separator. For this reason the height of the unit may be restricted, and if the plant layout requires a tall system, it must be comprised of units stacked on top of each other, with means for draining the water from each unit separately.

Finally, in horizontal flow systems steam may bypass the separator through the water-collecting trough at the base of the unit; careful design is needed in this region, because the higher velocity of the bypass flow may lower the critical reentrainment velocity. Sorokin, Popchenkov, and Burkat [7-14] state, for corrugated plates of a design to be detailed later, that this influence is obvious for plates 150 mm tall but is negligible for plates 400–600 mm tall.

7.1.3.1 Vertical Upward Steam Flow

There are two approaches to correlating results for vertical upward flow. The first derives from the flooding correlations (Sherwood et al. [7-15]) for dumped packings such as Raschig rings, as used in chemical technology. The rationalization of the correlation is to consider the surface area within the separating device to be redistributed as a set of vertical flow tubes, with water flowing down the walls and steam flowing up the core. A state will be reached as the steam flow is increased when the water can no longer drain; this defines the flooding point.

Bradie [7-16] and Bradie and Dickson [7-17] slightly rearranged the correlating groups of Sherwood et al. and obtained

$$\Pi_1 = \frac{s}{\epsilon^3} \frac{U_{fs}^2}{g} \left(\frac{\mu_f s}{\rho_f U_{fs}} \right)^{0.2} \qquad (7.1.12)$$

$$\Pi_2 = \frac{s}{\epsilon^3} \left(\frac{\rho_g}{\rho_f - \rho_g} \right) \frac{U_{gs}^2}{g} \qquad (7.1.13)$$

where s is the surface area of the packing per unit volume, ϵ is the voidage, and U_{fs} and U_{gs} are the superficial velocities of the water and the steam.

Bradie's correlations for some knitted wire mesh separators are given in Fig. 7.1.12. It is seen that the correlation is the same for all meshes of the same manufacture and the same wire diameter. There is however an appreciable variation of the critical velocity for meshes of the same manufacture but of different wire diameter, which emphasizes that testing should be carried out for specific meshes, unless a conservatively safe value is to be accepted for the steam loading.

The correlating line for Raschig rings is also given in Fig. 7.1.12, and some results of York and Poppele [7-24] for another knitted wire mesh give values of Π_2 approximately 80% of those for that line.

Corrugated plates are not usually employed with vertical upflow of steam, but Alen'kin et al. [7-18] carried out experiments with plate forms shown in Fig. 7.1.13. Their correlation is also shown in that figure and is represented by

$$\Pi_3 = 0.67 - 2.1\Pi_4 \qquad (7.1.14)$$

$$\Pi_3 = \log_{10} \left(\frac{\rho_g U^2}{2g\epsilon\rho_f G \cos \alpha} \mu_f^{0.16} \right) \qquad (7.1.15)$$

$$\Pi_4 = \left(\frac{W_f}{W_g} \right)^{1/4} \left(\frac{\rho_g}{\rho_f} \right)^{1/8} \qquad (7.1.16)$$

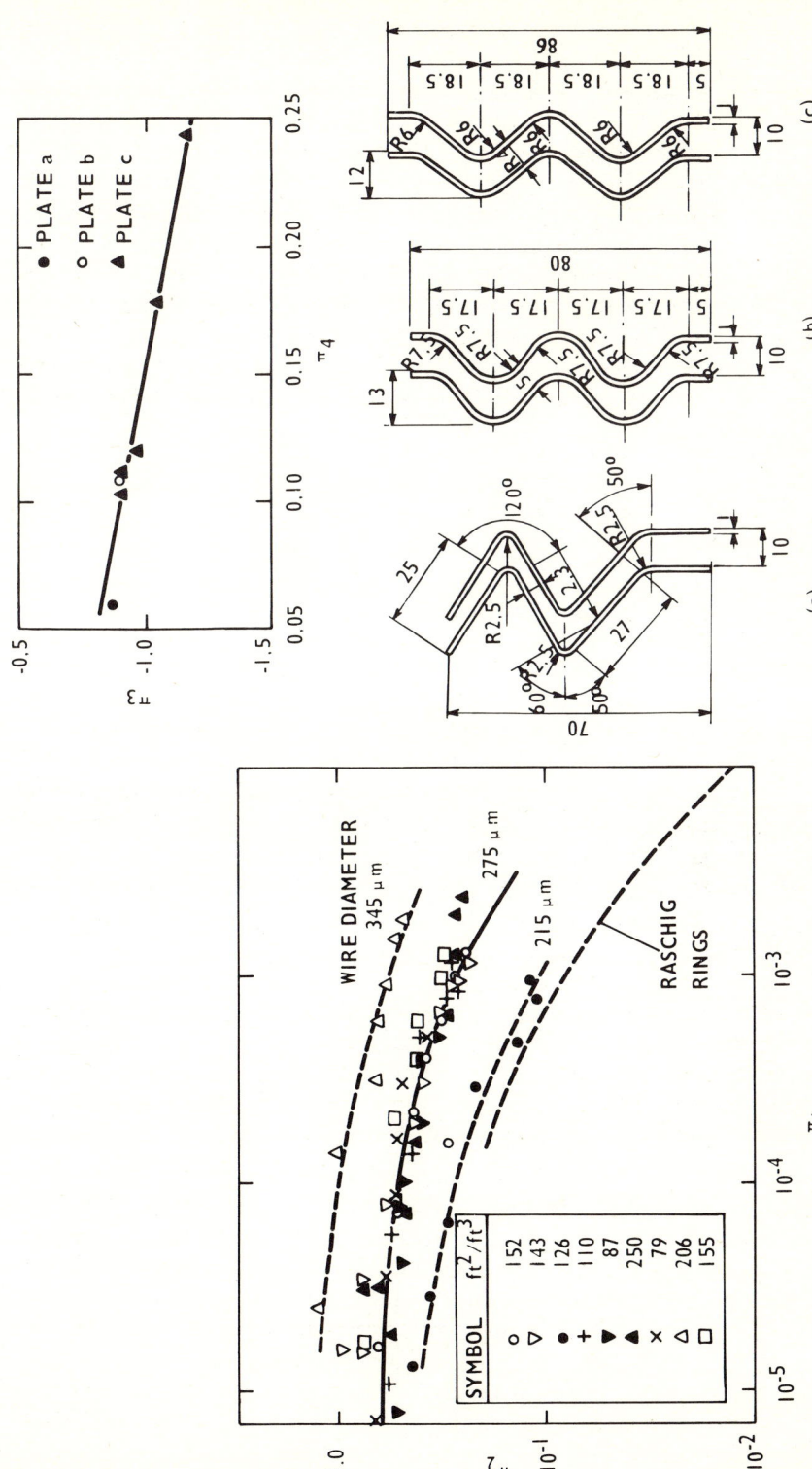

FIG. 7.1.12 Knitted wire mesh, critical velocities in upflow: Results of Bradie [7-16].

FIG. 7.1.13 Corrugated plates, critical velocity in upflow: Results of Alen'kin et al. [7-18].

where μ_f is in kg/m·s, U is the steam approach velocity, G is the gap between plates, α is the maximum angle the plate profile makes to the vertical, and W is the mass flow rate per unit total cross section.

The general form of the correlation was obtained in the manner described above with respect to Π_1 and Π_2, and the form of Eq. (7.1.14) is analogous to that found by Planowski and Kafarov [7-19] for dumped packings.

The second approach to correlation starts from the observation that the critical velocity is substantially independent of water loading, if that loading is small. We may then expect that the critical velocity is given in the form of a Weber number. However, it is also noted that the Laplace length scale is $[\sigma/g(\rho_f - \rho_g)]^{1/2}$ where σ is surface tension; thus we obtain a Weber number, which is called the Kutateladze number Ku in the Russian literature.

$$\mathrm{Ku} = \frac{\rho_g^{1/2} U}{[g\sigma(\rho_f - \rho_g)]^{1/4}} \tag{7.1.17}$$

However, we may also expect the critical velocity to be dependent upon some characteristic dimension L_0 of the separator, and thus we have the additional group

$$\Pi_5 = \frac{g(\rho_g - \rho_g)L_0^2}{\sigma} \tag{7.1.18}$$

Figure 7.1.14 shows some data of Wilson et al. [7-20] for a knitted wire mesh operated from 20 to 124 bar, plotted in the coordinates of Ku versus $g(\rho_f - \rho_g)/\sigma$. They indicated that for this particular mesh Ku is constant at 3.2 if the Laplace length scale is less than 1.5 mm. The arguments for this form of correlation are supported by this result because 1.5 mm is a reasonable magnitude for the characteristic dimension of the mesh. It is also to be expected that Ku would be of the order of magnitude of unity.

For large values of the Laplace length scale, Ku appears to vary as the inverse of the square of that scale, but the exact relationship is questionable on the basis of these limited data.

Values of Ku from Bradie's results for low liquid loadings [7-16] are also plotted in Fig. 7.1.14; they do not disagree with the results of Wilson et al., but it is obvious that there is a variation from mesh to mesh, perhaps because the characteristic mesh dimension is not included in the ordinate. Two points are also illustrated for Raschig rings in steam at 0.18 and 1 bar, as quoted by Sorokin et al. [7-21]. They lie somewhat low, perhaps because the tortuosity has not been taken into account, as with the cos α term in Eq. (7.1.15).

Finally, it is interesting to note that Sorokin et al. [7-21] put Ku in the range of 0.45-0.7, but Fig. 7.1.14 shows that this value can be greatly exceeded, at least in high-pressure steam/water systems. The general conclusion is that the critical velocity or breakthrough characteristics for

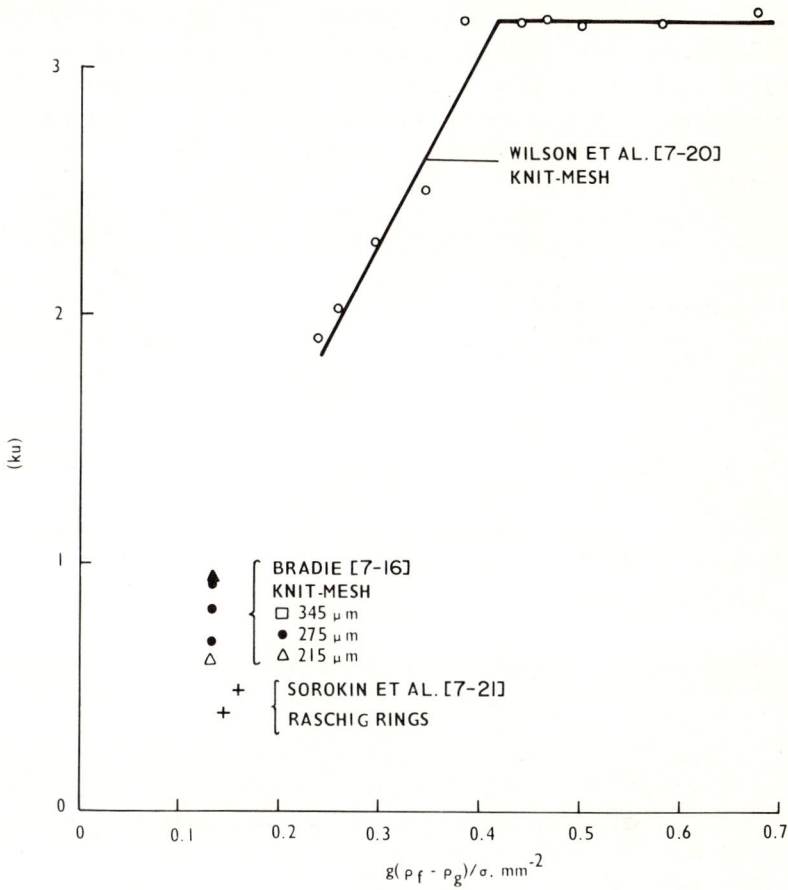

FIG. 7.1.14 Knitted wire mesh, critical velocities in upflow with negligible water loading.

systems with upflow of steam can be successfully correlated, but the correlation varies from design to design within a limited but significant range. This means that preliminary designs are warranted, but exact values of the critical velocity are subject to test. It also means that designs are open to improvement, with possible appreciable gains in reduction of equipment size.

7.1.3.2 Horizontal Steam Flow

Panasenko and Koslov [7-22] and Sorokin et al. [7-14] have shown that the breakthrough velocity is essentially independent of water loading for low water loadings when the steam flow is horizontal. Therefore two sets of results due to Wilson et al. [7-20] and Panasenko and Koslov [7-22], covering a wide pressure range, are represented in the coordinates of Ku and the inverse of the square of the Laplace length scale in Fig. 7.1.15. The plates of

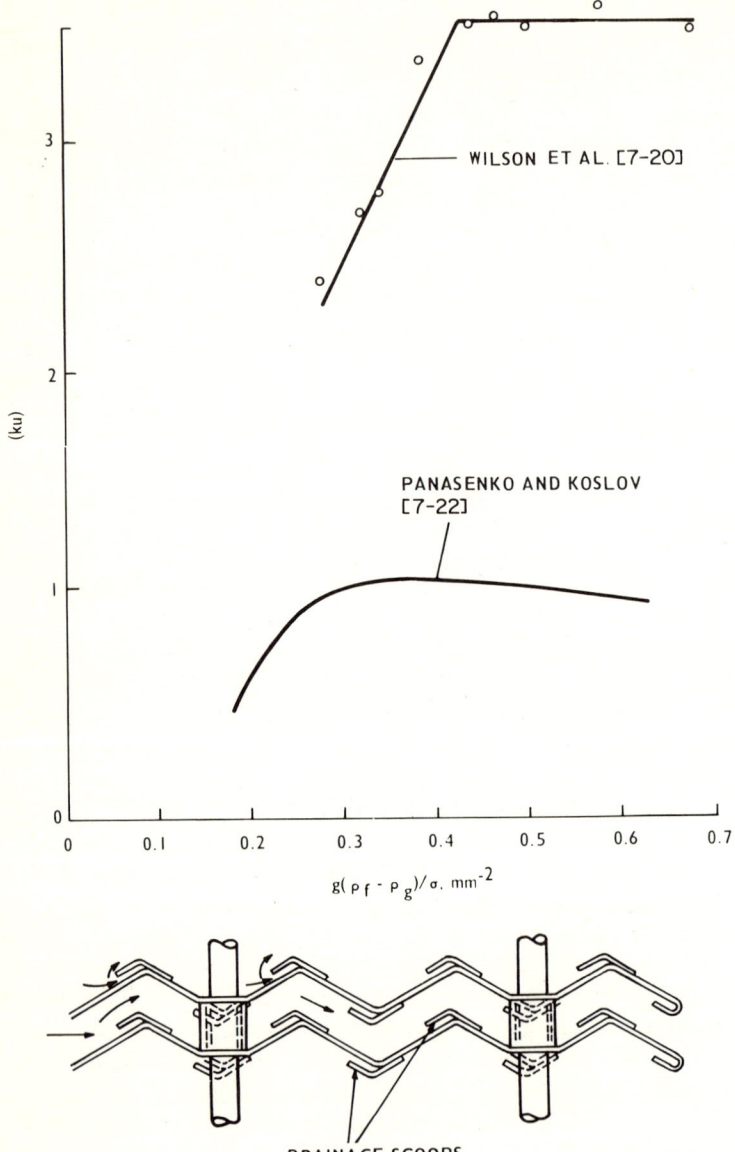

FIG. 7.1.15 Corrugated plates, critical velocities in horizontal flow with negligible water loading.

Wilson et al. are illustrated in the figure, but no dimensions are available. Panasenko and Koslov stated their active plate height to be 130 mm. The plates were 80 mm wide, spaced 5 mm, and with corrugations made up of 10 mm radius curves.

The results of Wilson et al. are close to their other results for knitted wire mesh given in Fig. 7.1.14, but with the limiting value of Ku equal to 3.5. The critical Laplace length is again about 1.5 mm, which suggests that the characteristic length scale of the corrugated plates is some much-smaller dimension than, say, the wave pitch. Panasenko and Koslov correlated their results by the empirical equation

$$U = [1.4 - 1.8(1 - x)] 10^{-0.0056p} \qquad (7.1.19)$$

where the critical steam velocity is in m/s, x is the weight fraction of steam at inlet, and p is the pressure in bars. The line drawn in Fig. 7.1.15 is for $x = 0$; considering that it is an empirical fit to data, it shows the same trends as the data of Wilson et al., but with the maximum value of Ku equal to unity and the critical Laplace length equal to 1.8 mm.

The plates of Wilson et al. with their special drainage scoops have a clear advantage over the simple plates of Panasenko and Koslov, though it must also be noted that the latter were not very tall and may be suffering from end effects, especially at the bottom. However, it is also noticed that water does not actually drain down the scoops illustrated in Fig. 7.1.15. A vortex is set up within the scoops, which displaces the water to drain as a ribbon in front of the scoops. Reentrainment, when it occurs, is from the ribbon, and this in itself suggests that improvements could be made. It is felt by the writer that there is scope for the development of other plate designs.

The only information on the influence of plate height is given by Sorokin et al. [7-14] for air and water at atmospheric pressure. They used plates (b) shown in Fig. 7.1.13 and their results are given in Fig. 7.1.16, where x is the steam quality at inlet. The performance with a plate height of 150 mm is much poorer than with a height of 440-600 mm, and they attributed the difference to end effects being dominant in the shorter plates. They found no

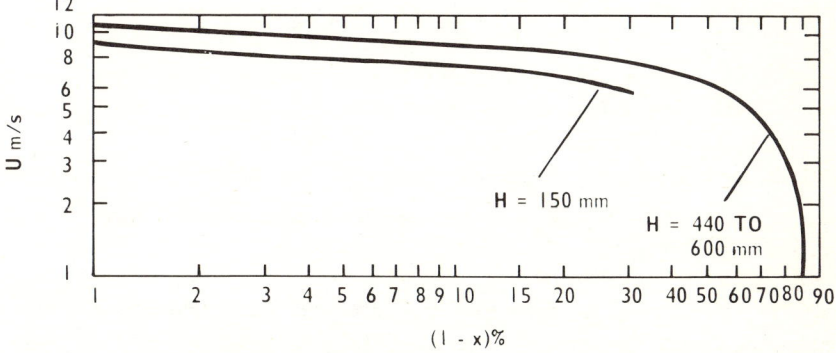

FIG. 7.1.16 Corrugated plates, critical velocity in horizontal flow; influence of water loading: Data of Sorokin et al. [7-21].

obvious influence of plate height H_p over the small range of 440–600 mm, but it should be noted that we would expect the product $H_p(1-x)$ to be a better correlating variable in Fig. 7.1.16 than $1-x$ by itself. It may be that the reduction in the influence of end effects as H_p increases is being balanced over the experimental range by the adverse influence of H_p, and it is not to be concluded that plates of indefinitely large height can be employed. Indeed, if a tall unit is required, the usual practice is to stack plate units on top of each other, with drainage devices at the bottom of each unit.

The influence of water loading is indicated by Eq. (7.1.19) and by the results given in Fig. 7.1.16. It is seen to be significant at the loadings pertinent to wet-steam turbine technology.

Finally, it should be noted that corrugated plates are rarely aligned such that the steam flow is horizontal. Usually the flow is inclined upward at an angle, which may be between 10° and 45° to the horizontal. The inclination influences the breakthrough velocity, often improving it, but that is subject to test. The only information on this subject known to the writer is that of

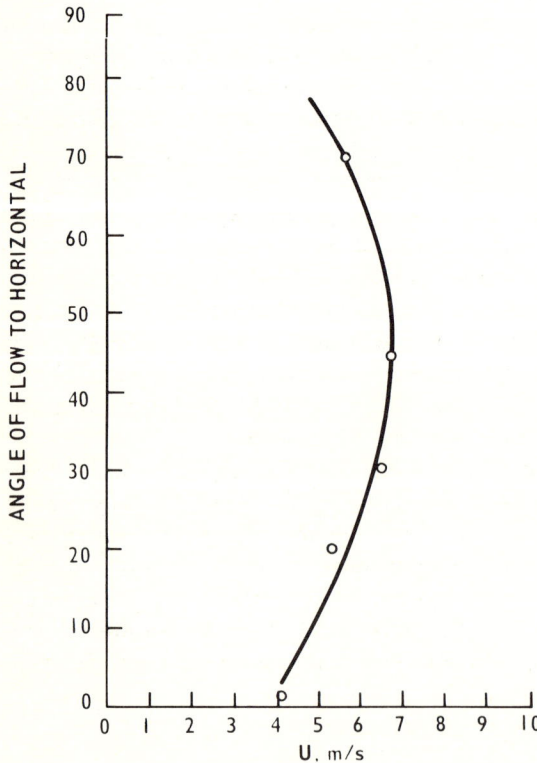

FIG. 7.1.17 Knitted wire mesh: Influence of angle of flow on critical velocity: Data of Dickson and Morrison [7-23].

Dickson and Morrison [7-23] for knitted wire mesh; it is given in Fig. 7.1.17. Here it is seen that performance is improved at first as the steam flow is varied from the vertical. The water drains to one side of the mesh and escapes more readily. The performance deteriorates as the inclination varies from 45° to 0° to the horizontal, but this same trend may not occur with corrugated plates.

7.2 EXTERNAL MOISTURE SEPARATOR-REHEATERS

R. L. Coit, P. D. Ritland, T. J. Rabas, and P. W. Viscovich

Moisture separator-reheaters (MSRs) (see Figs. 7.2.1 and 7.2.2) are a critical part of any nuclear power plant using pressurized water or boiling water reactors, and will become more of a critical part of our total energy picture as nuclear plants using light water reactors assume a larger role in power generation. The reasons MSRs are vital to nuclear plants with light-water reactors lie in the character of the steam produced; both types of reactor systems produce saturated steam at relatively low pressures that, if allowed to expand continuously through the turbine, would exhaust at 20–24% moisture. This excessive moisture would impair the efficiency of the turbine blades and cause them to erode at an excessive rate.

The reasons MSRs are so vital to our total energy picture can be found in their functions. In broadest terms, the functions of MSRs are to reduce the maintenance and increase the availability of large steam turbines, and to increase overall plant efficiency. More specifically, MSRs remove the moisture from and then superheat high-pressure turbine exhaust steam.

The number and size of MSRs shipped or placed in service have grown as the number and size of nuclear power plants have grown. Westinghouse shipped its first MSR in 1965 and placed it in service in 1967; since then the company has shipped 186 and placed another 87 in service.

While the number and size of MSRs have increased, their design has evolved in response to early operating experience and to the need for larger, more efficient units. At Westinghouse this evolutionary process has progressed through three well-defined generations until today its MSRs are highly reliable links in the complex chain of equipment needed to generate power in nuclear plants. Other manufacturers of MSRs—General Electric, Brown-Boveri, and Allis Chalmers (in association with Kraftwerk Union Aktiengesellschaft) in particular—have also followed evolutionary paths.

7.2.1 Physical Features of MSRs

An MSR is a pressure vessel with three basic parts: an outer shell (a cylinder, often with formed heads welded to each end), moisture separators (usually

FIG. 7.2.1 Four moisture separator-reheaters being installed on turbine floor of 700 MW plant.

chevrons but sometimes wire mesh), and one or more tube bundles to transfer heat from high-pressure steam to the partially spent cycle steam (Fig. 7.2.3). The first unit, installed in a 450 MWE plant, is 9 ft in dia., 40 ft long, and weighs 40 ton. In contrast, today's units are large: those designed for use in nuclear plants rated at 1200 to 1300 MWE are 12 ft 9 in. in dia., about 95 ft long, and weigh 212 ton.

FIG. 7.2.2 End view of moisture separator being installed in 700 MW plant; reheater drain tank shown on far right.

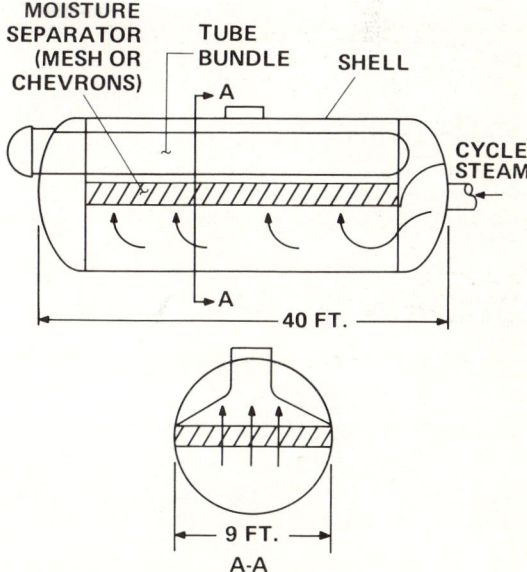

FIG. 7.2.3 Three basic parts of an MSR: outer shell, moisture separators, and one or more tube bundles.

In a nuclear plant, MSRs are normally placed horizontally and parallel to the turbine axis at or near the turbine elevation, and in the pipe system between the high-pressure and low-pressure turbines (Fig. 7.2.4).

The shell, heads, and internal supports of an MSR are fabricated of carbon steel plate; the tube sheets are usually clad carbon-steel forgings. Tubes can be 80-20 or 90-10 Cu-Ni, or seamless or welded carbon steel. Chevron moisture separators are typically carbon steel when used in PWR plants, and often stainless steel when used in BWR plants.

PWR and BWR reactors produce saturated steam at 900–1000 psig, or superheated steam with 50°F superheat. If the steam were allowed to expand normally through the turbine, the moisture content at the turbine exhaust would be 20–24%, a level so high that excessive blade erosion and losses in blade efficiency (and corresponding losses in cycle thermal efficiency) would result.

To minimize blade erosion and losses in efficiency, cycle steam at about 175 psig, 350°F, and with 12–14% moisture is piped from the high-pressure turbine exhaust through the crossunder piping system to a steam-inlet manifold in the MSR. From the manifold, the steam passes through the chevron moisture separators, where moisture is removed and the steam is restored to a dry and saturated state (Fig. 7.2.5).

The chevron moisture separators (Fig. 7.2.6) remove moisture from the wet steam by forcing the direction of flow to change. As the flow direction changes, the moisture particles drop out of the flow and impinge on the surface of the chevron-shaped steel. The impinged moisture then moves down

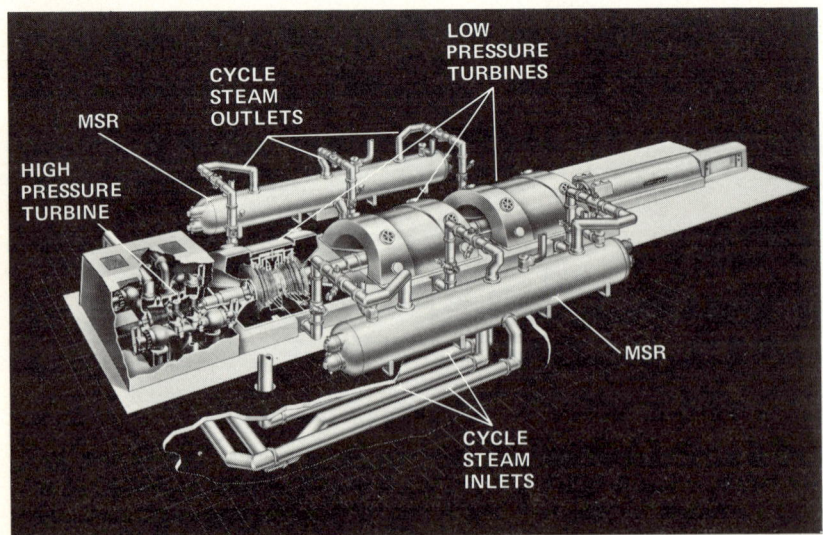

FIG. 7.2.4 MSRs are normally placed horizontally and parallel to the turbine axis, at or near the turbine elevation, and in the pipe system between HP and LP turbines.

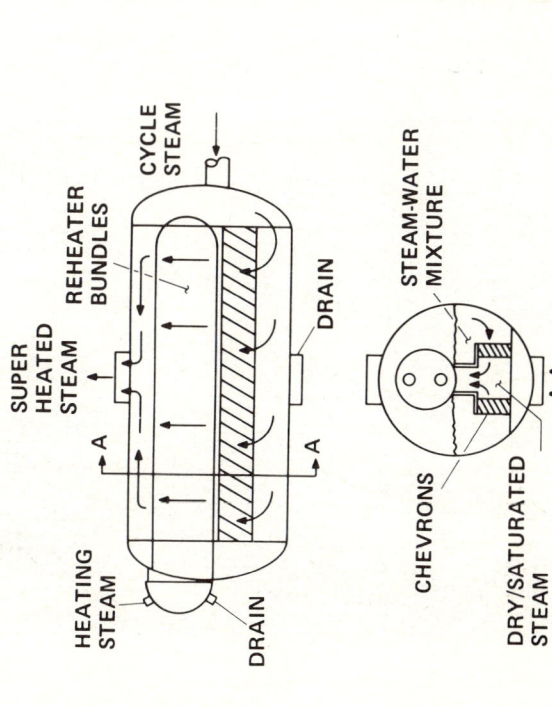

FIG. 7.2.6 Chevron moisture separators remove moisture from wet steam by forcing direction of flow to change.

FIG. 7.2.5 Cycle steam enters MSR through inlet manifold, passes through moisture separators, then passes over outside of a tube bundle; tube bundle is heated by steam taken from upstream of HP turbine main stop valve.

the steel surfaces and is returned to the feedwater steam system through drains.

The dried-cycle steam passes over the outside of low-profile, integrally finned tubes (Figs. 7.2.7 and 7.2.8), which are heated by steam taken from upstream of the HP turbine main stop valve. As it passes over the tubes, the cycle steam picks up about 125°F of superheat, then leaves the MSR through outlets on the top of the shell and flows to the LP turbine.

There are two basic reasons why MSRs are needed in nuclear power plants. First, cycle efficiency is increased $1\frac{1}{2}$-2% over a comparable nonreheat cycle, and electrical output is increased proportionally. Second, resulting moisture levels are lower at the turbine exhaust, 11% at 1.5 in.

Cycle efficiency and electrical output increase since it is more efficient to dry and superheat saturated steam outside the LP turbine than it is to handle high moisture levels inside the turbine; in short, turbine internal moisture separation is less efficient than external separation.

While increases in cycle efficiency can be empirically demonstrated in the power plant, it is difficult to prove in thermodynamic terms. On the surface it appears self-defeating to use high-pressure steam to reheat lower-pressure cycle steam. However, MSRs are able to increase the plant's overall efficiency by increasing the turbine's efficiency more than they decrease the plant's steam-use efficiency.

FIG. 7.2.7 Shop view of two tube bundles.

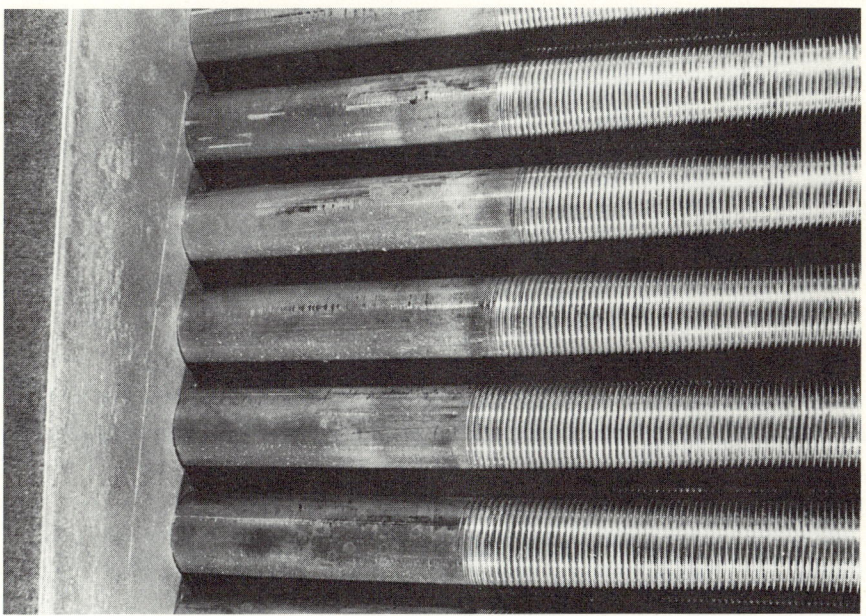

FIG. 7.2.8 Tubes in bundles are low profile, integrally finned, and heated by steam taken from upstream of HP turbine main stop valve.

To further improve cycle and MSR efficiency, the reheater portion of the MSR is divided into two pressure stages (Fig. 7.2.9), gaining an additional .3–.5% in turbine heat rate and kW over single-stage reheat. To illustrate, use of two-stage reheat increases cycle-steam temperature from about 350 to

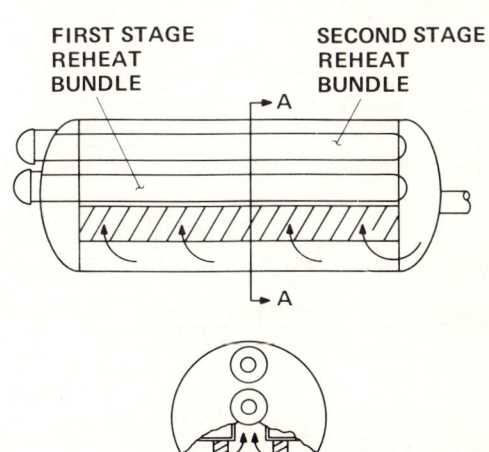

FIG. 7.2.9 Dividing reheater portion of MSR into two pressure stages improves cycle and MSR efficiency.

420°F as it passes over the first reheater tube bundle, and from about 420 to 500°F as it passes over the second tube bundle.

Steam taken from upstream of the throttle provides about half the heat needed when two-stage reheat is used; the remainder is supplied from steam taken from a high-pressure turbine extraction zone.

An added benefit from the second reason to use MSRs—to reduce moisture levels at the turbine exhaust—is reduced maintenance and increased availability of the turbine. Wet saturated steam erodes turbine blades and rotors more rapidly than dry steam, causing increased maintenance.

7.2.2 The Evolution of MSR Design

Westinghouse MSRs have gone through three design generations. The first unit (Fig. 7.2.10) of the first generation was shipped in 1965 and placed into service in 1967 at Southern California Edison's San Onofre unit no. 1. This first unit had a demister section of stainless-steel knitted wire mesh, integrally supported by a lightweight grid, and one stage of reheat. The cycle steam entered the unit through a horizontal straight duct located at one end of the shell, discharged into the shell through slots in the upper surface of the duct, then flowed vertically across the tube bundle.

Mesh demisters were used in the first-generation design: while mesh is able to effectively remove moisture, large areas placed in a horizontal configuration

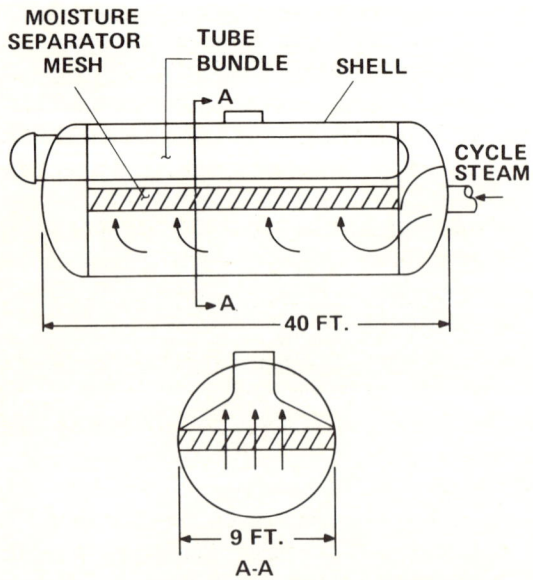

FIG. 7.2.10 Westinghouse first-generation design has demister section of stainless-steel knitted wire mesh.

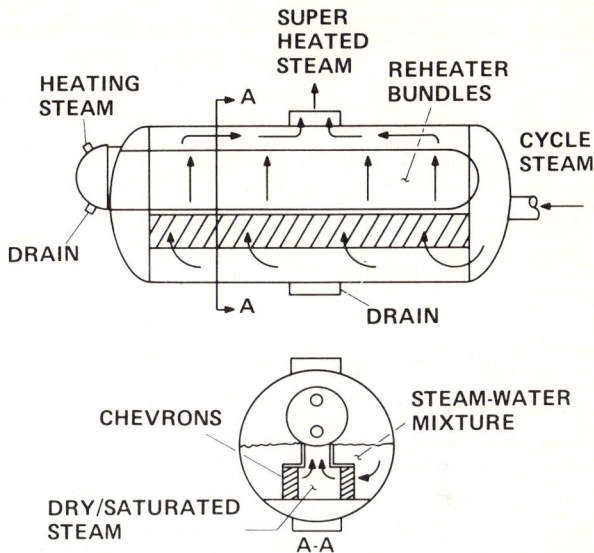

FIG. 7.2.11 Westinghouse second-generation design uses vertical chevrons instead of mesh to remove moisture, reducing size of unit.

are needed. Because of this need for a large horizontal area, vertical chevrons replaced mesh in the second-generation design, introduced in 1970 (Fig. 7.2.11). Second-generation MSRs allow 45% more steam to pass through a given-size unit, reducing the space needed in the plant by a comparable percentage. This reduced need for space per pound of steam is a primary reason why it is feasible to build and ship MSRs for the larger nuclear plants.

Third-generation MSRs (Figs. 7.2.12–7.2.15), introduced in 1972, are actually two second-generation MSRs connected by a common inlet section. This design, which allows use of one tank on each side of the turbine even when three double-flow LP tandem turbines are used, has three primary advantages over the second-generation design: the width of the turbine floor is reduced 12–24 ft depending on plant layout, external piping is simplified, and the effects of testing the interceptor valve are reduced.

Other manufacturers of MSRs have gone through similar design stages. Some of their first-generation designs had individual vessels for separation and reheat: separators were vane-type; heaters were provided as needed and utilized carbon-steel tubes with welded fins.

General Electric's second-generation MSR, and current offering, uses chevron separators. Tubes are low-finned Cu–Ni, carbon, or low-alloy steel.

An early European offering is somewhat different in that covers on each end of the pressure vessel are flat instead of formed. A current European offering has hemispherical heads. Regardless of head shape, the moisture separator is one level of 4-in.-deep wire mesh placed horizontally, and each of

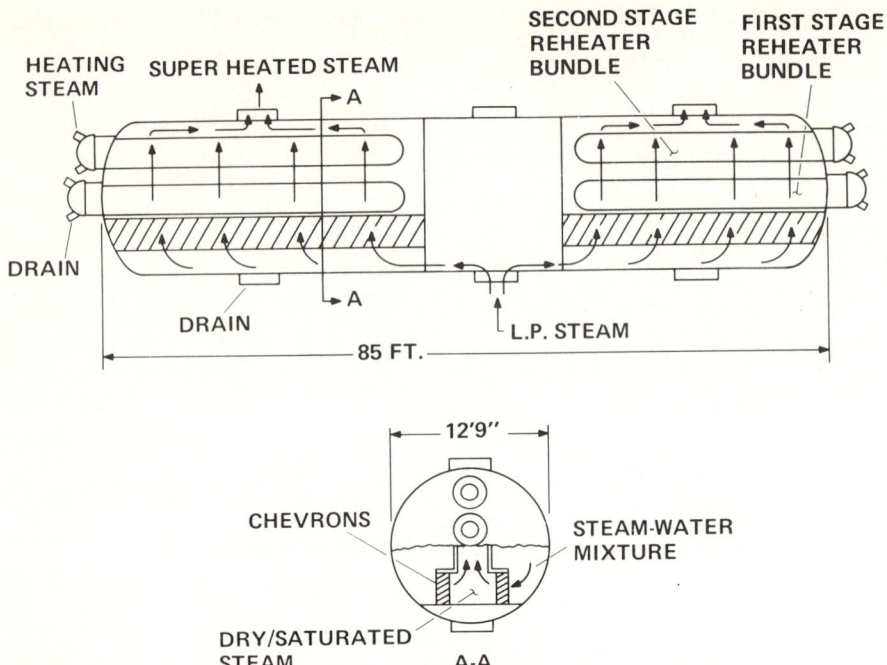

FIG. 7.2.12 Westinghouse third-generation MSRs are actually two second-generation MSRs connected by common inlet section.

FIG. 7.2.13 MSRs being assembled in Westinghouse plant at Lester, Pennsylvania.

FIG. 7.2.14 Internal supports are fabricated outside MSR shell, then inserted and welded in place.

FIG. 7.2.15 Cross-sectional view of third-generation MSR. In central portion is channel for tube bundles; left and right above center line are steam water distributing passages.

the two stages of reheat consists of two sets of three identical bundles placed in the same horizontal plane. U-tubes are $\frac{3}{4}$ in. OD, integrally finned, and fabricated of seamless carbon or welded stainless steel.

Allis-Chalmers Company and Kraftwerk Union Aktiengesellschaft are jointly offering two types of MSRs. Their vertical design utilizes two-stage separation: the steam first passes through a cyclone, then through vane-type separators similar to those used in early designs. The turbine steam is then reheated by being passed over U-tubes arranged vertically. Their horizontal units utilize vane-type separators and U-tube reheaters.

7.2.3 Operating Experience

An understanding of operating experiences is helpful to a general understanding of MSRs. These operating experiences are related to reheater startup, the steam-inlet manifold, tube vibration, reheater drainage, and individual tube drainage. The cause, effects, and analyses of these experiences provide a major part of the rationale for today's design.

7.2.3.1 Reheater Startup

In late 1966, during hot functional testing of the first operating unit at San Onofre and when the MSR shell was at atmospheric pressure and temperature, steam was allowed to enter U-tubes in the reheater, displacing the air in the tubes and causing the legs to heat at different and unpredictable rates. The result was that the upper leg of the tube became hot much faster than the lower, causing the upper leg to expand an inch or so more than the lower by the time the steam had reached the U-bend. This differential expansion caused the tube to bind with a cocking action in its supports located near the U-bend. As the steam turned around the U-bend and began to travel down the bottom leg the binding intensified, causing the bottom leg to expand and the tube to buckle (Fig. 7.2.16).

Test apparatus Two test models were constructed: The first model had the same bundle height as the San Onofre unit but was only 1 ft wide, and included 75 tubes (Figs. 7.2.17 and 7.2.18). The second model had

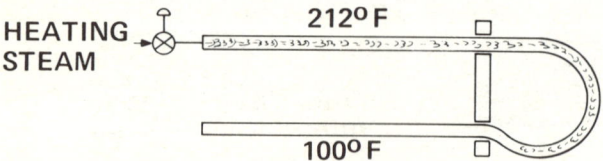

FIG. 7.2.16 Steam entering U-tubes displaces air in tubes and forces legs of tube to heat and expand at different and unpredictable rates. At San Onofre, this differential expansion caused tube to bind and cock in its supports.

FIG. 7.2.18 Tube bundle in model MSR shown in Fig. 7.2.17 includes 75 tubes and is full height, but only 1 ft wide.

FIG. 7.2.17 Test model MSR used to simulate conditions found during reheater startup.

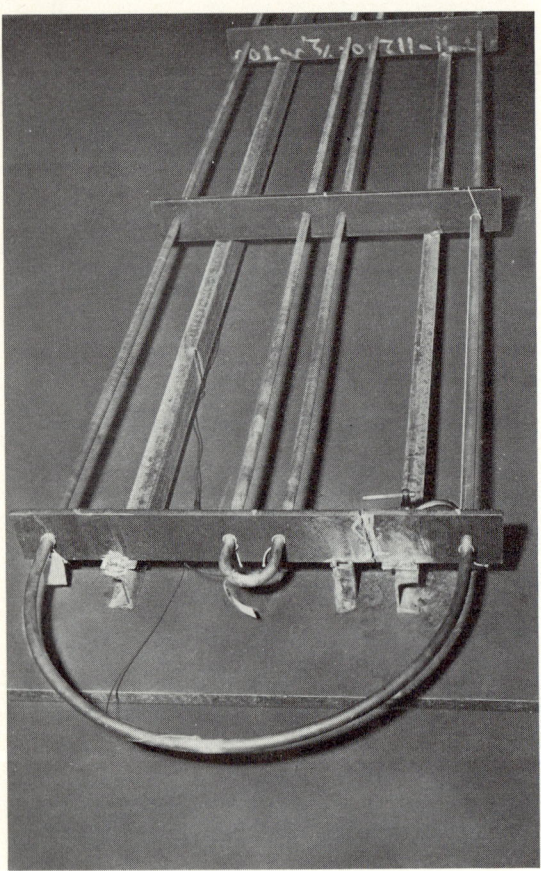

FIG. 7.2.19 Two-tube model used to test reheater start-up conditions included one tube with longest and another with shortest radius bend used in San Onofre MSR.

two full-scale tubes, one with the longest radius bend, the other with the shortest (Fig. 7.2.19).

Test procedures and results Tests on the full-height model were begun before startup at San Onofre, and initially focused on the unit's abilities to function within performance specifications. When the tube buckling at San Onofre was discovered, the full-height model was dismantled and a similar condition uncovered.

1 Tests on the two-tube model were designed to determine how introducing steam into the tubes at various rates, and the location of tube supports, affect tube expansion and buckling. To determine tube behavior over a wide range of operating conditions, steam was suddenly introduced to one

side of the cold U-bend, then to both sides, then to U-tubes uniformly preheated to 200°F. Tube supports were placed at various angles to the tube and purposely misaligned to demonstrate the tube's ability to expand freely when properly heated. The legs expand equally if leg temperature and movement of the air pocket through the legs is controlled by carefully injecting steam simultaneously into each leg. This procedure however is so difficult to control that it could not be reliably used in an operating plant. Any significant difference in expansion, and all bowing and buckling, is eliminated if the tubes are uniformly preheated to 200°F by the turbine cycle steam.

2 The full-height model was refitted with a new bundle and, as subsequent tests were run, the results of the tests on the two-tube model were verified. All tests used startup procedures developed during the two-tube tests, and sidewalls were removed and the tubes inspected after each test. No buckling occurred.

Application of test results to MSR design and operation The test results and conclusions led to a twofold recommendation on how to initially operate MSRs: First, air must be removed from the tubes and steam must not enter the tubes until the turbine load has reached 35%, effectively eliminating the heat wave that develops as the steam pushes air ahead of it in the tubes. Second, the tubes are preheated to approximately 200°F by the 35%-load LP steam, and heating steam is allowed to enter the tubes at a controlled rate, precluding condensation on the preheated tube walls. This procedure reduces the initial transfer of heat from the steam to the tubes and maintains tube temperature at 200°F, the shell-side temperature. When the steam valves are opened further, the steam already in the tubes will condense uniformly, allowing both legs to expand evenly and eliminating any bind in the supports.

The solution to reheater startup problems might have been to substitute U-tubes with straight tubes and a floating reverse chamber. However, continued use of U-tubes is justified when their freedom to expand and their abilities to assume configurations that provide a more closely balanced steam flow are considered.

7.2.3.2 Steam-Inlet Manifold

The wire mesh demister in the San Onofre MSR wore through locally and broke after operating only three months. The breaks in the wire were concentrated at the frame-support grids, although some breaks also occurred in other areas. The breaks were analyzed and their cause attributed to abrasion, in turn caused by the impact of the entering steam. The investigation centered on the inlet-steam manifold design.

Test apparatus Two models were constructed to examine the effects of steam distribution on demisters. An air model with a half-size lower chute and a 28% open turnaround was used to analyze how high jet velocities can be broken by horizontal baffles or "hats." Results of the air-model tests led to tests with a steam model able to manifold 70% of full flow and bypass 30% through the original turnaround. The manifold for the steam-test model was designed to handle steam with an inlet chute velocity of 150 ft/s, a volume flow rate of 1000 cfs, and a jet absolute velocity of 156 ft/s for all portions (the corresponding jet static loss is .75 psi).

Test procedures Four series of tests were run. The first focused on manifolding the steam flow along the lower chute of the MSR, thereby lowering the steam velocity as it impinges on the mesh. The second tests were run with a closed turnaround end and manifold with increased flow area near the turnaround end. To run the third tests, the number of vanes was increased and a perforated plate added above the mesh to force more even distribution of the incoming steam. To define the effects of the perforated plate on performance, the fourth series of tests was run without manifold vanes and hats. The flow of cycle and heating steam, reheater outlet temperature, and moisture carryover at 10 points along the length of the mesh were continuously measured during all tests.

Test conclusions During the first series of tests, calorimeter readings indicated carryover at the turnaround end, extending one-third of the way to the steam inlet. Carryover was also indicated by an increase in steam consumption from the 0-10% moisture runs. The mesh was overloaded because of the 28% bypass entering the upper chute through the original turnaround.

Calorimeter readings indicated relatively little carryover during the second series, although there was a marked decrease in outlet temperature and increase in steam consumption. The gross carryover was localized along the wall and primarily above the demister elements.

The third series of tests showed only slight carryover at the turnaround end, and no carryover elsewhere. Heating-steam balances showed no carryover. Calorimeter readings taken during the fourth test series indicated some carryover at the turnaround end, but insignificant carryover elsewhere.

Application of test conclusions to MSR design Laboratory tests have shown that when steam upflow through mesh is uniform and below a threshold velocity of approximately 5 ft/s, moisture removal could be 100%. More specifically, the manifold air tests demonstrated that uniform velocity was attainable, and the large-scale steam tests demonstrated that properly designed manifolds eliminated moisture carryover.

The steam-inlet duct in the first-generation design included a turnaround but not manifolds. All units of the first-generation design (with mesh demisters), whether in operation or on order, were quickly retrofitted or redesigned according to principles set forth by Dittrich and Graves [7-30]. The physical design was patterned after the test model and is shown in Fig. 7.2.20.

Field and laboratory tests have demonstrated that chevrons, used to replace mesh in second- and third-generation designs, also have a threshold velocity

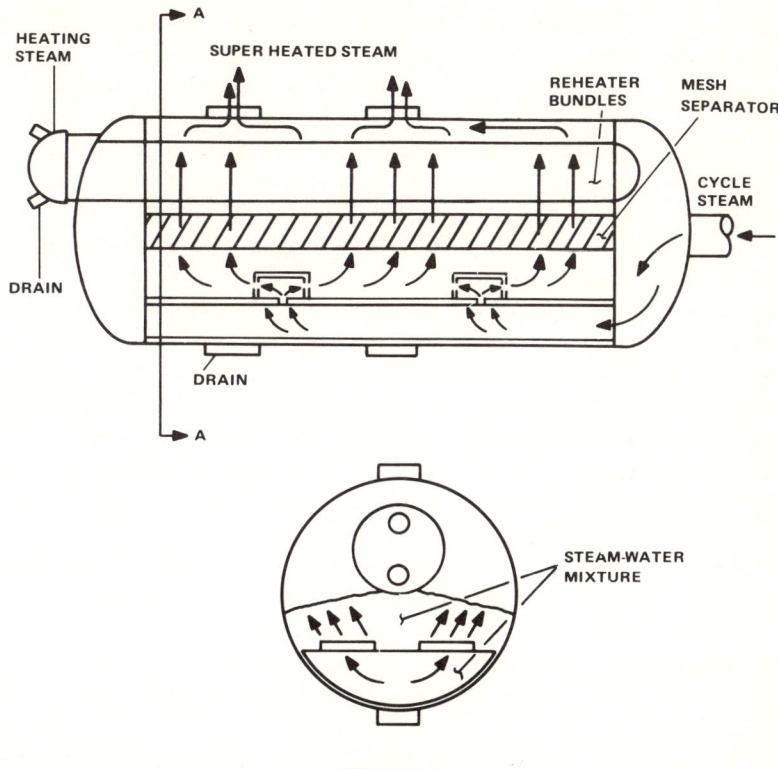

FIG. 7.2.20 As a result of steam distribution tests, operating MSRs of first-generation design and all units on order were retrofitted with manifolds.

below which moisture carryover is zero. Design of the manifolds used in these units is based on the conformal mapping technique developed by McKnown and En-run Hsu [7-28].

7.2.3.3 Tube Vibration

Tube vibration caused some tube wear and breakage at one 700 MW plant, and could have caused the same problem at similar units had they not been quickly retrofitted.

The phenomenon of tube vibration was carefully analyzed by the development engineers. The first task was to remove the bundles and carefully inspect the tubes for wear patterns. The tubes wore generally (and most failures occurred) in the upper portion of the bundle where the high specific volume of the reheated steam causes the crossflow velocity to be greatest. Some tubes on the sides of the bundle, where bundle-to-shell bypass leakage had caused local velocities much higher than normal, also wore and failed. Bottom tubes showed little wear.

Analysis of tube vibration centered on use of a cross-sectional model of an MSR tube bundle to derive data concerning peak-to-peak amplitude versus velocity response, correlation of amplitude with fluid dynamic head, and primarily the constant of proportionality, an experimental constant that varies with the type of tube and dimensional layout of the tube field.

These data were needed to determine ways to minimize damaging tube vibration, and in turn to use more efficiently the available heat transfer surface area of narrow tube bundles with high shell-side cross-flow velocity. In addition, the tests were needed to determine how to prevent tube and vibration failure when the normal steam flow through each MSR is nearly doubled, while testing the intercept and stop valves.

Tube vibration and resulting failure in MSRs is characterized by a large-amplitude whirling motion. This motion can be induced only by fluid-elastic-type flow; other types of flow, notably vortex shedding, have been ruled out as a cause of tube vibration in these units.

A fluid-elastic mechanism of flow is explained by a theory put forth by Connors [7-29]. The theory holds that there is a threshold (or critical) velocity above which large unstable vibrations occur. This threshold varies with the physical characteristics of the tube bundle and density of the fluid.

The relationship among critical velocity and other variables is

$$\frac{U_c}{f_0 D} = \beta \sqrt{\frac{m_0 \delta_0}{\rho D^2}}$$

where U_c = minimum free-flow area critical velocity
f_0 = natural frequency of individual tubes
D = diameter of tube
m_0 = weight per unit length of tube
δ_0 = logarithmic decrement of tube
ρ = density of fluid
β = experimental constant dependent on type of tube (finned, plain) and cross-sectioned geometry of bundle or tube pattern, particularly on T/D and b/D, where
T = tube transverse pitch
b = tube longitudinal pitch

Test apparatus Three design criteria were established for the basic test model (Fig. 7.2.21):

1 The model is able to evaluate the effects of flow on vibration across spans of 37.12 in. and 44 in.
2 The width is five tubes
3 The height allows a fully developed flow field and meets pressure-drop and flow limitations of the compressor

FIG. 7.2.21 Test model, five tubes wide, is able to evaluate effects of flow on vibration across spans of 37.12 in. and 44 in. Height of tube bundle allows fully developed flow field.

The first tests were made on a model with 11 vertical rows and 49 tubes. A Plexiglas plate, placed over the unit's entire face, allowed the test to be observed. To eliminate or minimize bypass, half-tubes were installed along the Plexiglas plate; the opposite face was placed within $\frac{1}{16}$ in. of the tubes.

Tube supports were made from actual units drilled in the shops to eliminate variance from improper tolerances. Tubes of 90-10 Cu-Ni were used, with 19 fins per in., each fin .012 in. wide and .050 in. high, and with a wall thickness of .049 in. under the fins. Their design is similar to the LP tube bundle in an operating MSR.

So that the worst possible vibratory conditions could be tested, it was necessary that the natural frequencies of all tubes be identical. To achieve this homogeneity, all tubes were cut to the same length and their frequency tested and adjusted by placing the tube in tension or compression. The average natural frequency of tubes in the model was 36.1 Hz and the average logarithmic decrement 0.0327. A specially designed end plug, fastened at both ends of the tubes to a 2 in. thick tube support (Fig. 7.2.22) was used to minimize damping and to control the final natural frequency of the tubes.

Test procedures Tests measured the amplitude of vibration as a function of ligament velocity. Amplitude of vibration was measured with accelerometers placed inside the tubes. Air flow began low, was increased in

(a)

(b)

FIG. 7.2.22 During testing of tube vibration, a specially designed end plug, fastened at both ends of the tubes to a 2-in.-thick tube support, was used to minimize damping and to control final natural frequency of the tubes.

predetermined steps, and amplitude readings were taken at each step. After maximum flow was reached, it was reduced in similar steps and corresponding measurements were taken. As a control, tests were made at several random flow points to uncover any inherent variables introduced by the step-by-step procedure. The ligament velocity was calculated by measuring the air flow through a calibrated nozzle located downstream of the test section.

Test conclusion *Qualitative results* were derived from viewing the tests and taking slow-motion films through the Plexiglas cover, and feeling the end plugs for vibration. Results were:

1. Tube vibration could be seen after accelerometer readings indicated peak-to-peak amplitudes of .05 in. or more.
2. Slight movement of the end plugs could be felt when displacement reached .05–.10 in.; movement was greater at the top of the section and the range from top to bottom was gradual.
3. Some tubes were more susceptible to large amplitude movement than others.
4. Dynamic forces overcame gravity and the tubes' natural stiffness to move each tube against the top of the tube hole before setting up a large vibratory motion.
5. A periodic loud shudder was heard at large accelerometer readings and high flow rates, presumably the result of a large number of tubes vibrating in unison. Steady-state accelerometer readings were not affected.

Quantitative results were derived from test data and fall into three primary areas: amplitude versus velocity response, correlation of amplitude with fluid dynamic head, and determination of the constant of proportionality β.

1. Peak-to-peak amplitude and its relationship to velocity were computed under incrementally increasing, decreasing, and random flow.
2. The correlation of amplitude and fluid dynamic head was computed from data derived from the tests. Correlation of amplitude with some parameter other than velocity is needed, since amplitude response is a function of fluid density, making it impossible to directly compare measured amplitude for various tubes at equal velocity. Correlating amplitude with fluid dynamic head is logical since (*a*) the amplitude of a vibrating tube depends fundamentally on the lift and drag forces acting on it, and (*b*) the lift and drag coefficients (the ratio of lift and drag forces to dynamic head) are dependent on the relative positions (amplitude) of the adjacent tubes (Fig. 7.2.23).
3. The constant of proportionality, the primary characteristic to be determined by the investigation of MSR tube vibration, was determined to be 10.9–12.4, and was calculated using the equation

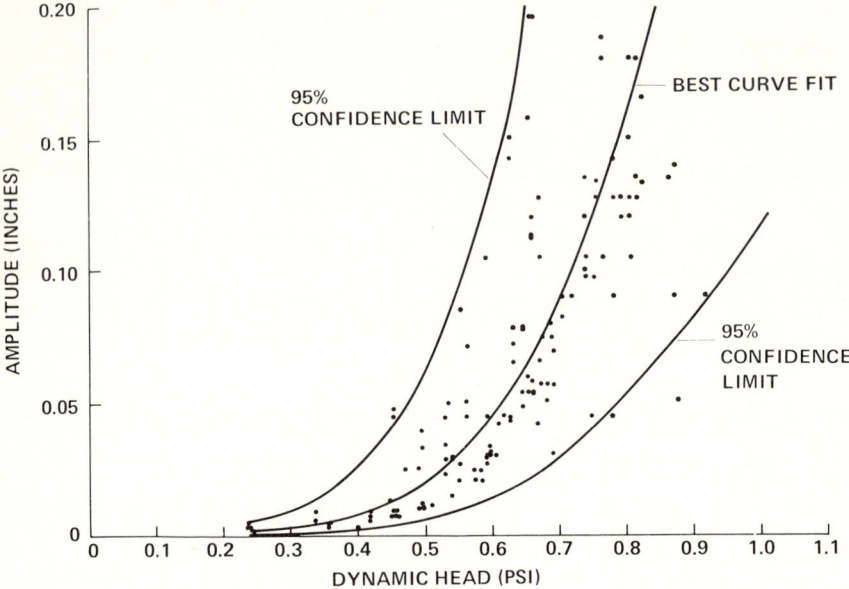

FIG. 7.2.23 Effect of flow on vibration amplitude.

$$\frac{U_c}{f_0 D} = \beta \sqrt{\frac{m_0 \delta_0}{\rho D^2}}$$

The determination of β depends strongly on the choice of nondamaging amplitude, a value that defines the acceptable velocity (or dynamic head) on which to base the velocity parameter. A more valid consideration is to find that point above which the amplitude increases very rapidly for a small change in flow. A conservative peak-to-peak amplitude of approximately .05 in. was chosen, since it is nondamaging and it lies close to the point on the velocity/amplitude curve where changes are rapid (Fig. 7.2.23). Based on the peak-to-peak amplitude of .05 in., β is 11.4. The lower limit for β is 10.9, the upper limit 12.4.

To apply the test data to an individual unit, it is necessary to define a quantity (Alpha) which would correspond to β. Those units with Alpha values above β are susceptible to tube vibration and failure, while those with Alpha values below β are not. From our field observations, units with intermittent Alpha values of 6.5 were not damaged, but units that ran steadily at 12.8 or higher, and/or intermittently at 22.0, suffered tube failures after several months. Test results correlated well with field observations. One unit in a 700 MW plant had Alpha values of 10.7 at the bottom of the bundle and 12.8 at the top of the bundle. Alpha reached 18.0 during testing of the intercept valve, slightly wearing the tubes in the bottom of the bundle (Fig. 7.2.24).

Application of test results to MSR design MSRs must operate under a wide range of flow conditions. Perhaps the most stringent of these conditions occurs during testing of the intercept valve, when velocity can rise to 180% and vibration to 300% of normal design.

The tests provided valuable data on the limits of tube vibration β and damping δ, ensuring that the bundles are designed to remain safely within these limits despite maximum shell-side mass-flow rates and pressure drops. More specifically, the tube bundles are now designed to withstand the maximum loading that occurs during testing of the intercept value, with a vibration peak-to-peak amplitude no greater than .05 in.

7.2.3.4 Reheater Drainage

Poor drainage of the reheater bundles has caused tube welds to fail and tubes to buckle in at least three MSRs of the first-generation design. The phenomenon was investigated in the field to simulate realistic conditions.

Test procedures and conclusions Sight glasses to indicate the water level in the drain tank (and the MSR) were installed in the drain lines of several MSRs. The water level and cycle-steam temperature were recorded, and it was found that as the water level rose the cycle-steam temperature dropped, indicating that the tubes were being intermittently flooded. This intermittent flooding led, in turn, to stresses caused by thermal transients and subsequent radial expansion and contraction.

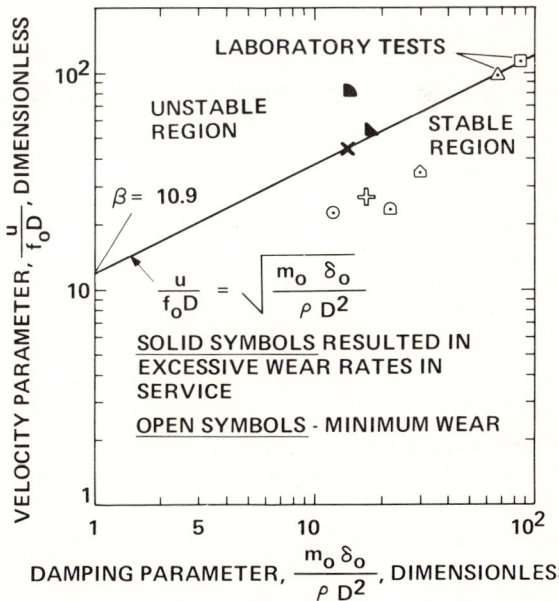

FIG. 7.2.24 Velocity parameters versus damping parameters for various MSR installations.

Application of test results to MSR design and operation The primary reason for the flooding was a "manometer effect" caused by draining two MSRs to a common tank. This arrangement might be adequate if the temperature control valves on the inlet to each reheater bundle operate in perfect unison and tube-side pressures remain constant under all operating conditions. But the control valves cannot operate in unison at all times, and the shell-side flow in one tank is often different than the other. The result is that drainage from the MSR with the lower inlet pressure is impeded and its tubes flood.

The solution is simple and effective: drain each reheater bundle into a separate tank (Fig. 7.2.25).

7.2.3.5 Individual Tube Flooding

A number of tube welds failed in several first-generation MSRs installed in 500 and 700 MW plants. The individual welds failed at a point approximately $\frac{1}{4}$ in. from the face of the tube plate and at the bottom and outside of the tubes (Fig. 7.2.26). Cracks in the welds showed considerable erosion, apparently along a leak path.

Test apparatus and procedures Over a three-year period, four MSRs in commerical operation were fitted with thermocouples at the top and bottom of tubes (Fig. 7.2.27) in selected vertical rows (Fig. 7.2.28). To determine the effects of varying loads, temperature at each location was continuously recorded.

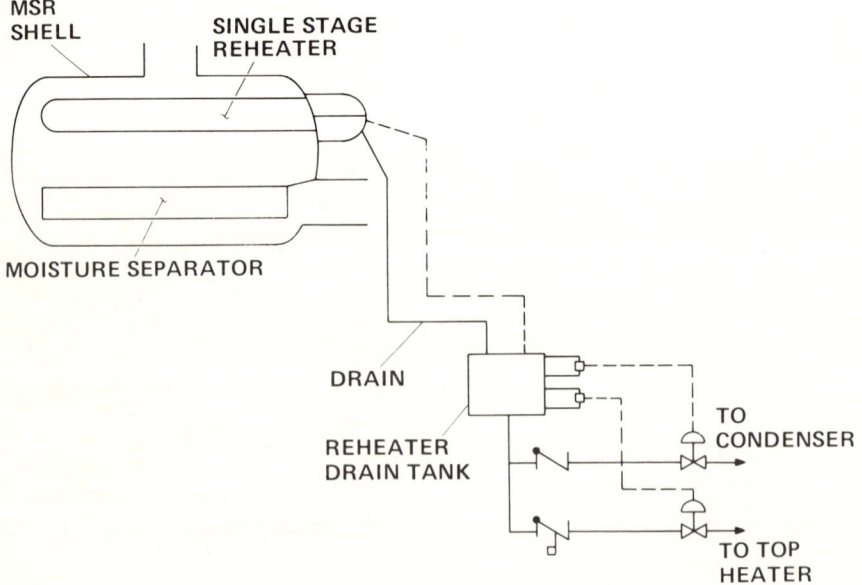

FIG. 7.2.25 Intermittent flooding of the reheater bundles has been eliminated by draining each bundle to a separate tank.

External Water Separators 361

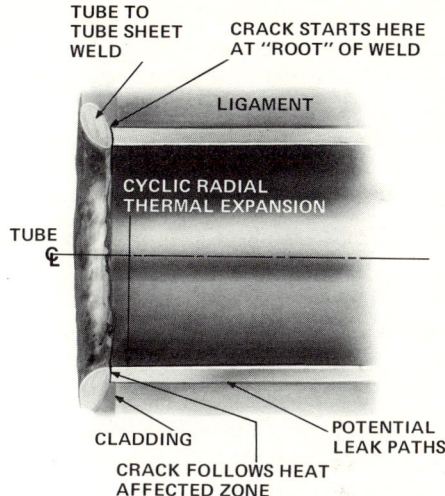

FIG. 7.2.26 In several first-generation MSRs, installed in 500 and 700 MW plants, individual welds failed at a point approximately $\frac{1}{4}$ in. from face of tube plate in bottom and outside tubes.

An extensive and detailed computer study was run concurrently with the field tests. The purpose of the study was to determine the amount of steam each row of tubes receives when operating at various loads, and with and without orifice plates.

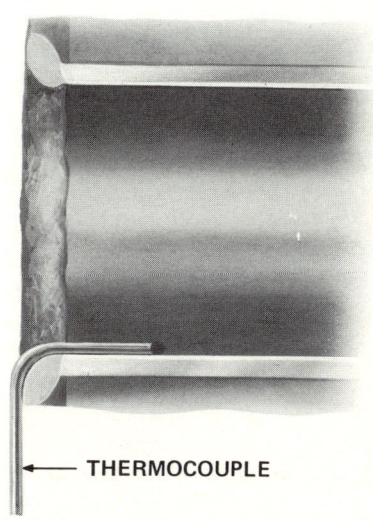

FIG. 7.2.27 To record temperature fluctuations, thermocouples were placed at top and bottom of some tubes.

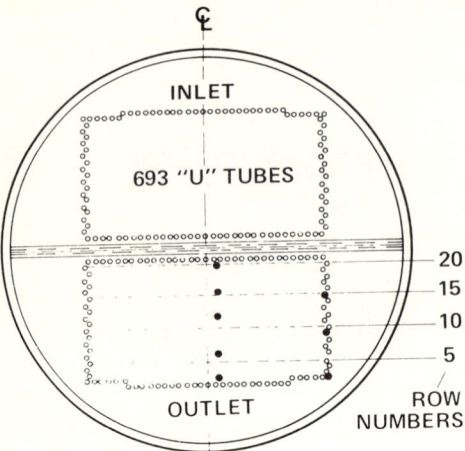

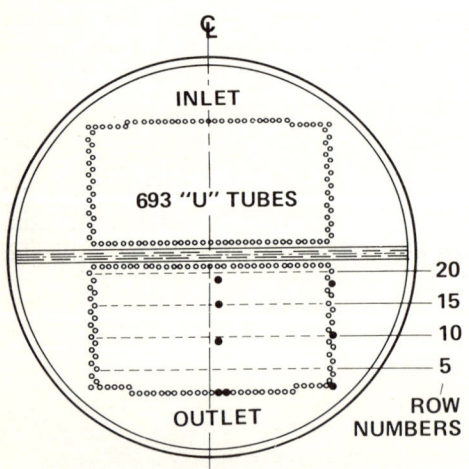

FIG. 7.2.28 The location of tubes fitted with thermocouples in two of the four units tested.

Test conclusions Thermocouple readings proved conclusively that some tubes were operating under unstable temperature conditions (Fig. 7.2.29). The instability reached such proportions that a cyclical temperature differential of 110°F between the large and small radius tubes was recorded in periods of 10–20 s. The cycles were slow enough for the tube to react by conduction and to cause radial expansion and contraction (Fig. 7.2.30). This expansion and contraction of the restrained tube ends eventually caused them to crack.

The large temperature differentials recorded were caused by a chilling mechanism hypothesized to be the result of flooding of individual tubes. The tubes flow in a plug flow or fill with water; the water flowing out of the

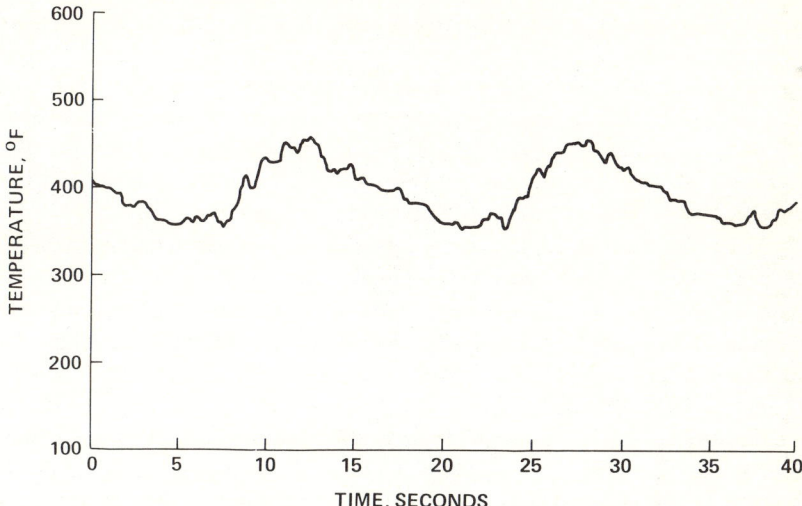

FIG. 7.2.29 Typical trace from thermocouple located in row 3 of MSR operating without orifice plate at 55% load.

tubes is then subcooled by the cycle steam, chilling and contracting the lower leg of the U-tube and potentially causing it to bind in its supports.

Application of test conclusions to MSR design While a detailed knowledge of the cause of operating instabilities in heat exchangers such as MSRs is not yet available, the solution is common practice: add a single-phase pressure

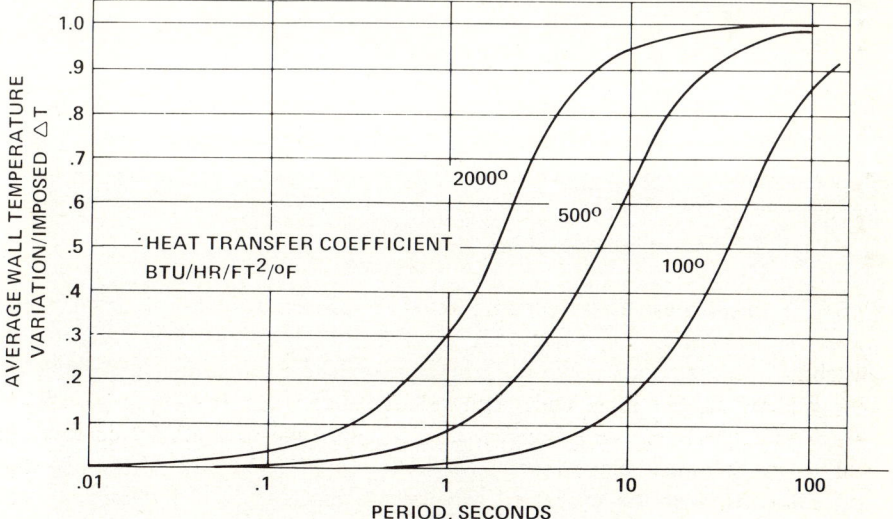

FIG. 7.2.30 Effects of periodic changes in heat-transfer conditions on tube wall temperatures; temperature cycles were slow enough to cause tube to expand and contract radially.

drop such as an orifice plate at the inlet to the tube bundle (Fig. 7.2.31).

Design of the orifice plate followed a three-step procedure. First, data from the field tests and computer study were compared, then adjusted as necessary. Second, a computer program based on the adjusted data was written. Third, the orifice plate was designed to provide acceptable limits of subcooling with the specified vent flow of 2.0% of heating-steam flow.

After the orifice plates were installed, their effectiveness in solving the problem was verified again by recording temperature at the tube discharge ends. The readings showed that cyclic-temperature differentials had been eliminated or greatly reduced; for example, a temperature change of 110°F before the orifice plate was installed was reduced to only 10°F by the plate (Tables 7.2.1 and 7.2.2).

Additional experiments have determined that loss of venting flow (also called "excess steam" or "scavenging steam") can initiate significant variations in cyclic temperature and oscillations in the flow of heating steam to the MSR. In several installations, the excess steam lines from four MSRs are vented through common pipes, manifolds composed of two to four feed vent lines.

Because these manifolds can restrict venting flow, their design has been revised in one of two possible ways: each excess steam line is separately piped to the extraction pipe or shell of the appropriate heater, or orifices are installed on each excess steam line and the line routed to piping large enough to ensure proper venting (2% of full-load steam flow) from each bundle under all operating conditions.

7.2.4 MSRs of the Future

The moisture separation and reheat function is needed in every nuclear power plant using light-water reactors. But, as PWR and BWR plants increase in size,

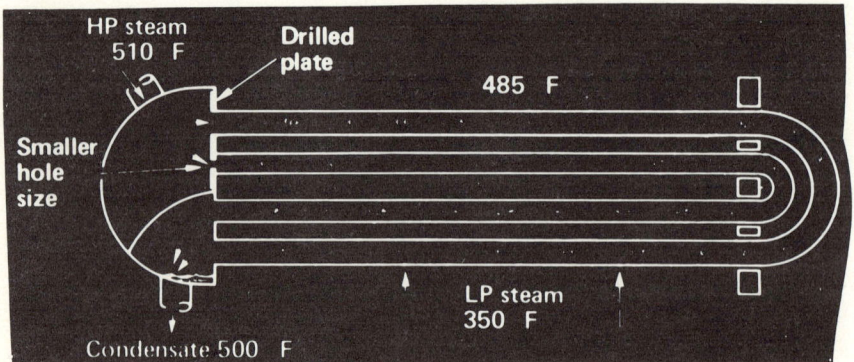

FIG. 7.2.31 Orifice plate on heating-steam inlet eliminated operating instabilities of tube bundle.

TABLE 7.2.1 Effect of orifice plates on condensate temperatures, 700 MW plant

% load	Row no.	Loc.	Orifice plate				No orifice plate			
			Temperature				Temperature			
			Max.	Min.	ΔT	Period	Max.	Min.	ΔT	Period
50	2	Center	436	426	10		518	408	110	10–20
	6	Center	510	504	6		507	439	68	10–20
	12	Center	520	520	0	0	527	521	6	
	16	Center	523	523	0	0	528	522	6	
	20	Center	526	526	0	0	523	511	12	
	2	Side	418	418	0	0				
	9	Side	405	405	0	0				
	15	Side	439	439	0	0				
67	2	Center	446	446	0	0	428	421	7	
	6	Center	514	507	7		465	459	6	
	12	Center	508	508	0		515	515	0	0
	16	Center	514	508	6		518	518	0	0
	20	Center	517	510	7		520	520	0	0
	2	Side	440	440	0	0				
	9	Side	420	420	0	0				
	15	Side	442	442	0	0				
97	2	Center	461	461	0	0	425	425	0	0
	6	Center	499	499	0	0	463	463	0	0
	12	Center	474	474	0	0	504	504	0	0
	16	Center	507	507	0	0	508	508	0	0
	20	Center	510	510	0	0	510	510	0	0
	2	Side	452	452	0	0				
	9	Side	427	427	0	0				
	15	Side	444	444	0	0				
97	2	Center	470	470	0	0				
	6	Center	508	501	7					
	12	Center	505	505	0	0				
	16	Center	514	514	0	0				
	20	Center	512	505	7					
	2	Side	449	449	0	0				
	9	Side	458	458	0	0				
	15	Side	441	441	0	0				

and if the function is to continue to be fulfilled in one vessel, MSRs will be needed that can dry and superheat more steam in a given shell volume. This in turn could reduce their equivalent number per LP turbine block and the space needed in the power plant.

To increase the space efficiency of an MSR it is probable that the tubes, tube bundle, and chevrons will all be modified to some extent, as well as their arrangement and location within the shell. The challenge is to make these modifications without decreasing the unit's thermal or separation performance.

TABLE 7.2.2 Effect of orifice plates on condensate temperatures, 500 MW plant

% load	Row no.	Loc.	Orifice plate				No orifice plate			
			Temperature				Temperature			
			Max.	Min.	ΔT	Period	Max.	Min.	ΔT	Period
56	1	Center	561	561	0	0	491	491	0	0
	1	Center					491	491	0	0
	9	Center	570	570	0	0	561	561	0	0
	15	Center	570	570	0	0	566	566	0	0
	19	Center	570	570	0	0	566	566	0	0
	2	Side					404	370	34	
	10	Side	570	570	0	0	449	387	62	
	18	Side	570	570	0	0	462	428	34	
78	1	Center	520	482	38		495	495	0	0
	1	Center					487	487	0	0
	9	Center	528	528	0	0	537	537	0	0
	15	Center	545	545	0	0	545	545	0	0
	19	Center	545	545	0	0	541	541	0	0
	2	Side					378	378	0	0
	10	Side	503	503	0	0	429	382	47	
	18	Side	537	537	0	0	416	416	0	0
60	1	Center	522	522	0	0	478	478	0	0
	1	Center					470	470	0	0
	9	Center	522	522	0	0	535	535	0	0
	15	Center	529	529	0	0	545	545	0	0
	19	Center	532	532	0	0	545	545	0	0
	2	Side					353	353	0	0
	10	Side	487	487	0	0	422	365	57	
	18	Side	529	529	0	0	445	408	37	

Development work on the tubes focuses on improving the efficiency of heat transfer. One promising possibility is use of more and higher fins; today's tubes have 19 fins per in., each .050 in. high. Increasing the number of fins to 26 per in. would increase the outside-to-inside ratio, and could add some 10% to present heat transfer capacity.

Tubes could be replaced with other types of heat exchangers. Plate fins, for example, could fulfill the same function as tubes in far less space. Because of material and design limitations, plate fins are manufactured only in small units; while the larger units needed in MSRs could be built, they are presently impractical.

Assuming that U-tubes will not soon be replaced, their present configuration is expected to remain for many years because of the need for drainage. And tube material, now typically Cu–Ni or carbon steel for use in PWR plants, and carbon steel for use in BWR plants, provides the optimum combination of heat transfer, corrosion resistance, and strength.

Changes in the tube-bundle configuration are contemplated; future bundles may be wider, shorter, and shallower, potentially decreasing the length of the MSR by a significant amount without increasing diameter.

Today's chevrons will likely keep their present shape for many years; they remove 100% of the free water particles over 15 μm, yet use less space than any other known mechanical separation. Although considerably more development work is needed, electrostatic separation offers the advantage of further reduction in space requirements, but has the disadvantage of needing considerable power to operate.

Offshore nuclear plants on large floating platforms in the Atlantic Ocean three miles from shore will be operating in only a few years. They pose unique but not necessarily difficult design problems. Because of wave action, one end of an MSR on an offshore platform could be 3 ft higher than the other at any given moment, and the unit could seesaw within this 3 ft arc every 13 s. Because this cycling could cause intermittent flooding of the tubes used in present designs, the tubes may be larger in diameter and shorter in length. In addition, the unit will be anchored to the platform in such a way that it is free to move for limited distances in any direction.

Plants with reactors cooled by liquid metal, probably sodium, may or may not need external moisture separation and reheat. Westinghouse is the lead contractor in The Project Management Group, a consortium formed to design and build the first prototype liquid-metal fast-breeder reactor, a 350–400 MW unit owned by the Tennessee Valley Authority. The present proposal is to provide steam at the turbine inlet at 1450 psi and 900–950°F, and not to use external moisture separation or reheat. This does not necessarily mean that MSRs are doomed to extinction, as liquid metal reactors displace present designs; a great deal of testing at the prototype plant will undoubtedly lead to a number of design changes before commercially sized LMFB reactors are technically or economically feasible, probably not before 1985 or 1990.

NOMENCLATURE

a	constant in Eq. (7.1.10)
d	drop diameter
D	wire diameter
g	acceleration due to gravity
G	gap between tubes or wires in a bundle
h	gap between corrugated plates
H	h/y_0
H_p	height of corrugated plate
H_k	depth of knitted wire mesh
l	stopping distance for drop
L	defined by Eq. (7.1.5)

m	number of wave cycles in corrugated plate
p	pressure
P	pitch of corrugations
s	surface area per unit volume of knitted wire mesh
U	steam velocity
U_s	superficial steam velocity
W	superficial mass flow rate per unit area
x	steam quality
y_0	corrugation amplitude
Ku	Kutateladze number, Eq. (7.1.17)
Re_d	drop Reynolds number, Eq. (7.1.3)
Stk	wire Stokes number, Eq. (7.1.1)
Stk_p	plate Stokes number, Eq. (7.1.9)
α	maximum angle of corrugated surface to the vertical
ϵ	voidage of knitted wire mesh
η_w	fractional separation efficiency of a wire
η_k	fractional separation efficiency of knitted wire mesh
η_p	fractional separation efficiency of one wave of a corrugated plate
η_{po}	overall fractional separation efficiency of corrugated plates
η_T	fractional separation efficiency of a tube or wire bundle
μ	viscosity
ν	kinematic viscosity
$\Pi_1, \Pi_2, \Pi_3, \Pi_4, \Pi_5$	dimensionless groups given by Eqs. (7.1.12), (7.1.13), (7.1.15), (7.1.16), and (7.1.18)
ρ	density
σ	surface tension

Subscripts

g	steam phase
l	water phase

REFERENCES

7-1 Gardner, G. C.: *Proc. Inst. Mech. Eng.*, vol. 178, p. 593, 1962.
7-2 Brun, R., and Negler, H.: *NACA Tech. Note No.* 2904, 1953.
7-3 Brun, R., Lewis, W., Perkins, P., and Serafini, J.: *NACA Rept. No.* 1215, 1955.
7-4 Fuchs, N. A.: "The Mechanics of Aerosols," Pergamon, Oxford, 1964.
7-5 Strauss, W.: "Industrial Gas Cleaning," Pergamon, Oxford, 1966.
7-6 Wong, J. B., and Johnstone, H. F.: *Univ. Ill. Eng. Exp. Station Tech. Rept. No.* 11, 1953.
7-7 May, K. R., and Clifford, R.: *Ann. Occup. Hyg.*, vol. 10, p. 83, 1967.

7-8 Stairmand, C. J.: *Trans. Inst. Chem. Eng.*, vol. 28, p. 130, 1950.
7-9 Carpenter, D. L., and Othmer, D. F.: *A.I.C.H.E.J.*, vol. 1, p. 549, 1955.
7-10 Bürkholz, A.: *Chem.-Ing.-Tech.*, vol. 42, p. 1314, 1970.
7-11 Davies, C. N., and Peetz, V.: *Proc. Roy. Soc.*, vol. A234, p. 269, 1956.
7-12 Katz, E. J.: "Study of Inertial Mist Collectors," M. Sc. thesis, Pennsylvania State University, State College, Pa., 1959.
7-13 Bürkholz, A., and Muschelknautz, E.: *Chem.-Ing.-Tech.*, vol. 44, p. 503, 1972.
7-14 Sorokin, Yu. L., Popchenkov, I. N., and Burkat, V. S.: *Chem. Petro. Eng.*, no. 12, p. 781, 1966.
7-15 Sherwood, T. K., Shipley, G. H., and Holloway, F. A. L.: *Ind. Eng. Chem.*, vol. 30, p. 765, 1938.
7-16 Bradie, J. K.: "Entrainment Studies," Ph.D. thesis, Heriot-Watt University, Edinburgh, 1969.
7-17 Bradie, J. K., and Dickson, A. N.: Fluid Mechanics and Measurements in Two-phase Flow Systems, *Symp. Inst. Mech. Eng., Leeds,* paper no. 24, 1969.
7-18 Alen'kin, N. F., Golub, S. I., Gol'dshtein, I. F., and Rosen, A. M.: *Khim. Prom.*, no. 10, p. 782 (C.E. Trans. 4920), 1967.
7-19 Planowski, A. N., and Kafarov, W. W.: *Khim. Prom.*, no. 4, p. 13, 1946.
7-20 Wilson, J. F., Grenda, R. J., Klumb, R. H., Littleton, W. E., Meyer, W. C., and Yent, H. W.: *A.C.N.P.* 65003, 1965.
7-21 Sorokin, Yu. L., Demidova, L. N., and Kuz'min, N. P.: *Chem. Petro. Eng.*, no. 8, p. 664, 1968.
7-22 Panasenko, M. D., and Koslov, Yu. V.: *Teploenergetika*, no. 8, p. 69, 1962.
7-23 Dickson, A. N., and Morrison, D. S.: Multiphase Flow Systems, *Inst. Chem. Eng. Symp., Glasgow,* series no. 38, paper no. K2, 1974.
7-24 York, O. H., and Poppele, E. W.: *Chem. Eng. Progr.*, vol. 59, no. 6, p. 45, 1963.
7-25 Lemezis, S., Bushey, J. R., and Rabas, T. J.: Progress in Moisture Separator Reheaters for Nuclear Power Plants, *Amer. Power Conf.*, April 22-24, 1969.
7-26 Artusa, F. A.: Turbine Cycles for Nuclear Power Plant Application, *Amer. Power Conf.*, April 25-27, 1967.
7-27 Discharge Coefficients for Combustor-Liner Air Entry Holes, *NASA Tech. Note* 8663.
7-28 McKnown, J. S., and Hsu, En-run: "Application of Conformal Mapping to Divided Flow," Mid-Western Conference on Fluid Dynamics, p. 153, Edwards, Ann Arbor, Mich., 1951.
7-29 Connors, H. J., Jr.: Fluidelastic Vibration of Tube Arrays Excited by Cross Flow, in "Flow-Induced Vibration in Heat Exchangers," pp. 42-56, ASME Publ., 1970.
7-30 Dittrich, R. T., and Graves, C. C.: Discharge Coefficients for Combustor-Liner Air Entry Holes. I. Circular Holes with Parallel Flow, *NASA Tech. Note* 3663, 1956.

Author Index

The bracketed number preceding a series of page numbers is the reference number by which the author is represented (with or without his name) on the pages cited. The last number in each series is usually the page on which the complete reference is found. Absence of a bracketed number indicates that the author's name appears without a reference number on the page(s) listed. An italic letter *n.* after a page number indicates a footnote.

Abraham, F. F., [1-26] : 43, 57
Aleksandrov, A. A., [2-5] : 60, 62, 124
Alen'kin, N. F., [7-18] : 330, 331, 369
Ames Research Staff, NACA, [3-2] : 132, 187
Anderson, B. W., [4-6] : 195, 258
Anderson, G. H., [4-16] : 201, 259
Angus, J. C., [4-20] : 208, 259
Armstrong, R. L., [5-29] : 286, 290
Artusa, F. A., [7-26] : 369

Babcock, W. R., [1-25] : 43, 57
Baehr, H. D., [1-16] : 23, 56
Baker, D. W. C., [5-27] : 285, 290
Bakhtar, F., [3-15] : 139, 188
Barschdorff, D., 128
  [2-21] : 89, 125
  [3-7] : 138–140, 143, 145, 147, 151, 172, 174, 176, 177, 188
  [3-17] : 142, 143, 188
Bauman, K., 2, 105
  [2-32] : 105, 126
  [5-1] : 261, 289
Becker, R., 128, 157
Beldecos, N., 105
Bergh, H., [4-7] : 195, 258
Binder, R. C., [4-3] : 195, 258
Binnie, A. M., 127
  [3-6] : 135, 136, 187
  [3-9] : 139, 188
  [3-14] : 140, 143, 188
Blowney, W., 105

Boltzmann, L., 128
Boyer, D. W., [2-18] : 86, 125
Bradie, J. K.:
  [7-16] : 330–333, 369
  [7-17] : 330, 331, 369
Brown, F. T., [4-5] : 195, 258
Brun, R.:
  [7-2] : 318, 319, 368
  [7-3] : 318, 319, 368
Brunton, J. H., [5-26] : 285, 290
Bryer, D. W., [4-15] : 201, 259
Buckle, E. R., [3-31] : 179, 189
Buff, F. P., 128
Burkat, V. S., [7-14] : 329, 333, 335, 369
Burkhard, H.-P., [3-10] : 139, 145, 146, 148, 150, 153–156, 188
Bürkholz, A.:
  [7-10] : 322, 323, 328, 369
  [7-13] : 323, 326, 328, 369
Bushey, J. R., [7-25] : 369

Callendar, H. L., [2-2] : 60, 124
Campbell, B. A., [3-15] : 139, 188
Carpenter, D. L., [7-9] : 320, 369
Carslaw, H. S., [1-24] : 42, 57
Cataldi, C., [5-19] : 283, 290
Chambré, P. L., [3-27a] : 164, 189
Chapin, J. H., [1-21] : 40, 56
Chareyre, R., [4-45] : 241, 259
Cheng, C. F., [5-19] : 283, 290
Christ, A., [4-35] : 225, 228, 259

Christie, D. G.:
 [5-12]: 271, 289
 [5-22]: 283, 290
Churchill, S. W., [2-31]: 103, 125
Clifford, R., [7-7]: 320, 321, 368
Cole, J. E., [2-11]: 94, 98, 125
Connors, H. J., Jr., [7-29]: 354, 369
Cox, H. J. A., [2-33]: 105, 126
Craig, H. R. M., [2-33]: 105, 126
Crane, R. I.:
 [2-16]: 81, 125
 [2-23]: 87, 125
 [2-38]: 113, 126
 [3-26]: 163, 172, 176, 178, 180, 182, 184, 188
 [4-18]: 201, 202, 259
 [4-25]: 212, 259
 [4-30]: 218, 259
 [5-7]: 268, 289
Crowe, C. T., [1-25]: 43, 57

Daneshyar, H., [4-8]: 197, 259
Davidson, B. J.:
 [2-17]: 86, 91, 125
 [2-23]: 87, 125
 [3-26]: 163, 172, 176, 178, 180, 182, 184, 188
 [5-7]: 268, 289
Davies, C. N.:
 [4-40]: 236, 247, 259
 [7-11]: 322, 323, 369
Deich, M. E., 128
 [2-25]: 94, 98, 125
 [3-21]: 151, 152, 176, 179, 188
 [3-25]: 163, 188
 [3-32]: 180, 181, 189
 [3-34]: 180, 181, 189
 [3-36]: 180, 182, 189
 [5-11]: 270, 289
Dejc, M. E., [1-17]: 35, 56, 135, 136, 180
Dejong, V. J., [2-28]: 94, 125
Demetri, E. P., [3-29]: 166, 167, 189
Demidova, L. N., [7-21]: 332, 333, 335, 369
Dickson, A. N.:
 [7-17]: 330, 331, 369
 [7-23]: 336, 337, 369
Ditrich, 352
Dobbins, R. A.:
 [2-11]: 94, 98, 125
 [4-46]: 243, 259
Döring, W., 128, 157

Drain, L. E., [4-22]: 209, 210, 259
Dukler, A. E., [4-37]: 235, 259
Dunning, W. J., 128
 [4-20]: 208, 259
Durst, F.:
 [4-21]: 208, 259
 [4-24]: 210, 211, 259
Dussourd, J. L., [4-17]: 201, 259
Dzung, L. S., [2-4]: 60-62, 124

Eaton, J. L., [5-27]: 285, 290
Einstein, A., 128
Elliott, D. E., [5-30]: 286, 290
Elzarov, 199
Eschenroeder, A. Q., [2-18]: 86, 125
Evans, H. D.:
 [2-15]: 79, 114, 125
 [4-50]: 245-260

Farkas, L., 128, 157, 160
Feder, J., [1-19]: 37, 56, 156, 162, 163
Filippov, G. A.:
 [3-21]: 151, 152, 176, 179, 188
 [3-31a]: 177, 189
 [3-36]: 180, 182, 189
 [3-37]: 183, 184, 189
 [5-11]: 270, 289
Finlayson, P. C., [4-12]: 198, 199, 259
Firey, J. C., [2-28]: 94, 125
Flatt, F., 105
French, M. J., [4-20]: 208, 259
Frenkel, J., 157
Freudenreich, J. V., [5-24]: 284, 290
Freudenrich, I., 105
Fuchs, N. A.:
 [1-29]: 44, 47, 57
 [7-4]: 318-320, 368

Gallant, H., 127n.
Gardner, F. W., [5-20]: 283, 290
Gardner, G. C., 326, 327
 [1-2]: 3, 56
 [1-37]: 52, 53, 57
 [2-36]: 105, 126
 [5-21]: 283, 284, 290
 [7-1]: 317, 368
Gardzilewicz, A., [3-23]: 154, 188
Gibbs, J. W., 128
Glauz, R. D., [2-19]: 86, 125
Gloger, M., [6-2]: 299, 315

Gol'dstein, I. F., [7-18] : 330, 331, 369
Golovin, M. N., [1-40] : 46, 57
Golub, S. I., [7-18] : 330, 331, 369
Graves, 352
Green, J. R., [3-14] : 140, 143, 188
Gregory, B., [5-16] : 289
Grenda, R. J., [7-20] : 332–325, 369
Groh, G., [1-15] : 27, 56
Guy, H., 105
Gyarmathy, G., 128, 184, 243
 [1-18] : 38, 40, 43, 48, 53, 56, 154, 164, 165, 176, 179
 [2-9] : 67, 68, 105, 106, 118–120, 122, 125
 [3-3] : 132, 134, 138–140, 143–145, 148, 151, 154, 156, 174, 176, 187
 [3-10] : 139, 145, 146, 148, 150, 153–156, 188
 [3-13] : 139, 148, 149, 181, 188
 [3-28] : 164, 189
 [3-39] : 184, 185, 189
 [4-43] : 241, 259
 [5-6] : 267, 289
 [5-18] : 278, 289

Hässler, G., [1-34] : 49, 50, 57
Hall, G. J., [2-18] : 86, 125
Hammitt, F. G., [5-25] : 285, 290
Hancox, N. L., [5-26] : 285, 290
Handbook of Chemistry and Physics, [1-14] : 28, 56
Harris, F. R., [5-10] : 269, 289
Hawthorne, W. R., [2-12] : 88, 125
Hayward, G. W.:
 [5-12] : 271, 289
 [5-22] : 283, 290
Hegetschweiler, H., [1-8] : 13, 16, 18, 56
Heller, W., [4-49] : 245, 260
Helmholtz, H. von, 35, 127, 128
Hermann, R., 127
Hill, P. G., 128
 [1-12a] : 22, 32, 56
 [3-29] : 166, 167, 189
 [3-30a] : 189
Hodkinson, J. R., [4-40] : 236, 247, 259
Holloway, F. A. L., [7-15] : 330, 369
Hopkins, D. F., [4-19] : 203, 259
Horlock, J. H.:
 [2-3] : 60, 124
 [2-37] : 106, 126
 [4-8] : 197, 259
Hossli, W., [1-9] : 13–15, 19, 56

Hsu, En-Run, [7-28] : 353, 369
Huang, Y. C., [5-25] : 285, 290

Iberall, A. S., [4-4] : 195, 258
Ignatevskii, E. A., [3-37] : 183, 184, 189
Ikeda, T., [3-35] : 180, 189

Jacobs, D., [2-24] : 94, 125
Jaeger, J. C., [1-24] : 42, 57
Jankowski, T., [1-23] : 41, 57
Jizmagian, G. S., [4-46] : 243, 259
Johnstone, H. F.:
 [1-21] : 40, 56
 [7-6] : 320, 321, 368
Joos, G., [3-24] : 157, 188

Kafarov, W. W., [7-19] : 332, 369
Kalinin, A. V., [2-26] : 94, 125
Kang, S.-W., 128
 [1-20] : 40, 56, 164
Kantrowitz, A., 128
Kasprzyk, S. von, [4-34] : 224, 259
Katz, E. J., [7-12] : 322, 323, 325, 369
Keenan, J. H., [1-12a] : 22, 32, 56
Keller, H., [6-1] : 293n., 315
Kelvin, Lord (see Thomson, W.)
Kent, R. P., [5-29] : 286, 290
Keyes, F. G., [1-12a] : 22, 32, 56
Kirillov, I. I.:
 [1-39] : 53, 57
 [2-10] : 105, 106, 125
 [5-14] : 271, 272, 289
Kirkwood, J. G., 128
Kistler, A. L., [4-28] : 216, 217, 259
Klumb, R. H., [7-20] : 332–335, 369
Konorski, A., [1-23] : 41, 57
Koslov, Yu. V., [7-22] : 333–335, 369
Kraftwerk-Union, [1-7] : 7, 10, 11, 56
Król, T.:
 [3-20] : 148, 150, 151, 154, 188
 [3-23] : 154, 188
Krzeczkowski, S.:
 [1-31] : 57
 [2-40] : 114, 126
Kuhrt, F., 128
Kurshakov, A. V., [3-34] : 180, 181, 189
Kuz'min, N. P., [7-21] : 332, 335, 369

Lagun, V. M., [4-14] : 198, 199, 259

Author Index

Langford, R. W.:
 [1-35]: 49, 51, 57
 [2-39]: 114, 126
 [4-1]: 194, 258
 [4-36]: 228, 229, 259
Laws, J. O., [2-41]: 115, 126
Lemezis, S., [7-25]: 369
Lesch, F.:
 [3-10]: 139, 145, 146, 148, 150, 153–156, 188
 [3-13]: 139, 148, 149, 181, 188
 [4-43]: 241, 259
 [5-6]: 267, 289
Lewis, W., [7-3]: 318, 319, 368
Liepmann, H. W., [2-22]: 90, 99, 100, 125
Lin, S., [3-18]: 143, 188
Lindley, 199
Littleton, W. E., [7-20]: 332–335, 369
Loewenstein, L. C., [2-35]: 105, 126
Lord, M., [4-29]: 217, 259
Lord Kelvin (see Thomson, W.)
Lothe, J., 128
Luck, W., [4-33]: 221, 259
Ludewig, M., [1-4]: 9, 11, 56

McAdams, W. H., [1-22]: 40, 57
McAllister, D. H.:
 [1-33]: 48, 57
 [5-15]: 271, 282, 284, 289
MacDonald, A. N., [5-22]: 283, 290
McKown, J. S., [7-28]: 353, 369
Manas, B., [4-45]: 241, 259
Mantzouranis, B. G., [4-16]: 201, 259
Marriott, J. B., [5-30]: 286, 290
Marx, P. P., [4-51]: 249, 260
May, K. R.:
 [4-38]: 236, 259
 [7-7]: 320, 321, 368
Mayhew, Y. R., [2-13]: 76, 85, 125
Melling, A.:
 [4-21]: 208, 259
 [4-25]: 212, 259
Meyer, H., 128
 [3-3]: 132, 134, 138–140, 143–145, 148, 151, 154, 156, 174, 176, 187
Meyer, W. C., [7-20]: 332–335, 369
Miller, E. H.:
 [1-10]: 18, 20, 56
 [2-34]: 105, 106, 122, 126
 [5-5]: 265–267, 269, 289
Mills, A. F., [3-30]: 167, 172, 176, 189
Moore, C. T., [5-15]: 271, 282, 284, 289

Moore, J. G.:
 [1-12a]: 22, 32, 56
Moore, M. J.:
 [1-35]: 49, 51, 57
 [1-36]: 51, 52, 57
 [2-16]: 81, 125
 [2-23]: 87, 125
 [2-39]: 114, 126
 [3-26]: 163, 172, 176, 178, 180, 182, 184, 188
 [4-1]: 194, 258
 [4-11]: 198, 259
 [4-18]: 201, 259
 [4-36]: 228, 229, 259
 [5-7]: 268, 289
Morkovin, M. V., [4-27]: 216, 217, 259
Morrison, D. S., [7-23]: 336, 337, 369
Morrow, D. L., [4-20]: 208, 259
Moses, C. A., [3-22]: 151, 153, 181, 188
Mühlhäuser, H., [1-6]: 8, 13, 18, 19, 56
Mugele, R. A.:
 [2-15]: 79, 114, 125
 [4-50]: 245, 260
Muschelknautz, E., [7-13]: 323, 326, 328, 369
Musick, V. S., [5-19]: 283, 290

National Advisory Committee for Aeronautics (NACA), [3-2]: 132, 187
National Aeronautics and Space Administration (NASA) Technical Notes, [7-27]: 369
National Engineering Laboratory Tables, [2-1]: 59, 124
National Physical Laboratory, [2-14]: 78, 125
Naumchik, B. V., [5-14]: 271, 272, 289
Negler, H., [7-2]: 318, 319, 368
Neller, 326, 327
Nicola, M. C., [2-6]: 63, 125
Nosovitskii, A. I., [5-14]: 271, 272, 289

Oswatitsch, K., 127, 128
 [3-4]: 134, 135, 187
 [3-8]: 139, 164, 169, 188
Othmer, D. F., [7-9]: 320, 369

Panasenko, M. D., [7-22]: 333–335, 369
Pankhurst, R. C., [4-15]: 201, 259

Parker, G. J.:
  [1-32]: 48, 57
  [5-13]: 271, 289
Parsons, N. C., [5-17]: 274, 289
Parsons, R. H., [5-9]: 269, 270, 289
Peetz, V., [7-11]: 322, 323, 369
Penndorf, R. B., [4-41]: 239, 259
Perelman, A. Ya., [4-47]: 244, 259
Perkins, P., [7-3]: 318, 319, 368
Petr, V., 128
  [2-29]: 94, 96, 97, 98, 125
  [3-11]: 139, 143, 144, 155, 156, 188
  [4-42]: 241, 259
Peuster, K., [1-11]: 15, 56
Pigford, R. L., [1-21]: 40, 56
Pink, H.-R., [1-3]: 4, 5, 56
Planowski, A. N., [7-19]: 332, 369
Popchenkov, I. N., [7-14]: 329, 333, 335, 369
Poppele, E. W., [7-24]: 330, 369
Pound, G. M., 128
Pouring, A. A., 128
  [2-20]: 88, 125
  [3-16]: 140, 176, 177, 188
Povarov, O. A.:
  [3-31a]: 177, 189
  [5-11]: 270, 289
Prandtl, L., 127
Priahin (Pryakhin), V. V.:
  [3-31a]: 177, 189
  [5-11]: 270, 289
Probstein, R. F., 128
Prokopovicz, J., [1-23]: 41, 57
Pryakhin, V. V. (*see* Priahin, V. V.)
Putnam, A. A., [1-40]: 46, 57
Puzyrewski, R.:
  [2-40]: 114, 126
  [3-12]: 139, 140, 156, 166, 172, 176, 188

Rabas, T. J., [7-25]: 369
Ralph, W. J., [1-28]: 44, 57
Rankine, W. J. M., 81
  [2-7]: 63, 125
Ringeis, W. K., [1-11]: 15, 56
Riollet, G., [1-5]: 8, 16, 18, 56
Roberts, A. G., [4-12]: 198, 199, 259
Robertson, J. M., [4-19]: 203, 259
Rogers, G. F. C., [2-13]: 76, 85, 125
Rosen, A. M., [7-18]: 330, 331, 369
Roshko, A., [2-22]: 90, 99, 100, 125
Rudinger, G., [2-27]: 94, 125

Russell, K. C., [1-19]: 37, 56, 156, 162, 163
Rutz, J., [4-32]: 221, 222, 259
Ryley, D. J.:
  [1-27]: 44, 57
  [1-28]: 44, 57
  [1-30]: 47, 57, 163
  [1-38]: 45, 52, 57
  [5-13]: 271, 289

Saltanov, G. A.:
  [3-19]: 145, 147, 148, 151, 153, 154, 188
  [3-25]: 163, 188
  [3-32]: 180, 181, 189
  [3-34]: 180, 181, 189
  [3-36]: 180, 182, 189
  [3-37]: 183, 184, 189
Samoilovich, G. S., [4-9]: 197, 259
Schaaf, S. A., [3-27a]: 164, 189
Schmidt, E., 128
  [1-12]: 22, 25, 26, 32, 56, 161
Schodl, R., [4-10]: 197, 198, 259
Schofield, P.:
  [2-34]: 105, 106, 122, 126
  [5-5]: 265–267, 269, 289
Schuder, C. B., [4-3]: 195, 258
Sculpher, P.:
  [1-36]: 51, 52, 57
  [4-11]: 198, 259
Seban, R. A., [3-30]: 167, 172, 176, 189
Seleznev, L. J., [3-19]: 145, 147, 148, 151, 153, 154, 188
Serafini, J., [7-3]: 318, 319, 368
Shapiro, A. H.:
  [2-12]: 88, 125
  [2-30]: 100, 125
  [4-17]: 201, 259
Sherwood, T. K., [7-15]: 330, 369
Shifrin, K. S., [4-47]: 244, 259
Shipley, G. H., [7-15]: 330, 369
Shpenzer, G. G., [5-14]: 271, 272, 289
Siegenthaler, A.:
  [3-10]: 139, 145, 146, 148, 150, 153–156, 188
  [5-6]: 267, 289
Simoyu, L. L., [4-14]: 198, 199, 259
Small, J., [1-38]: 45, 52, 57
Smith, A.:
  [4-2]: 195, 258
  [5-8]: 268, 269, 289
  [5-28]: 286, 290

Smith, A. (*continued*):
  [5-29]: 286, 290
  [5-30]: 286, 290
Smith, D., 105
Smith, K., 105
Smith, L. T., [3-33]: 180, 181, 189
Soderberg, C. R., [5-3]: 262, 263, 289
Sorokin, Yu. L.:
  [7-14]: 329, 333, 335, 369
  [7-21]: 332, 333, 335, 369
Spencer, R. C., [1-10]: 18, 20, 56
Spengler, P., [3-39]: 184, 185, 189
Stairmand, C. J., [7-8]: 320, 369
Stein, G. D., [3-22]: 151, 153, 181, 188
Stever, H. G., [3-5]: 136, 137, 187
Stepanchuk, V. F.:
  [3-25]: 163, 188
  [3-32]: 180, 181, 189
Stodola, A., 127
  [1-1]: 3, 56, 164
  [2-35]: 105, 126
  [5-2]: 261, 289
Strasser, W., [1-11]: 15, 56
Strauss, W., [7-5]: 320, 368
Suzuki, A., [3-35]: 180, 189
Szilárd, L., 128

Thomson, W. (Lord Kelvin), 35, 128
Thorpe, A. D.:
  [4-31]: 219, 259
  [5-23]: 284, 290
Tijdemann, H., [4-7]: 195, 258
Tipping, J. C.:
  [1-35]: 49, 51, 57
  [2-39]: 114, 126
  [4-1]: 194, 258
Todd, K. W.:
  [4-13]: 198, 199, 259
  [5-16]: 289
Tolman, R. C., 128
Trakjtengerts, M. S., [2-5]: 60, 62, 124
Traupel, W., 105
  [1-13]: 24, 25, 48, 56
  [5-4]: 263, 289
  [6-3]: 313, 315
Trojanovskij, B. M., [1-17]: 35, 56, 135, 136, 180
Tsiklauri, G. V.:
  [3-19]: 145, 147, 148, 151, 153, 154, 188
  [3-32]: 180, 181, 189
Tubman, K. A., [1-28]: 44, 57

Usagi, R., [2-31]: 103, 125

Vander Hulst, H. C., [4-39]: 236, 259
Volmer, M., 128, 156, 157
von Helmholtz (*see* Helmholtz)
von Kasprzyk (*see* Kasprzyk)
Vukalovich, M. P., [2-5]: 60, 62, 124

Wakeshima, H., 128, 162
Wallach, M. L., [4-49]: 245, 260
Walters, P. T.:
  [2-23]: 87, 125
  [3-26]: 163, 172, 176, 178, 180, 182, 184, 188
  [4-44]: 241, 244, 247, 259
  [4-48]: 245, 260
  [5-7]: 268, 289
Wang, C. P., [4-23]: 209, 259
Warren, H., 105
Weber, A., 128, 156, 157
Wegener, P. P.:
  [1-18a]: 36n., 56
  [2-27]: 94, 125
  [3-1]: 128, 156, 157, 163, 187
Weyer, H., [4-10]: 197, 198, 259
Whitelaw, J. H.:
  [4-21]: 208, 259
  [4-24]: 210, 211, 259
Whybrew, 199
Wicks, M., [4-37]: 235, 259
Wieselberger, 127
Williams, G., [4-29]: 217, 259
Willoughby, P. G., [1-25]: 43, 57
Wilson, C. T. R., 127
Wilson, J. F., [7-20]: 332–335, 369
Witting, H., [3-29]: 166, 167, 189
Wong, J. B., [7-6]: 320, 321, 368
Wood, M. R.:
  [1-27]: 44, 57
  [4-31]: 219, 259
  [5-23]: 284, 290
Wood, N. B.:
  [3-38]: 184, 185, 189
  [4-26]: 212, 214, 259
Woods, M. W., [3-9]: 139, 188
Wu, B. J.-Ch., [3-27]: 163, 189
Wulff, W., [4-35]: 225, 228, 259

Yablonik, R. M.:
  [1-39]: 53, 57

[2-10]: 105, 106, 125
Yatcheni, I. A.:
  [3-34]: 180, 181, 189
  [3-36]: 180, 182, 189
Yellot, J. I., 127
Yent, H. W., [7-20]: 332–335, 369

York, O. H., [7-24]: 330, 369
Yousif, F. H., [3-15]: 139, 188

Zeldovich, Y. B., 157
Zeuner, G., [2-8]: 64, 125
Zierep, J., [3-18]: 143, 188

Subject Index

Absorption coefficient per wavelength, 96, 98
Access holes, in turbines, 198–199
Acoustic equations (linearized small perturbation equations), 95–99
Acoustic velocity:
  "equilibrium," 96, 97
  factors affecting, 97–98
  "frozen," 96
  in nozzles, equation, 133
  in wet steam, 94–99
Adiabatic equilibrium, 33–34
Adiabatic reversion, 33–35
  entropy rise, 35
Aerodynamic efficiency, 2
Agglomeration, of droplets, 47
  (See also Droplets)
Air, condensation at very low temperatures, 128
  (See also Humid air)
"Alkazene," as manometer fluid, 192
Allis-Chalmers moisture separator-reheaters, 337, 348
Alpha (see Baumann factor)
Alsdorf (W. Germany) steam/gas-turbine power plant, 251
Anemometers and Anemometry, 207–217
  fluctuation equation, 215
  (See also Hot-wire anemometers; Laser-Doppler anemometers, etc.)
Angle of impingement, droplets on blade tips, 284
Angular momentum, of gas phase, 117
Angular scattering, 237
Antierosion devices, 9
  (See also Baumann stage; Erosion protection, etc.)
Atomization, 263
  in LP turbines, 49–51
  relation to droplet size, 29

Avogadro number, of water, 21

Back-pressure, trends in U.S. and W. Germany, 7
Balzer's Ltd. interference filter data, 246
Baumann factor (α), 105, 261, 269
  mean values for different turbines, 105
Baumann rate, 121
Baumann rule, 2, 106
Baumann stage, 9, 12
Bessel functions, 196
"Bess" 500 MW experimental low-pressure turbine, 271–273
  water content, 284
Bidispersions, drop size distribution, 244
Blade carriers, erosion-corrosion, 291, 292
Blade tips:
  droplet-size distributions, 251, 253
  effect of geometry on wetness losses, 265
  erosion, 304–305
  wetness values, 251, 253
Blade-to-blade flow, in wet-steam turbines, 71
Blade roots:
  droplet-size distributions, 251, 253
  wetness values, 251, 253
Blades:
  design provisions, 303–305
  droplet impingement, 284
  efficiency, effect of moisture, 261–269
  erosion, 283–288
    reducing design, 303–305
  length effect on last-stage isentropic heat drop, 8–9
  protection, 283–288
  suction slots, 11
  surface frictional loss, 118
  thermal relaxation, 70
  wakes, 194

Blades (*continued*):
  water re-entrainment from, 114
  wet-steam flow, 69-70
  (*See also* Fixed blades; Moving blades, etc.)
Boiler drums, 317
Boiling-water reactor (BWR) plants, 11-13
  erosion, 283
  HP turbine water extraction, 273
  MSRs, 337-367
  stainless steel uses, 340
Boiling-water reactors (BWR), 11-13
Boltzmann constant, 37, 157
Boltzmann's distribution law, 37, 156
Bouguer law, 240, 241
Boundary layers, fluctuation modes, 216
Breakthrough characteristics, external WS, 329-337
Breakthrough velocity, of separators, 318, 336-337
Break-up, of droplets, 49-51
Brown-Boveri Co.–Goerz laser anemometers, 208
Brown-Boveri moisture-separator-reheaters, 337
Brownian coagulation, 47
Brownian motion (*see* Thermal motion)
BWR (*see* Boiling-water reactors)

Cadmium Blue cell, Hayakawa type 5 BC-102, 247
Calorimeters, for psychometry, 225
Carbon steel:
  erosion-corrosion, 19
  in HP turbine flange, 19
  in MSR parts, 340
  for MSR tubes in future PWR and BWR plants, 366
Carnot efficiency, 1, 2, 3
Cascade flow, condensation in, 180-181
Cascades, condensation in, 179-186
Cavitation boiling (*see* Flashing)
CEGB (*see* Central Electricity Generating Board)
Central Electricity Generating Board (CEGB) turbines, 269, 286
Centrifugal force, effect on turbine flow, 53
Centrifugal separators, 271
Centrifuging loss, 121
  coefficient, 117
CERL:
  turbine probes, 200
  steam tunnels, 214
  wetness probe, 227-234
Channel walls, liquid motion on, 52-53
Chevron plate separators (*see* Corrugated plate separators)
Choked flow, 85
Choking flow, Mach number dependence, 99
Chromium steel, 18, 19, 283
  in erosion-corrosion prevention, 291-293
  in erosion protection, 19
  in wet-steam turbines and piping, 296-297
Clapeyron equation, 76, 95-96
Clausius-Clapeyron equation, 22
Cloud chamber expansion, 127, 131
Coagulation:
  of droplets, 44-48
  in steam-droplet mixtures, 44-48
Coarse water:
  definition, 106, 217
  droplets collection efficiency, 218
  effect on temperature measurement, 201
  extraction, 122
  mechanical separation, 217
  tracers, 221
  in wet-steam turbine stages, 107
Coarse water losses, 112-117, 121
  for fixed blade, 113-116
  for moving blade, 116-117
Coarse water measurement:
  absolute methods, 217-222
  mechanical separation methods, 217
  tracer methods, 221
  in turbine interstages, 217-220
  in wet steam, 217-222
Compressibility factor (*see* Real-gas factor)
Compression processes, equations, 60
Computational methods, 25, 60, 62, 103, 172-176
  for condensing flow, 171
  for convergent-divergent nozzles, 87-88
  for nozzle flow, 172-176
  Runga-Kutta method, 78, 103, 172
  for transonic flow, 85-88
  for two-phase flow, 169-174
  for wet-steam flows, 91-94
Computer programs:
  for analyzing light-scattering probe results, 250
  involving equations of state, 25
  in MSR design, 364
  for orifice plate design, 364
Computer studies, 60, 62, 172-176
  of MSRs, 361-364

Condensation:
  basic concepts, 129-131
  in cascades, 179-186
  effect of foreign nuclei, 177-179
  expanding vapors, 129-131
  in flowing steam, 127-189
  heterogeneous (*see* Heterogeneous condensation)
  historical aspects, 127-128
  homogeneous (*see* Nucleation)
  at HP, 128
  at low supersonic Mach numbers, 128
  in multistage HP turbine, 91
  nozzle vs. turbine flows, 184-186
  in nozzles, 140-145
  in overexpanded nozzles, 181
  spontaneous (*see* Nucleation)
  in turbines, 179-186
  in two-dimensional steam-flow field, 92-94
Condensation fronts, 161
Condensation shock (*see* Shocks)
Condensing flow:
  effect of inlet conditions, 141-143
  equations, 173
Condensing method:
  accuracy, 226-228
  for wetness measurement, 223-224, 226-228
Conformal mapping technique, for manifold design, 353
Conrad-type probes, 200
Conservation of energy equations (*see* Energy equations)
Conservation of mass (continuity) equations, 70-73, 95, 102
  for nozzle flow, 133
  for one-dimensional flow with nucleation, 84
  in terms of wetness fraction, 84
  for wet-steam flow, 71-73
Conservation of momentum equations (*see* Momentum equations)
Constant of proportionality, in tube vibration, 357-358
Continuity equations (*see* Conservation of mass equations)
Control volume equations, 71-72
Cooling, through-flow (TFC), 7
Cooling towers (CT), 7
  drift measurement probes, 253-257
  droplet size distribution, 253-257
  liquid flow distribution, 256

Copper-nickel alloys (Monel):
  for BWR turbine cylinders, 283
  for MSR tubes, 340, 345, 355
  in future PWR plants, 366
Coriolis force, effect on turbine flows, 53
Corrosion, 18-19
  in nuclear power plants, 13
  (*See also* Erosion-corrosion)
Corrosion-erosion (*see* Erosion-corrosion)
Corrugated plate separators (demisters, chevrons):
  basic concepts, 317-318
  breakthrough characteristics, 329-337
  breakthrough velocity, 318
  critical velocity, 318, 331-332, 334
    in upflow, 330-331
  drainage scoops, 334, 335
  effect of:
    inclination, 336-337
    plate height, 335-336
    water loading, 335-336
  efficiency, 318, 324-328
  horizontal steam flow, 333-337
  impaction-type, 282
  for MSRs, 352-353
    in future, 367
  in nuclear power plants, 338, 340-341, 345, 346
  performance in wet-steam turbines, 317-337
  plate form, 324-326
  steam velocity in, 318
  theory and parameters, 324-328
  vertical upward steam flow, 330-333
  wavelength, 324
  for wet-steam turbines, 326-327
Critical droplet size, 35-37, 157-158
  relation to supersaturation ratio, 36
Critical (throat) speed, equation for nozzles, 133
Critical velocity (*see under* Corrugated plate separators; Knitted wire mesh separators, etc.)
Critical velocity for vibration (*see* Threshold velocity for vibration)
CT (*see* Cooling towers)
Cycle efficiency, improvement in nuclear power plants by MSRs, 342-344
Cyclone separators, 282
Cylinders:
  erosion protection, 282-288
  water extraction, 272, 276, 277

Demisters (*see* Knitted wire mesh separators)
Dibromoethylbenzene, as manometer fluid, 192
Dielectric methods, for wetness measurement, 221–222
Dimensionless equations, for wet-steam steady flow, 77–78
DISA laser anemometers, 208, 214
Displacements, system responses, 65
Dissociation, of water, 21*n*.
Dissymmetry, in light-scattering by fogs, 237
Doppler frequency detection:
  by fringe (noncoherent) method, 211–213
  by reference beam (coherent) method, 209–211
  (*See also* Laser-Doppler anemometry)
Drag coefficient, 43, 68, 115
Drag force:
  equations, 77, 103
  large droplets, 48
  in steam-droplet mixtures, 43–44
  of steam on water drops, 68
  Stokes law, 319, 320, 324
Drainage devices, 11
  for HP turbines, 17
  (*See also* Slits; Suction slits, etc.)
Drift measurement probes, 253–257
Drop size:
  effect on:
    acoustic velocity, 97
    expansion, 83
    relaxation zones, 103–104
  (*See also* Droplet size)
Drop trajectories, 205
Droplet number, 80–81
  conservation equations, 71, 72
  distribution calculation, 244
  measurement:
    by light-scattering probe, 249–257
    in two-phase flows, 249–257
  in wet steam, 67, 249–257
Droplet size:
  categories vs. various processes, 29
  critical (*see* Critical droplet size)
  growth by agglomeration, 47
  measurement (*see* Droplet size measurement)
Droplet size distribution, 79–81, 176–177, 243
  in CT, 253–257
  of droplets torn from blade trailing edges, 49–51
  in fogs, 245
  in LP turbines, 251, 253
  measurement:
    by light-scattering probe, 249–257
    in two-phase flows, 249–257
Droplet size measurement, 139–140
  coarse water, 234–236
  focused direct shadowgraph, 234–235
  fog, 236–241
  large droplets, 234–236
  light-scattering methods, 236–241
  magnesium oxide slide method, 236
  needle probes, 235–236
  optical techniques, 236–241
  photographic method, 234–235
Droplet trajectories, 204–205
Droplets:
  break-up, 49–51
  coagulation, 48
  collisions, 45
  critical size (*see* Critical droplet size)
  deformation, 49–51
  density distribution measurement, 139
  deposition, 48
  diffusion, 48
  distance-to-diameter ratio, 31
  drag, 43–44, 48, 68
  equilibrium number, 158
  growth, 128, 165, 168, 176–177
    relation to droplet size, 29
  growth rate equations, 163–168
  impact, 48
  impingement, 48
  large, drag, 48
  number concentration, 30
  number of molecules contained in, 31, 36
  rebounding, 48
  size distribution (*see* Droplet size distribution)
  specific number, 31
  surface temperature, 37
  surface tension, 128, 162
  terminal settling speed, 44
  velocity measurements, 249–257
  in wet steam, 28
  (*See also* Drops; Fog droplets; Water droplets, etc.)
Drops:
  effective diameters, at HP and LP, 81
  large:
    acceleration, 115
    dimensionless velocity, 116
    equation of motion, 115–116
    Stokes law resistance, 319

Subject Index

stopping distance, 319
"Dry" stage work-loss coefficient, 121–122
Dry steam flow, 194
Dust, effect on condensation, 129, 177–179
Dynamic viscosity, 26

Economics:
  nuclear turbines, 5, 13
    European vs. U.S., 13
  power plant turbines, 6
  power production, 4–5
  steam turbines, 4–5
  turbine-generator unit size, W. German vs. U.S., 5
Effective drop diameter, at HP and LP, 81
Efficiency:
  change under transient conditions, 313
  constant isentropic, 60
  effect of water in turbines, 104–122
  of HP stages, 17
  losses (*see* Energy losses; Losses; Wetness losses, etc.)
  of nuclear turbines, 5
  overall, 106–107
  and overspeed, 313
  of separators, 318–321
  in steam turbines, 34–35
  (*See also* Aerodynamic efficiency; Carnot efficiency; Fractional separation efficiency; Thermodynamic efficiency, etc.)
Electric power:
  consumption, 4
  plants (*see* Nuclear power plants; Power plants; etc.)
  production, improvement by MSRs, 342–344
Electrical conductivity:
  effect of salts on, 27–28
  feedwater in HP power plants, 28
  live steam, 28
  water, 27–28
End-blade erosion:
  effects of:
    annulus area, 303
    CT, 303
    design of blade and casing, 302
    end-blade height, 303
    low back-pressure, 303
    moisture content of steam, 299–300
    tip velocity of blades, 300–301
    two- vs. four-flow turbines, 303
  vs. erosion-corrosion, 291

Energy deficit, 118
Energy (conservation) equations:
  for acoustic velocity in wet steam, 95
  for nozzle flow, 133
  for one-dimensional flow:
    with nucleation, 84
    without nucleation, 70, 74–75
  for stream tube, 75
  for steady flow, 75
  for wet steam flow, 74–75
    control volume, 74
    shock waves, 102
    two-phase mixture, 74–75
  for zero slip, 84
Energy losses:
  from:
    blade-surface friction, 119
    centrifuging of fog or coarse water, 119
    fog-droplet transport, 118
    fog nucleation, 119
    fog supersaturation, 119
    turbulence, 119
  in wet-steam turbines, 59–126
Enthalpy, 75
  change, 60, 67
  for liquid phase, 76
  fluctuations in LP turbines, 185
  Gibbs free, 37, 158
Enthalpy-entropy (Mollier) charts, 17, 22–23, 223
  for isentropic stage, 107
  for single-stage condition line, 107, 108
  steam, 23
  subcooled dry steam, 32
  in subcooling definition, 32
Enthalpy loss coefficients:
  moving blades, 108
  nozzles, 108
Enthalpy losses, 117
Entropy:
  equation, 133
  real-gas factor dependence on, 24–25
  (*See also* Enthalpy-entropy charts; Temperature-entropy diagrams, and isentropic entries)
Equations of state, 95, 102
  graphical representation, for water, 23
  for nozzle flow:
    caloric, 133
    entropy, 133
    thermal, 133
  steam, 59–64
    subcooled (supersaturated), 59

Equations of state, steam (*continued*):
   superheated, 59–63
   wet, 62–64
   for wet-steam flow, 75–76
Equilibrium number:
   of clusters in supersaturated states, 158–160
   of droplets, 158
Equilibrium population, of critical droplets, 158
Equivalent wetness fraction, 33
Erosion, 1–3, 18, 19
   in blades, 10, 52
   vs. droplet size, 29
   in end-blade tips, 304
   from high-speed droplet impact, 48, 50
   in HP blading, 252
   in light-water nuclear power plants, 252
   in LP turbines and blades, 9, 10
   in moving blades, 105
   in steam pipes, 252
   in steam turbines, 127
   worming, 281, 283
Erosion coefficient, 299–303
   in blade design, 303–304
   as function of:
      load type, 300
      steam cycle type, 300
Erosion-corrosion:
   causes, 291–293
   evaluation, 293–296
   effects of:
      moisture, 293, 295, 296
      pH of feedwater, 292, 293
      steam velocity, 293, 295
      temperature, 291–294
   predicted vs. measured values, 298
   prediction, 296–298
   from steam leakage in HP turbine, 19
Erosion protection:
   by:
      drainage slits, 303, 307, 310
      flame hardening, 312
      steam-heating, 305–311
      Stellite, 312
   of turbines, 282–288
Erosion resistance, 285–288
Erosion shields, 286–288
Erosion testing machines, 286
Error propagation factors, in nozzle experiments, 139
Euler equation, 117

Euler turbine equation, 110
Excess steam (*see* Venting flow)
Exhaust wetness, 105
Exit (cross-over) pressure, 17–18
Exit wetness, 1
Expansion:
   one-dimensional wet-steam, 83
   in steam-engine cylinders, 81
   of wet steam through HP turbine, 92
Expansion index:
   for equilibrium flow with frictional losses, 79
   for equilibrium isentropic flow, 79
   for gaseous phase of wet steam, 79
   for nonequilibrium flow, 79
Expansion lines, 6–10
   of HP turbines, 17–18
   in LP turbines, 8–10
Expansion processes, equations, 60
Expansion rates, 145, 155
   effects on:
      condensation, 127–131, 138
      stroke time, 81
   in steam engines, 81–82
External water separators, 317–369
   breakthrough characteristics, 329–337
Extinction coefficient, 238–240
   mean, 243
   vs. particle size, 240, 245–249
   for size distribution of small particles, 243
Extinction curves:
   Mie theory, 240
   for water droplets, 240
Extinction method:
   apparatus, 239
   of fog drop size measurement, 238–249
Extraction belts, efficiencies of open vs. louvred, 279–281

Feedwater:
   electrical conductivity, 28
   effect of pH on erosion-corrosion, 292–293
Ferrous ions, erosion-corrosion by, 291
Ferrybridge turbines, erosion control, 273
Films:
   in wet steam, 28
   in wet-steam turbines, 52–53
Filters (*see* Interference filters, etc.)
Fins, for MSR tubes, 366
First Law of Thermodynamics, 74
Fixed (stator) blades (stationary blades):
   coarse water losses, 113–116

design provisions, 303–307
erosion control:
  by steam heating, 305–311
  by suction slits, 303–307
  slits, 303–307, 311
  steam-heated, 305–311
  water flow patterns, 52
  (See also Hollow stationary turbine blades)
Flame hardening:
  automatic, 312
  for erosion protection, 312
Flashing, 38, 39
Flooding, 330
Flooding point, 330
Flow (see Axisymmetric flow; Gas flow; Steam flow; etc.)
Flow equations:
  effect of nucleation, 83–91
  matrix vector form, 85
  for one-dimensional flow with nucleation, 83–85
Flow models, vapor phase, 203
Flow patterns:
  Pitot tubes, 204
  steam, 297
  water on blades, 52–53
Flowing steam, condensation, 127–189
  (See also Steam flow)
Fluctuating pressures, in pressure measurement, 195–198
Fluctuation equation, for anemometers, 215–217
Fluctuation sensitivity, of anemometers, 215–217
Fluid flow, elastic-type, effect on tube vibration, 354
Flux ratio, 240
Fog drop size measurement:
  angular scattering method, 241
  equipment, 246
  extinction method, 238–249
Fog droplet size:
  distribution, 245
  effect on wetness losses, 122
  in diverse nozzles, 148–151
  measurement, 145, 148–151, 154–156, 181–183
Fog droplets:
  drag, 121
  flow, 111–117
  number determination, 145, 148–151
  transport losses, 118

Fog drops, fractional deposition on blading, 114, 115
Fog wetness fraction measurement:
  absolute methods, 222–228
  absorption methods, 221
  accuracy, 225–228
  condensing method, 223–224, 226–228
  dielectric methods, 221–222
  errors of different methods, 225–228
  heating method, 222–224, 226, 228–234
  methods requiring calibration, 221–222
  psychrometric method, 224–225, 227, 228
  throttling method, 222, 223, 226, 228
  in wet steam, 221–228
Fogs:
  coarse, 29
  effect on turbine efficiency, 106
  fine, 29
  formation, 145, 148–151, 155
  in wet-steam turbine stage, 107
  work-loss coefficients, 112
Foreign nuclei, effect on condensation, 129, 177–179
Fourier integral, 99
Fourier series, for flow field disturbance, 98–99
Fractional separation efficiency, 317–328
  corrugated plates, 324–328
  knitted wire mesh, 318–323
  for potential flow around a wire, 319
  single wire, 321
Frequency response, of pressure transducers, 197–198
Friction:
  in Laval nozzles, 132–134
  losses due to, 79
  on blade surfaces, 118
  from coarse water, 118

Gas constant, 75
  calculation, 62
  effective, 60
  vs. temperature, 61
Gas dynamic equations
  for: one-phase (monophase) nozzle flow, 133
    shock-free flow, 132–133
    two-phase flow, 170
Gas dynamics, wet steam, 59–126
  (See also Nonideal gases)

GEC turbines, 269
  erosion control, 273
General Electric moisture separator-reheaters, 337, 345
Geothermal plants, erosion problems, 3
Gibbs free enthalpy, 37
  change, 158
Glass beads, separation efficiency, 326–327
Gundremmingen nuclear power plant, 15

Heat, effect on gas flow, 136–137, 145
  (See also Enthalpy; Latent heat release; Latent heat transfer; Temperature)
Heat balance, 215, 230
Heat conduction, 76
  nonstationary, within droplets, 42–43
  (See also Thermal conductivity)
Heat convection, 76, 214–215
Heat exchange, during adiabatic reversion, 34
Heat flow, 214–215
Heat release, on condensation, 88–91
Heat transfer, 31
  in condensation, 163–165
  convective, 214–215
  in droplet dispersions, 80
  from droplets, 39–41, 103, 166
  equations, 103
  improvement in future MSR tubes, 366
  latent, 119–120
  losses, 120
  in MSR, 338
  in MSR tube bundles, 363
  rates, from droplets, 166
  between vapor and droplets, 39–41
  to walls, 41
  in wet steam, 38–43
Heat transfer coefficient, 40, 76, 80, 164, 166
  for droplets in steam, 67
Heating method, for wetness measurement, 222–224, 226–234
  accuracy, 226–228
Heaton Works turbines, 263–266, 269, 286
Helmholtz equation (see Kelvin-Helmholtz equation)
Heterogeneous condensation (on foreign nuclei), 129, 177–179
High-pressure (HP)-high-temperature power plants:
  electrical conductivity of feedwater, 28
High-pressure (HP) steam turbines, 3, 6, 35

condensation in, 91
design, 15, 16
double-flow, 15–16
erosion, 19, 283
  (See also Erosion-corrosion)
expansion in, 17, 18
extraction belts, 281
feedwater electric conductivity, 28
moisture, 18–20
MSRs, 343
in nuclear power plants, 15
  (See also Nuclear turbines)
relaxation time, 70
scavenging process, 47
section, 70
separators, 282
thermal coagulation in steam, 47
wet steam, 283
  "Kate" experimental, 263–264
  "Mary" experimental, 266–267, 278–281
  models, 35
  wetness loss factors, 269
  wetness losses, 265–267
  (See also Multistage high-pressure steam turbines)
High-pressure (HP) steam-water systems, 332–333
Hollow stationary turbine blades:
  drainage through slits, 303–307, 310
  steam-heating, 305–311
Homogeneous condensation (see Nucleation)
Horizontal steam flow, in separators, 333–337
Hot-wire anemometers (and anemometry), for wet steam, 212–217
HP steam turbines, etc. (see High-pressure steam turbines, etc.)
Humid air, condensation in nozzles, 140–145
Hydrazine, in feedwater, 293

Ideal gas laws, 59
Idealized flow in, Laval nozzles, 132
Impact, of droplets, 48
Impaction (collision) efficiency, 44, 46, 47
Impulse blade section, 94
Impulse cascade, calculated 2-dimensional flow, 94
Inertial effects, vs. droplet size, 29
Inertial impaction parameter, 44
Inertial relaxation:

Subject Index

calculation at Pitot tube mouth, 81
  of water droplets, 68–69
Inertial relaxation time, 69, 70
Inertial separators, 317–318
  (*See also* Corrugated plate separators;
  Knitted wire mesh separators)
Influence coefficients, for wet steam, 88–89
Inlet:
  pressure, 17
  stagnation state in nozzle flow, 133
  superheat, 17, 20
  wetness, 17
Institut für Dampf- und Gasturbinen der
  RWTH (Rheinisch-Westfälische
  Technische Hoch-schule) Aachen, light
  scattering probe, 249
Instrumentation, for testing wet steam, 191–260
Interference filters, for drop size measurement, 246–247
Interferograms, of condensing flow, 142–143
Interferometers, 139
Intermediate-pressure (IP) steam turbines, 6
Intermediate water separation, 4
Intermittent purging, 195–196
International Formulation ("IFC 1967")
  equation of state, 22
Influence coefficients, for wet steam, 88–89
"Inveresk" paper-mill turbine, 274–277, 279–280, 282
Ions, effect on condensation, 129, 177–179
  (*See also* Salts)
IP steam turbines (*see* Intermediate-pressure
  steam turbines)
Iron hydroxide, in erosion-corrosion, 291
Isentropic expansion index, 60, 63–64, 79, 97
Isentropic exponent k, 26
Isentropic flow, 79
Isentropic heat release, 261
Isentropic output, for wet steam, 120
Isentropic wet-steam stage, 106–111, 120
Isokinetic sampling, 231–234
Isothermal compressibility, 96
Iteration methods, 86–94, 207

"Kate" experimental high-pressure turbine, 263–264
Kelvin-Helmholtz equation, 35
Kiel probes, 191, 200
KN (*see* Knudsen number)

Knitted wire mesh separators (demisters):
  basic concepts, 317–318
  breakage, 351–352
  breakthrough characteristics, 329–337
  breakthrough velocity, 318, 336–337
  critical velocity, 318, 330–333, 336–337
  effects of:
    steam distribution, 351–353
    inclination, 336–337
  efficiency, 318–323, 328
  in MSRs, 338, 344, 351–353
  performance in wet-steam turbines, 317–337
  sinusoidal flow stream, 328
  steam velocity in, 318
  testing with H_2SO_4 drops in air, 323
Knudsen number (Kn), 39, 43, 45, 67, 77, 214
Kraftwerk Union AG moisture separator-reheaters, 337, 348
Kutsteladze number (Ku), 332–335
  (*See also* Weber number)

Laminar diffusion, of droplets, 48
Laplace length scale, 332–335
Large droplet drag, 121
Laser-Doppler anemometers (and anemometry), 207–212
  optical system, 209–212
Lasers, for transmittance measurements, 241–249
Latent heat release:
  in droplet dispersions, 80
  to droplets by condensing vapor, 76
  rate, 164
Latent heat transfer, to steam flow, 119–120
Latex hydrosols, for calibration of light-scattering probes, 250
Laval nozzles, 131–137
  condensation in, 141, 180–183
  condensation shocks, 135–136
  fog droplet size, 154
  friction, 132–134
  idealized flow, 132
  Mach number profiles, 134
  one-dimensional flow with condensation, 169–177
  pressure, 134
  shock-free flow, 132–133
  shock waves, 134
  shocks with heat release, 136–137, 145
  steady gas flow in, 130–137

Laval nozzles (*continued*):
  steam flow, 127
  testing equipment, 137–140
  two-dimensional effects, 134–136
Ligament velocity, 355, 357
Light attenuation probes, 139–140, 143, 144, 183
Light scattering, of monochromatic light by spheres, 238
Light-scattering probes:
  applications, 250–257
  calibration, 250
  for drift measurement at CT, 253–257
  for droplet size measurement, 249–257
  for HP applications, 252–253
  for live-steam lines, 252–254
  for LP applications, 250–252
  principles, 249–250
  for wetness measurement, 249–257
  windows, 247, 252–253
Light sources, for extinction measurements, 246–247
Light-water-cooled reactor nuclear power plants:
  erosion-corrosion, 3, 291, 296
  MSRs, 337–367
  wetness probes, 252–253
  (*See also* Nuclear power plants)
Linearized small perturbation equations (*see* Acoustic equations)
Liners:
  erosion protection, 282–288
  louvred cylindrical, 281–282
Liquid-metal fast-breeder reactor (LMFBR) plants, MSR absence in proposed designs, 367
Live steam:
  electrical conductivity, 28
  lines, droplet measurements, 252–254
  pressure, 1, 2
  temperature, 1
LMFBR plants (*see* Liquid-metal fast-breeder reactor plants)
Losses (in efficiency):
  calculation, 120
  of gaseous phase, 117–119
  from steam expansion to supercooled condition, 120
  (*See also* Aerodynamic losses; Stage-work losses; Wetness losses, etc.)
Louvre plate separators (*see* Corrugated plate separators)
Low-alloy steel:
  for BWR turbine cylinders, 283
  in MSR tubes, 345
Low-pressure (LP) steam:
  gas dynamics, 69
  pressure measurements, 192
Low-pressure (LP) steam flow:
  coarse water wetness fraction, 218–220
  computations, 81
  vs. condenser pressure and turbine power output, 7
  effective drop diameter, 81
Low-pressure (LP) steam nozzles, erosion prevention, 285
Low-pressure (LP) steam-turbine end-blades:
  erosion, 298–312
    causes, 298–299
    prevention by steam heating, 307–311
    prevention by suction slits, 303–305, 310
  tip speeds, 288
Low-pressure (LP) steam-turbine stages:
  design parameters, 302
  loss components, 121
Low-pressure (LP) steam turbines, 3, 35
  blades (*see* Blades; Fixed blades; Low-pressure steam-turbine end-blades; Moving blades)
  centrifuging loss coefficient, 117
  condensing, 6–11
  design, 8–9
  erosion, 288
  exhaust condensation, 183
  final stage:
    loss calculations, 121–122
    typical data, 8–9
  fog wetness measurement, 227–234
  moisture, 9, 10
  moisture influence on blading efficiency, 261
  moving blade flow calculations, 91–92
  in nuclear power plants, 15
  (*See also* Nuclear turbines)
  relaxation time, 69–70
  scavenging process, 45, 47
  section through wet-steam stages, 70
  steam reheating, 14–15
  thermal coagulation in, 47–48
  unsteady flow measurement, 212–214
  water flow patterns, 53
  wet-steam:
    "Bess" experimental, 271–273, 284
    models, 35
  wetness-loss factors, 269

wetness measurement by heating method, 227–234
LP steam turbines, etc. (*see* Low-pressure steam turbines, etc.)

Mach number (M), 40, 43, 85, 214
  as design parameter, 94–99
  effect on condensation, 127, 128, 134
  for nozzle flow, 133
  vs. pressure, 200
  profiles in Laval nozzles, 134
  in shocks with heat addition, 136–137
Manifolds, for MSRs, 364
  (*See also* Steam-inlet manifolds)
"Manometer effect," in MSR, 360
Manometers:
  fluids for, 192
  for wet-steam flows, 191–192
Marine propulsion, saturated steam turbines, 2–3
"Mary" experimental high-pressure wet-steam turbine, 266–267, 278–281
  droplet-size measurements, 267
  wetness-loss factors, 267
Mass balance, of droplets, 167
Mass flow rate, 223
Matrix vector form, of flow equations, 85
Maxwell equations, for basic thermodynamic relationships, 76
Maxwell's law, 37
Mean free path, steam, 26, 27
Mellins transform, 244
Mercury, as manometer fluid, 192
Microballoons, separation efficiency, 326–327
Mie curves, 241
Mie theory, 237–239
  comparison with experiment, 248–249
"MIT 1969" equation of state, 22
Moisture:
  effect on:
    blading efficiency, 261–269
    end-blade erosion, 299–300
  effects and handling in:
    HP turbines, 18
    LP turbines, 9, 10
  (*See also* Water; Wetness; etc.)
Moisture removal devices, 269–282
Moisture separator-reheaters (MSR):
  basic parts, 337–339
  design, 344–348, 351, 359–367
  division into two pressure stages, 343–344

drainage, 359–360
effect of orifice plates on condensate temperatures, 365, 366
effect on overspeed, 315
external, 337–367
future modifications for large nuclear power plants, 364–367
head shapes, 345
individual tube flooding, 360
for liquid metal reactors, 367
manifolds, steam-inlet, 351–353
"Manometer effect," 360
for nuclear power plants, 337–367
  placement, 340
operation, 348, 351
physical features, 337–344
startup, 348–351
  simulation by test model, 349–350
testing, 348–353
tubes (*see* Tube banks; Tube bundles)
velocity vs. damping parameters, 359
Westinghouse, 337, 344–347
Molecular clusters:
  equilibrium concentration vs. cluster size, 37, 38
  in vapor phase, 37
Mollier charts (*see* Enthalpy-entropy charts)
Momentum (conservation) equations, 70, 73–74, 95, 102, 133
  for gas phase of wet steam, 74
  for liquid phase of wet steam, 73–74
  for nozzle flow, 133
  for one-dimensional flow with nucleation, 84
  for zero slip, 84
Momentum input, of coarse water entering fixed blades, 119
Momentum thickness, 118
Monel metal (*see* Copper-nickel alloys)
Monodispersions, 70
  drop size measurement, 241–249
  of drops, 79–81
  transmittance, 241–243
Monophase flow, gas dynamic equations, 133
Moving (rotor) blades, 1–3
  coarse water losses, 116–117
  enthalpy loss coefficients, 108
  erosion, 1, 2, 105
    prevention by steam heating of associated stationary blades, 307–311
    prevention by suction slits, 310
  flow calculations for LP turbines, 91–92

Moving (rotor) blades (*continued*):
  hardening, 2
  protection of leading edge, 311–312
  shielding, 2
  tip section flow, 93
  tip speed, 3
  water flow patterns, 53
MSR (*see* Moisture separator-reheaters)
Multistage high-pressure (HP) steam turbines:
  condensation in, 91
  efficiency losses, 121

Neurath, RWE, power plant, 257
Newton's Second Law, 73
Nitrogen, condensation at very low temperatures, 128
Nomenclature, chapter lists, 54–56, 122–124, 186–187, 257–258, 288–289, 367–368
Nonequilibrium processes, 31
  expansion in steam engines, 82
Nonideal gases:
  classification parameters, 61
  equations of state, 75–76
Nonisokinetic sampling, 231–234
Nozzles:
  area, effect on gas flow, 138
  area relation, 133
  condensation in, 140–145
  convergent-divergent, 87–88
  enthalpy loss coefficient, 108
  experimental results, 137–156
  fluid states at various distances from wall, 135
  flow:
    with initially superheated steam, 87
    two-dimensional effects, 134–136
  flow equations, 133
  nucleation, 88–91
  pressure distributions, 145–148
    undulating, 135–136
  profiles, optical axis, 246, 247
  testing equipment, 137–140
  (*See also* Laval nozzles)
Nu (*see* Nusselt number)
Nuclear power plants:
  condensation in, 128
    HP turbines, 91
  cycle efficiency, improvement by MSR, 342–344
  electric output, improvement by MSR, 342–344
  exhaust moisture reduction by MSR, 342–344
  erosion problems, 9
  HP wet-steam turbines, 11–20
  LP condensing turbines, 6
  multistage HP turbines, 91
  offshore, MSR problems, 367
  Rolphton (Ontario), 269
  turbine water extraction, 273–274
  turbines, 9
    wet-steam, 105
  wet-steam problems, 3
  (*See also* Boiling-water reactor plants; Light-water-cooled reactor nuclear power plants; Liquid-metal fast-breeder reactor plants; Pressurized-water reactor plants; Steam-generating heavy-water reactor plants)
Nuclear reactors:
  steam wetness problems, 3
  (*See also* Boiling-water reactors; Pressurized-water reactors; Steam-generating heavy-water reactors, etc.)
Nuclear turbines, 15, 105
  cycles, temperature-entropy diagram, 4
  efficiency, 5
  erosion coefficient, 297
  erosion-corrosion, 291, 292, 295
  parameters in Europe vs. U.S., 13
Nucleation (spontaneous condensation, homogeneous condensation), 37, 70–71, 129, 156–177
  vs. droplet size, 29
  effect on:
    drop-size distribution, 244
    flow equations, 71, 83–85
    steam flow, 83–85, 88–91
  effects of:
    foreign nuclei, 177–179
    salts, 28
    time lag, 128
    wake eddies, 184
  equations, 93–94
  in expanding steam flow, 105
  losses due to, 119–120
  in nozzles, 88–91
    supersonic, 120
  premature, 184, 185
  in rotor tip sections of LP turbines, 92
  secondary, 244
  temperature effects, 184–186

theories (*see* Nucleation theories)
time delay, 162
Nucleation rate, 157–162, 176
  effects of:
    pressure, 161
    subcooling, 161
    and surface tension, 161–162
Nucleation theories:
  classical, 128, 159–161
  droplet growth, 163–168
  modified, 162–163
  revised, 128
Nuclei, effect on condensation, 177–179
Numerical methods (*see* Computational methods)
Nusselt number (Nu), 214

Offshore nuclear power plants, 367
Oil drops, separation, 325
One-dimensional equations, for wet-steam flow, 70–83, 169–172
One-dimensional flow, 71
  with nucleation, 83–91
    equations, 169–172
    experiment vs. theory, 176–177
    in Laval nozzles, 169–177
    numerical solution, 172–176
    with shock waves present, 177
    typical results, 174–176
One-phase flow (*see* Monophase flow)
Ontario Hydro power plant, 282
Optical axis, in nozzle profiles, 246, 247
Optical instruments, 139–143
Optical interferometry, 142–143
Optical ports:
  fog-excluder, 247
  for transmittance measurements, 247
  window materials, 247, 252–253
Optical probes, 248–249
  (*See also* Light-scattering probes)
Optical refractive index:
  saturated steam, 16
  saturated water, 28
  water, 27–28
Orifice plates:
  effect on condensate temperatures:
    in 700 MW plant, 365
    in 500 MW plant, 366
  for MSR tube bundles, 364
Overall efficiency, 106–107
Overall loss coefficient, 110–111
Overpressure calculation, 207

Overspeed:
  calculations, 313, 314
  after major load rejection, 312–315
  prevention, 313–315
    by bypass and intercept valves, 313, 315
    by controlled drainage, 315
    by trip devices, 313

Parsons turbines, 269, 274
Particle size:
  determination from extinction measurements, 241–249
  relation to mean extinction coefficient, 245
Passivation, 296
pH, effect on erosion-corrosion, 293
Photocells, for extinction measurement equipment, 247
Photography:
  of droplets, 234–235
  slow-motion, of tube vibration, 357
Pickering turbine, 274, 275, 278, 280, 281, 283
Pipes, erosion-corrosion, 291, 295–296
Pitot pressure, 201–207
Pitot probes, 140
  heads, 198–201
  (*See also* Pitot tubes)
Pitot tubes, 194, 197–198, 201–207
  droplet trajectories, 82, 206
  flow patterns, 204
  hooded, 198
  inertial relaxation calculation, 81
  two-dimensional, 204
Plate fins, for future MSRs, 366
Poisson relations, for nozzle flow, 133
Polarization ratio, 237–238
Polydisperse liquid phase, 79–81
Polydispersions:
  drop size measurements, 241
  spectral turbidity, 245
  transmittance, 241–242
Polytropic change, 61–62
Polytropic expansion index, 60
Polytropic vapors, 61
Power plants:
  economic aspects, 4–5
  fossil fueled, 6, 269
  manufacturing cost trends, 4–5
  Southern California Edison, San Onofre, 344, 348, 350, 351
  "Stella" Southpower station, 269–271
  unit size trends, 4–5
  (*See also* Nuclear power plants, etc.)

Pr (*see* Prandtl number)
Prandtl-Meyer expansion, 92
Prandtl number (Pr), 26, 40, 70
Prandtl tubes, 191
Premature nucleation, 184, 185
Pressure:
  effect on:
    acoustic velocity, 97
    condensation, 127, 131, 134, 143, 144
    gas flow, 138
    mean free path of steam molecules, 26, 27
    nucleation rate, 161
    saturation temperature, 76
  measurement (*see* Pressure measurement)
  (*See also* high-pressure, intermediate-pressure, low-pressure, and pressure entries)
Pressure distributions:
  in nozzles, 145–148, 176–177
  in turbine cascade, 180
Pressure lines, response to fluctuating pressure, 195–198
Pressure measurement:
  with continuously purged air lines, 192–195
  instrumentation, 138–139
  with intermittently purged air lines, 195
  by laser-Doppler anemometry, 207–212
  response to fluctuating pressures, 195–198
  subatmospheric, 193
  in turbines, 198–201
  in wet-steam flows, 191–201
Pressure transducers, 191, 195–198
Pressurized-water reactor (PWR) plants, 12
  carbon steel for MSRs, 340
  erosion, 283
  HP turbine water extraction, 273
  MSRs, 337–367
Pressurized-water reactors (PWR), 11–13
Probability-size distribution of droplets (*see* Droplet size distribution)
Probes (*see* Light-scattering probes; Optical probes; Pitot probes, etc.)
Project Management Group, LMFBR, 367
Psychometry:
  for wetness measurement, 224–225, 227, 228
  accuracy, 226, 228
Pulsating phenomena, in condensation, 128, 137
Pulse-height analyzers, 251
Purge flow, 193–195

PWR (*see* Pressurized-water reactors)

Quartz windows, for optical ports, 247

Radioactive contamination, of steam, 13
Rankine-Hugoniot equations, 90
Raschig rings, 330, 332, 333
Rayleigh region, 237
Rayleigh scattering, 237
Re (*see* Reynolds number)
Real-gas (compressibility) factor (Z), 24–26
  as function of entropy, 24–25
  saturated steam, 26
Rebounding, of droplets, 48
Receiving systems, for extinction measurement, 247–249
Refractive index, optical (*see* Optical refractive index)
Reheat, 2, 3
Reheaters, moisture separator- (*see* Moisture separator-reheaters)
Relaxation, 64–70, 79
  behind oblique shock, 103–104
  influence on flow equations, 71
  in wet-steam flow, 79
    calculations, 81–83
  (*See also* Inertial relaxation; Thermal relaxation)
Relaxation time, 64–66, 69–70
  physical significance, 64–65
  (*See also* Inertial relaxation time; Thermal relaxation time)
Relaxing medium, 64, 66
Remixing factor, 327
Resistance thermometers, 140, 143, 144
Response times, for air-purged lines, 195, 197
Reynolds number (Re), 40, 43, 45, 103, 214
  of droplets, 40, 320
RH (*see* Steam reheaters)
Rivulets, in wet-steam turbines, 52–53
Rolphton (Ontario) nuclear power plant, 269
Rotor blades (*see* Moving blades)
Runga-Kutta numerical procedures, 78, 103, 172

Salts:
  effects on:
    condensation, 178–179

electrical conductivity of steam and water, 27–28
San Onofre power plant (Southern California Edison), 344, 348, 350, 351
Sapphire windows, for light-scattering probes, 252–253
Saturated steam, 2–3, 11, 31
  erosion-corrosion, 291–298
  mean free path of molecules, 26, 27
  property data, 26–28
  real-gas factor, 26
  temperature measurement, 201
Saturated water, 31
  erosion-corrosion, 291–298
  property data, 25–28
Saturation, 20–21
  temperature, 21
  temperature vs. pressure, 24, 76
Sauter mean diameter, 243
Scattering angle, 250
Scattering patterns, 237
Scattering volume, 250, 251
Scavenging process, 45, 47
Scavenging steam (see Venting flow)
Secondary steam inlet, 17
Semipolytropic vapors, 62
Sensing heads, for pressure, 198–201
Sensitivity (see Fluctuation sensitivity; Steady-flow sensitivity, etc.)
Separation efficiency:
  parallel wires, 323
  single-wire, 320–321
  (See also Fractional separation efficiency)
Separators:
  centrifugal, 271
  depth, 329
  height, 329
  horizontal steam flow, 333–337
  vertical upward steam flow, 330–333
  (See also Corrugated plate separators; Cyclone separators; External water separators; Knitted wire mesh separators; Moisture separator-reheaters; Water separators, etc.)
Shock, 180, 181
  "aerodynamic" normal, 89–91
  due to condensation, 120, 180, 181
  in Laval nozzle, 135–136
  V-shaped, 135–136
  with heat release, effect on nozzle flow, 136–137, 145
  oblique, 99, 100

Shock-free flow:
  gas dynamic equations, 132–133
  in Laval nozzles, 132–133
Shock waves:
  in condensation, 128
  effect on condensation, 143, 144, 177
  energy losses, 120
  in Laval nozzles, 134
  Mach number dependence, 99
  thickness, 101
  in wet steam, 99–104
Single-stage condition lines, 108
Singularities:
  in integration of ordinary differential equations, 86–87
  saddle-point, 87
Sinusoidal oscillations, system responses, 65, 195–198, 328–329
Slits:
  in blades, 303–307, 310
  suction, 303–307, 310
Slow-motion photography, of tube vibration, 337
Smoke particles, effect on condensation of humid air, 179
Smoluchowski equation, 47
Sonic flow, 85, 87
Southern California Edison, San Onofre power plant, 344, 348, 350, 351
Space flight, condensation problems, 128
Spatial dispersion (spacing) of wetness, 29–30
Specific droplet number, 30–31
Specific heat, 26
  at constant pressure, 60
  of droplets, 29–30
  ratio for nozzle flow, 133
  vs. temperature, 61
Specific volume, 78–79
  ratio, steam/water, 26, 27
Spectra of droplet size (see Droplet size distribution)
Spectral turbidity, of polydispersions, 245
Stage efficiency, 107–108
Stage work-loss coefficients, 110–111
Stage work losses:
  components, 111–119
  due to nucleation, 119–120
  due to supercooling, 119–120
Stainless steel:
  in knitted wire mesh separators, 344
  in MSRs, 344
Static pressure distribution, measurement in nozzles, 139, 140, 145–148

Stationary blades (see Fixed blades)
Stator blades (see Fixed blades)
Steady flow:
  equations, 102
    energy, 74–75
    momentum, 74
Steady-flow sensitivity:
  of anemometers, 214–215
  in wet steam, 214–215
Steam:
  condensation in nozzles, 140–145
  equations of state, 59–64
  physical properties, 21–28
  supersaturation, 31–36
  (See also Flowing steam; Wet steam, etc.)
Steam distribution tests, 352–353
Steam/droplet mixtures:
  mechanics, 43
  thermodynamics, 35–38
Steam engines:
  gas dynamics, 63
  nonequilibrium expansion, 82
  simulated expansion in cylinders, 81
  stroke time, 81
Steam flow (see Dry steam flow; Horizontal steam flow; Two-phase steam flow; Vertical upward steam flow; Wet-steam flow; and as subhead under Laval nozzles, etc.)
Steam/gas-turbine power plants:
  Alsdorf (W. Germany), 251
  light-scattering probe installation, 251, 252
Steam-generating heavy-water reactor (SGHWR) plants:
  Winfrith, 269
Steam-generating heavy-water reactors (SGHWR):
  erosion, 283
  HP turbine water extraction, 273
  moisture removal, 269
Steam-heated fixed blades, 305–311
Steam-inlet manifolds:
  design for eliminating moisture carryover, 351–353
  for MSRs, 351–353
  testing, 351–353
Steam-path configuration, effect on erosion-corrosion, 295–297
Steam reheaters (RH), 318
  for nuclear steam turbine plants, 14
  size, effect on overspeed, 315
Steam superheat, 4
Steam tables, 59
Steam tappings, 17

Steam turbines:
  blades (see Blades; Fixed blades; Moving blades)
  historical aspects, 1
  (See also High-pressure steam turbines; Intermediate-pressure steam turbines; Low-pressure steam turbines; Wet-steam turbines)
Steam velocity:
  effect on erosion-corrosion, 294, 295
  in separators, 318
Steam wetness:
  history of problems from, 1–5
  (See also Wet steam; Wetness)
Steel:
  erosion-corrosion, 291–296
  for erosion protection, 283
  (See also Carbon steel; Chromium steel; Low-alloy steel; Stainless steel)
"Stella" Southpower station, 269–271
Stellites, 2, 283, 287–288
  for erosion protection, 312
  erosion resistance, 285–288
  for shielding blade tips, 9
Stokes drag law, 320
Stokes law resistance of drops:
  to motion, 319
  to relative motion with steam, 324
Stream tube equations:
  control volume, 71–72
  energy, 75
  flow, 70–83
    one-dimensional, 72
  momentum, 74
Stream tubes, calculated condensation in, 93–94
Streamline-curvature techniques, 71
Stroke time, in steam engines, 81
Subatmospheric pressure, measurement in wet steam, 193
Subcooled dry steam, 32
Subcooled (supersaturated) steam, 31, 91
  equations of state, 59
  gas dynamics, 61
  potential work capacity, 120
  stage efficiency, 120
  temperature measurement, 201
Subcooled vapors, 80
  temperature fields, 164
Subcooled (supersaturated) water, 31
Subcooling (supersaturation), 31–36
  and condensation, 128, 129
  effect:

Subject Index

on nucleation rate, 161
on turbine efficiency, 105–106
in steam, 31, 32, 91, 127
Subsonic conditions, in nozzles, 89
Subsonic flow, 88, 89
Subcritical heat addition, in nozzles, 89
Suction slits:
  in blades for erosion control, 303–307, 310, 311
  improper shape, 306
  proper shape, 307
Sulfur droplet dispersions, as testing systems, 247–249
Sulfuric acid drops, separation, 323
Supercooling, 91
  losses due to, 119–120
Supercritical heat addition, 88–91
Superheated steam, 6, 31, 105
  equations of state, 59–62
  gas dynamics, 61
  temperature measurement, 201
Superheated water, 31
Supersaturated expansion, 120–121
Supersaturated steam, etc. (see Subcooled steam, etc.)
Supersaturation (see Subcooling)
Supersonic cascade flow, 180
Supersonic conditions, in nozzles, 89
Supersonic expansion fans, 180–181
Supersonic flow:
  blade tips, 3
  fluctuation modes, 216–217
  in nozzles, 88
  in turbines, 94
Supersonic outlet velocity, of nozzles, 180
Surface (capillary) energy, of droplets in flow, 119
Surface tension:
  of droplets, 128, 162
  effect on pressure transmittal, 192
  relation to nucleation, 161
  between water and saturated steam, 26
Symbols (see Nomenclature)

Temperature:
  within droplets, 38–39
  effects on:
    erosion-corrosion, 291, 293–294
    gas constant, 61
    nucleation, 184–186
    refractive index of saturated water, 28
    specific heat, 61

measurement in wet-steam flows, 201
ranges for erosion-corrosion, 291
(See also Live-steam temperature, etc.)
Temperature-entropy diagrams, 2, 4
  nuclear steam turbine cycles, 4
  (See also Enthalpy-entropy (Mollier) charts)
Temperature fields, in subcooled vapor, 164
Tennessee Valley Authority (TVA) prototypes:
  LMFBR, 367
Terminal settling speed:
  droplets in turbine flow, 44
  steam/droplet mixtures, 43–44
Testing equipment:
  for nozzles, 137–140
  for wet steam, 191–260
  (See also names of specific instruments)
TFC (see Through-flow cooling)
Thermal accommodation coefficient, 166
Thermal choking, 88, 137
Thermal coagulation, 47
  vs. droplet size, 29
Thermal coefficient of expansion, 96
Thermal conductivity, 26
  (See also Heat conduction)
Thermal effects:
  lag, in wet steam, 41–43
  time constants, 42
  (See also Temperature)
Thermal (Brownian) motion, 47
Thermal relaxation, in wet steam, 67–68
Thermal relaxation time, 67
  for wet steam, 81
Thermocouples:
  probes, 201
  in tube bundle testing, 361–363
Thermodynamic efficiency, 1, 2
  (See also Carnot efficiency)
Thermodynamic equilibrium, 20
Thermodynamic loss, in wet-steam turbines, 34–35
Thermodynamics, of steam/droplet mixtures, 35–38
  (See also Enthalpy; Entropy, etc.)
Thin-wire resistance thermometer probes, 140, 143, 144
Three-dimensional flows, pressure measurements, 198–199
Threshold (critical) velocity for vibration, 354
Throat speed (see Critical speed)

Throttling method:
  for wetness measurement, 222, 223, 226, 228
  accuracy, 226, 228
Through-flow cooling (TFC), 7
Time constants, 64
  of thermal effects, 42
Titanium, in turbine blades, 13
Tool steels, erosion resistance, 287-288
Tracers, for coarse water, 221
Transducers, pressure (*see* Pressure transducers)
Transmission curves, for interference filters, 246-247
Transmittance, of suspensions, 240
Transmittance measurement, 241-242
  equipment, 246-249
  for fog drop size measurement, 246-249
Transonic flow:
  in nozzles, 85-86
  in turbines, 94
Transonic solution, of flow equations, 85-88
Tube banks:
  efficiency, 328
  sinusoidal flow stream, 328-329
Tube bundles:
  breakage, 338-343, 349, 350, 353
  buckling, 350-351, 359
  design, 359
  drainage, 359-360
  flooding, 359-363
  in MSRs, 353-359
  of future, 367
  temperature fluctuations, 361-363
  vibration (*see* Tube vibration)
  wear, 353
  weld failure, 360-361
Tube vibration, 353-359
  amplitude, 355-359
  amplitude vs. ligament velocity, 355-357
  analysis, 354
  effect of flow, 354-359
  models, 354-359
  in MSRs, 353-359
  optimum amplitude, 359
  test apparatus, 354-355
  testing, 354-359
Tubes:
  future modifications, 366
  Pitot (*see* Pitot tubes)
  (*See also* Stream tubes; Tube banks; Tube bundles; Tube vibration; U-tubes)

Tungsten wires:
  for anemometers, 212, 214
  gold-plated, 214
Turbidity:
  of bidispersions, 244
  integral equation, 244
  of suspensions, 240
Turbine blades (*see* Blades; Fixed blades; Moving blades)
Turbine/generator unit size trends, W. Germany vs. U.S., 5
Turbines (*see* High-pressure steam turbines; Intermediate-pressure steam turbines; Low-pressure steam turbines; Nuclear turbines; Steam turbines)
Turbulence:
  effect on condensation in nozzles, 180, 182
  in separators, 327
  from unfavorable slot shapes, 306
  in water film on pipes, 313
Turbulent diffusion, of droplets, 48
TVA (*see* Tennessee Valley Authority)
Two-dimensional effects:
  in Laval nozzles, 134-136
  in nozzle condensation, 140
Two-dimensional flow fields:
  condensation in, 92-94
  shock waves, 99-104
Two-phase steam flow:
  basic concepts, 1-57
  condensation in, 127-189 (*See also* Condensation; Nucleation)
  droplet size measurement, 249-257
  gas dynamics, 59-126
  equations, 170
  instrumentation, 191-260
  in separators, 317-369
  in turbines (*see* High-pressure steam turbines; Intermediate-pressure steam turbines; Low-pressure steam turbines; Nuclear turbines; Wet-steam turbines)
  wetness fraction measurement, 249-257
  (*See also* Horizontal steam flow; Vertical upward steam flow; Wet-steam flow)

U-tube reheaters, 348, 351, 363
U-tubes:
  in future, 366
  in MSRs, 348
Ultrasonic probes, for water thickness on pipe walls, 313

Subject Index

Unsteady flow:
  analysis, 218
  equations, 94–99
  measurement in wet steam, 212–217

VAK-Kahl, steam lines, 253
Valves:
  bypass, 315
  effect on overspeed, 313, 315
  intercept, 315
Vane-type separators, 348
Vapor pressure, 24
Vapor pressure function, 21
Vapor velocity, 118
Velocity, critical (*see* critical velocity *subhead under* Corrugated plate separators; Knitted wire mesh separators, etc.)
Velocity equations, 85
Velocity measurement:
  in droplets, 249–257
  effect of wetness fraction, 208
  by hot-wire anemometry, 212–217
  by laser-Doppler anemometry, 207–212
  light-scattering probes, 249–257
  in wet-steam flows, 201–217
Velocity probes, 249
Venetion-blind separators (*see* Corrugated plate separators)
Venting flow, in MSRs, 364
Vertical upward steam flow, 330–333
  in corrugated plate separators, 330–331
  in knitted wire mesh separators, 330–333
  in separators, 330–333
Vibration, in tubes (*see* Tube vibration)
Vortex shedding, effect on tube vibration, 354

Wakes:
  in blades, 194
  eddies effect on nucleation, 184
  effect on condensation, 180
  in LP turbines, 49–50
Walls:
  heat transfer to, 41
  (*See also* Channel walls)
Water:
  atomized, for light scattering probe calibration, 250
  Avogadro number, 21
  chemical formula, 21
  critical point, 21
  electrical conductivity, 27–28
  equations of state, 22–25
  gas constant, 21
  forms in wet-steam turbine stage, 107
  heavy (*see* Steam-generating heavy-water reactors)
  mass of one molecule, 21
  molecular radius, 21
  molecular weight, 21
  physical properties, 21–28
  triple point, 21
  (*See also* Coarse water; Moisture; Steam; Wetness, etc.)
Water droplets:
  in erosion-corrosion, 298–299
  inertial relaxation, 68–69
Water erosion protection, of turbines, 282–288
Water extraction:
  belts, 270
  from cylinders, 272, 276, 277
Water-filled lines, for pressure measurement, 191–192
Water film evaporation, effect on overspeed, 313
Water-hammer relationship, 285
Water-moderated reactor systems:
  influence of moisture on blading efficiency, 261
  (*See also* Boiling-water reactors; Pressurized-water reactors, etc.)
Water separators (WS), 3, 18
  in nuclear steam turbine plants, 14
  (*See also* External water separators)
Watson pressure-jet atomizers, 263
Weber number (We), 49, 332
  of droplets, 114
  (*See also* Kutateladze number)
Westinghouse liquid-metal fast-breeder reactor, 367
Westinghouse moisture separator-reheaters, 337, 344–347
Wet steam, 6, 31
  acoustic velocity in, 94–99
  behavior, 35–53
  calculated expansion, 92
  calculated condensation in stream tubes, 93–94
  composition, 20, 21
  drop size measurement, 245–247
  equations of state, 22–25, 62–64
  fog wetness fraction, 221–222
  gas dynamics, 59–126

Wet steam (*continued*):
  heat exchange with walls, 28
  heat transfer, 38–42
  influence coefficients, 88–89
  motion between phases, 28
  nucleation, 28
  physical data, 21–28
  pressure measurement, 191–201
  properties, 20–35
  shock waves in, 99–104
  state parameters, 28
  steam wetness, 28
  structure, 28–31
  subatmospheric pressure measurement, 193
  subcooling (supersaturation), 28, 31–35
  temperature measurement, 201
  testing equipment, 191–260
  thermal lag effects, 41–43
  thermal relaxation, 67–68
  unsteady-flow measurement, 212–217
  vapor pressure, 20, 21
  velocity measurement, 201–217
  wetness fraction measurement, 217–234
  (*See also* Steam wetness)
Wet-steam flow:
  calculations, 81–83, 91–94
  continuity equation, 71–73
  through convergent nozzle, 83
  dimensionless equations for steady flow, 77–78
  energy equations, 74–75
  equations of state, 75–76
  inequilibrium, 68
  momentum equation for gaseous phase, 74
  momentum equation for liquid phase, 73–74
  one-dimensional equations, 70–83
  one-dimensional expansion, 83
  supplementary equations, 76–77
  in turbines, 59–126
  (*See also* Horizontal steam flow; Two-phase steam flow; Vertical upward steam flow)
Wet-steam region, enthalpy-entropy chart, 24
Wet-steam turbines:
  basic concepts, 1–57
  blades (*see* Blades; Fixed blades; Moving blades)
  condensation in, 179–186
  cylinder protection, 283
  design for minimum loss, 122
  droplets analysis, 251–253
  efficiency, empirical corrections, 104–105
  energy losses, 59–126
  erosion-corrosion (*see* Erosion-corrosion)
  erosion-protection, 282–288 (*See also* Erosion protection)
  exhaust moisture level in nuclear power plants, 342
  experimental development, 261–290
  flow, condensation in, 181–183
  fog drop size measurement, 241
  fossil fueled, erosion coefficient as function of load, 299
  joints, erosion-corrosion, 291, 292
  nuclear (*see* Nuclear turbines)
  operating experience, 291–315
  optical probes, 248–249
  overspeed, calculation vs. test results, 314
  pressure measurement, 195, 198–201
  pressure fluctuations, 195–198
  relaxation effects, 70
  water extraction for BWR, PWR, and SCHWR, 273
  wetness losses, 104–122
  (*See also* High-pressure steam turbines; Intermediate-pressure steam turbines; Low-pressure steam turbines; Nuclear turbines)
Wetness:
  at control surfaces, 118
  detection, 140
  exit, 1
  inhibiting effect on nucleation, 178–179
  perturbation due to change in saturation pressure, 95–96
  (*See also* Moisture; Steam wetness; Water)
Wetness fraction, 29
  effect on acoustic velocity, 97
  of wet steam, 78–79
Wetness fraction measurement:
  light-scattering probe, 249–257
  in steam flows, 217–234
  in two-phase flows, 249–257
  (*See also* Fog wetness fraction measurement)
Wetness loss factors, 262–264
  HP, 267, 269
  LP, 268, 269
  vs. velocity ratio, 262–264
Wetness losses, 121
  calculations, 122
  due to fog-droplet transport, 118
  effect of fog droplet size, 121–122

as function of inlet superheat in HP turbines, 20
in laboratory turbines, 106
in turbines, 104–122, 261–269
Wetness probes:
thermodynamics, 228–234
using heating method, 228–234
Wever-type centrifugal separators, 271
Wilson limit, 267, 268
Wilson lines, 127–130, 145, 148, 151–155, 183–185
Wilson points, 129, 130, 138, 145, 148, 151–155, 175
of steam at HP, 153–154
of steam at LP, 151–153
Wilson zones, 129
Wind tunnels, condensation in, 127–128
Winfrith steam-generating heavy-water reactor, 269

Wire banks, sinusoidal flow streams, 328–329
Work-loss coefficients, 118
for control surface, 118
control volume, 108
fog droplets, 112
surfaces of rotor control volume, 116
coarse water, 116–117
Worming erosion, 281–283
Woven-wire mesh separators, 282
(*See also* Knitted wire mesh separators)
WS (*see* Water separators)

Xenon-arc lamps, as light source for extinction measurements, 246, 247

Z value (*see* Real-gas factor)